Uncertainty Quantification

Interdisciplinary Applied Mathematics

Volume 47

More information about this series at http://www.springer.com/series/1390

Christian Soize

Uncertainty Quantification

An Accelerated Course with Advanced Applications in Computational Engineering

 Springer

Christian Soize
Laboratoire Modélisation et Simulation
  Multi-Echelle (MSME)
Université Paris-Est Marne-la-Vallée
  (UPEM)
Marne-la-Vallée, France

ISSN 0939-6047   ISSN 2196-9973   (electronic)
Interdisciplinary Applied Mathematics
ISBN 978-3-319-85372-7   ISBN 978-3-319-54339-0   (eBook)
DOI 10.1007/978-3-319-54339-0

Mathematics Subject Classification (2010): 15B52, 35Q62, 35Q74, 35R30, 35R60, 37N15, 47H40, 60G60, 60H25, 60J22, 62B10, 62C10, 62F15

Printed on acid-free paper

This Springer imprint is published by Springer Nature
The registered company is Springer International Publishing AG
The registered company address is: Gewerbestrasse 11, 6330 Cham, Switzerland

Foreword

In numerical modeling and simulation, some degree of uncertainty is inevitable in the ability of the model to truly describe the physics of interest and/or in the data this model uses to assist in describing these physics. For this reason, and because numerical predictions are often the basis of engineering decisions, uncertainty quantification has been a subject of concern for many years. With the advent of multi-scale and multiphysics modeling, finding practical and yet rigorous ways to interpret uncertainty and characterize its impact on the assessment of probable outcomes has become ever more challenging. This book contributes to paving the way for meeting this objective one day.

The book focuses primarily on fundamental notions in the stochastic modeling of uncertainties and their quantification in computational models. Mechanical systems are the privileged applications because of the additional expertise of the author in this field. All of aleatory, epistemic, parametric, and nonparametric uncertainties are addressed.

The book begins with a description of those fundamental mathematical tools of probability and statistics that are directly useful for uncertainty quantification. It proceeds with a well-carried out description of some basic and a few more advanced methods for constructing stochastic models of uncertainties. It pays particular attention to the problem of calibrating and identifying a stochastic model of uncertainties when experimental data is available. For completeness, it also overviews the main approaches that have been developed for analyzing the propagation of uncertainties in computational models. Covering all of these topics in less than 330 pages, making them interesting to faculty, students, researchers, and practitioners, is a daunting challenge. The author has overcome this difficulty by skillfully organizing the content of this book and referring to mathematical proofs rather than deriving them in the middle of high-level descriptions.

I have been privileged to read and study the lecture notes on which this book is based. I have found them to be among the most useful ones for my understanding of the topic. Consequently, I invited the author to deliver in February 2015 at Stanford University an accelerated course on *Uncertainty Quantification in Computational Mechanics* based on these lecture notes. Given the success of this course,

I invited the author again in June 2016 to give a shorter version at the Aberdeen Proving Ground site of the Army Research Laboratory (ARL). The course was well attended by scientists and engineers from ARL representing five different laboratory divisions and several graduate students. For all these reasons, I expect this book to be a useful reference and a source of inspiration.

Stanford University   Charbel Farhat
Stanford, CA, USA
November 23, 2016

Preface

This book results from a course developed by the author and reflects both his own and collaborative research regarding the development and implementation of uncertainty quantification (UQ) techniques for large-scale applications over the last two decades.

The objectives of this book are to present fundamental notions for the stochastic modeling of uncertainties and their quantification in computational models encountered in computational sciences and engineering. The text covers basic methods and novel advanced techniques to quantify uncertainties in large-scale engineering and science models. It focuses on aleatory and epistemic uncertainties, parametric uncertainties (associated with the parameters of computational models), and non-parametric uncertainties (induced by modeling errors). The text covers mainly the basic methods and novel advanced methodologies for constructing the stochastic modeling of uncertainties. To this effect, it presents only the fundamental mathematical tools of probability and statistics that are directly useful for uncertainty quantification and overviews the main approaches for studying the propagation of uncertainties in computational models. Important methods are also presented for performing robust analysis of computational models with respect to uncertainties, robust updating of such computational models, robust optimization, and design under uncertainties. It also carries out the calibration and identification of the stochastic model of uncertainties when experimental data is available. The methods are illustrated on advanced applications in computational engineering, such as in computational structural dynamics and vibroacoustics of complex mechanical systems and in micromechanics and multiscale mechanics of heterogeneous materials.

This book is intended to be a graduate-type textbook for graduate students, engineers, doctoral students, postdocs, researchers, assistant professors, professors, etc. For learning a difficult interdisciplinary domain, such as the UQ domain, the fundamental difficulty is not the access to the knowledge but is related to the understanding of the knowledge organization, to the level of expertise that is required for solving a complex problem, and to the use of methods that are scientifically validated. Such an expertise is obtained by thinking and not by carrying out exercises.

The aim of the author is to propose a graduate-type textbook that leads the reader to think and not to train him/her with exercises.

The writing style has deliberately been chosen as short and direct, avoiding unnecessary mathematical details that may prevent access to a quick understanding of the concepts, ideas, and methods. Nevertheless, all the mathematics tools that are presented are absolutely correct and scientifically rigorous. All the required hypotheses for using them are given and explained. Any approximation that is introduced is commented and the limitations are specified. The useful references are given for a reader who is interested in finding the mathematical details for proving some mathematical results. A short paragraph is written at the beginning of each chapter, summarizing the objectives of the chapter and explaining the interconnections between the chapters. The chapters have a large number of subsections which will allow readers to clearly find and learn about specific topics that are useful to fulfill the broader chapter objectives.

An objective of this book is to present constructive methods for solving advanced applications and not to re-explain in detail the classical statistical tools, which are already developed in excellent textbooks in which there are academic examples that allow for training the undergraduate students.

The main objective is effectively to not recopy one more time the classical statistical methods for solving academic problems in small dimension with a very small number of scalar random variables. The book presents classical and advanced mathematical tools from the probability theory and novel methodologies in computational statistics that are necessary for solving the large-scale computational models with uncertainty quantification in high dimension. The book presents a set of mathematical tools, their organization, and their interconnections, which allow for constructing efficient methodologies that are required for solving challenging large-scale problems that are encountered in computational engineering. The book proposes to the readers a clear and identified strategy for solving complex problems.

The book includes several topics that are not covered in published research monographs or texts on UQ. This includes random matrix theory for uncertainty quantification, a significant part of the theory on stochastic differential equations, the use of maximum entropy and polynomial chaos techniques to construct prior probability distributions, the identification of non-Gaussian tensor-valued random fields in high stochastic dimension by solving statistical inverse problem related to stochastic boundary value problem, and the robust analysis for design and optimization. Many of these topics appear in research papers, but this is the first time that all of these topics are presented in a unified manner. Another novel aspect of this book is the fact that it addresses uncertainty quantification for large-scale engineering applications, which differentiates the methods proposed in this book from classical approaches. The theory, the methodologies, their implementation, and their experimental validations are illustrated by large-scale applications, including computational structural dynamics and vibroacoustics and solid mechanics of continuum media.

Université Paris-Est Marne-la-Vallée   Christian Soize
Marne-la-Vallée, France
November 24, 2016

Acknowledgments

I thank Professor Charbel Farhat from Stanford University for inviting me to give this accelerated course that I have especially written for this occasion and giving me the opportunity to write such an accelerated course on the basis of research years that I have carried out in the area of uncertainty quantification. Thanks to his invitation, I spent a very exciting period for teaching and research at Stanford University.

I thank all the colleagues, researchers, doctoral students, and postdocs who have worked with me since year 2 000 on important subjects related to uncertainty quantification and who are my coauthors in the papers referenced in this book:
Allain JM, Arnoux A, Arnst M, Avalos J, Batou A, Capiez-Lernout E, Capillon R, Cataldo E, Chebli H, Chen C, Clément A, Clouteau D, Cottereau R, Desceliers C, Duchereau J, Duhamel D, Durand JF, Ezvan O, Farhat C, Fernandez C, Funfschilling C, Gagliardini L, Ghanem R, Gharbi H, Grimal Q, Guilleminot J, Heck JV, Kassem M, Le TT, Leissing T, Lestoille N, Macocco K, Mbaye L, Mignolet MP, Naili S, Nguyen MT, Nouy A, Ohayon R, Pellissetti M, Perrin G, Poloskov IE, Pradlwarter H, Rochinha FA, Ritto TG, Sakji S, Sampaio R, Schueller GI, and Talmant M.

I would like to particularly thank Professor Roger Ghanem from the University of Southern California and Professor Marc Mignolet from Arizona State University, with which we have always carried out exciting scientific collaborations.

I thank my great friend, Professor Roger Ohayon, from CNAM in France who, in addition of our fructuous and intensive scientific collaborations, has carefully reread the manuscript of this book.

Contents

Acronyms

dB	decibels
dBA	A-weighted decibels
a.s.	almost surely
l.i.m	limit in mean square
ml.i.m	limit in mean square
pdf	probability density function
r.v	random variable
w.r.t	with respect to
APSM	algebraic prior stochastic model
BVP	boundary value problem
CPC	cardboard-plaster-cardboard
CPC	cardboard-plaster-cardboard
DAF	dynamic amplification factor
DOF	degree of freedom
DOFs	degrees of freedom
FKP	Fokker-Planck equation
FRF	frequency response function
GOE	Gaussian orthogonal ensemble
ISDE	Itô stochastic differential equation
KL	Karhunen-Loève expansion
MaxEnt	Maximum entropy
MATLAB	language of technical computing from MathWorks
MCMC	Markov Chain Monte Carlo
MSC-NASTRAN	computational mechanics software commercialized by MSC Software Corporation
PAR	prior algebraic representation
PC	polynomial chaos
PCA	principal component analysis
PCE	polynomial chaos expansion
POD	proper orthogonal decomposition

QoI	quantity of interest
ROB	reduced-order basis
ROM	reduced-order model
RVE	representative volume element
SROM	stochastic reduced-order model
SVD	singular value decomposition
UQ	uncertainty quantification

Chapter 1
Fundamental Notions in Stochastic Modeling of Uncertainties and Their Propagation in Computational Models

Abstract *This first chapter deals with a very short overview of fundamental notions, of concepts, and of vocabulary, concerning the aleatory uncertainties and the epistemic uncertainties, the sources of uncertainties and the variabilities (that are illustrated by showing experimental measurements for a real system), the role played by the model-parameter uncertainties and by the modeling errors in a nominal computational model, the major challenges for the computational models, and finally, concerning the fundamental methodologies.*

1.1 Aleatory and Epistemic Uncertainties

Aleatory uncertainties concern physical phenomena, which are random by nature such as:

— the pressure field in a fully developed turbulent boundary layer,
— the geometrical distribution of the inclusions inside the matrix of a biphasic microstructure.

Epistemic uncertainties concern the *parameters* of a computational model, for which there is a lack of knowledge, and also the *modeling errors*, which cannot be described in terms of the parameters of the computational model, as for instance:

© Springer International Publishing AG 2017
C. Soize, *Uncertainty Quantification*, Interdisciplinary Applied Mathematics 47,
DOI 10.1007/978-3-319-54339-0_1

- the lack of knowledge in the mechanical description of a boundary condition in a structure,
- geometrical tolerances induced by the manufacturing process in a structure,
- mechanical properties of materials resulting from a change of scale (without scale separation),
- the existence of hidden degrees of freedom in a complex mechanical system either induced by the use of some reduced kinematics in a computational model (beam theory instead of the 3D elasticity) or associated with secondary dynamical subsystems that are not modeled in the computational model.

In the framework of the *probability theory* and *mathematical statistics* of *uncertainty* quantification, there is no need to distinguish these two types of uncertainty (and we simply say *Uncertainties*) because the tools are exactly the same,

- for the stochastic modeling of uncertainties,
- for analyzing the propagation of uncertainties in the computational model,
- for the identification of the stochastic models by solving statistical inverse problems.

1.2 Sources of Uncertainties and Variabilities

The *designed system* is used for manufacturing the *real system* and for constructing a high dimensional computational model by using a mathematical-mechanical modeling process for which the main objective is the prediction of the responses of the real system in its environment. It should be noted that the high dimensional computational model is also called the high-fidelity computational model or simply, the *computational model* and sometimes, the *nominal computational model* or the *mean computational model*. The real system, submitted to a given environment, can exhibit a variability in its responses due to fluctuations in the manufacturing process and due to small variations of the configuration around a nominal configuration associated with the designed system. The computational model, which results from a mathematical-mechanical modeling process of the design system, has parameters which can be uncertain. In this case, there are *uncertainties on the computational model parameters*. In addition, the modeling process induces some modeling errors defined as the *model uncertainties*. Figure 1.1 summarizes the two types of uncertainties in a computational model and the variabilities of a real system, in which $\mathbf{U}$ and $\mathbf{U}^{nobs}$ are the observed and the nonobserved quantities, and where $\mathbf{u}^{exp}$ and $\mathbf{u}^{nobs,exp}$ are the corresponding quantities for the real system (only the observed quantity $\mathbf{u}^{exp}$ is assumed to be measured). It should be noted the following important items.

- The errors related to the method used for the construction of an approximate solution $(\mathbf{U}_n, \mathbf{U}_n^{nobs})$ of $(\mathbf{U}, \mathbf{U}^{nobs})$ (such as the finite element method that is frequently used for constructing the computational model) have to be reduced and controlled, and should not be considered as uncertainties. However, the errors

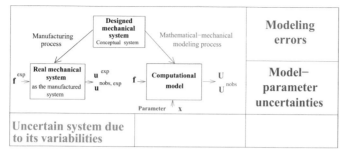

Fig. 1.1 Designed system, real system, computational model, sources of uncertainties, and variabilities.

induced by the use of a reduced-order computational model instead of the high dimensional computational model can be viewed as uncertainties.

– There are two types of uncertainties: the uncertainties on the model parameters called the *model-parameter uncertainties* and the *model uncertainties* induced by the *modeling errors*.

– There are some *variabilities* in the real system, due to the *manufacturing process* and due to *small differences in the configurations*: an experimental configuration of a complex mechanical system differs from the designed mechanical system and is never perfectly known.

1.3 Experimental Illustration of Variabilities in a Real System

Figure 1.2 shows an illustration of the variabilities in a complex real system, which is a car. This figure displays the experimental vibroacoustic measurements for 20 cars of the same type with optional extras [72]. Each curve represents the graph of the modulus of the measured frequency response function (FRF) for the internal

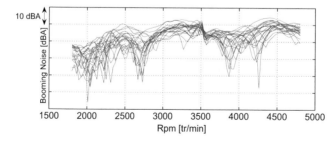

Fig. 1.2 Measurements of the FRF for the internal noise (acoustic pressure) in 20 cars induced by engine excitations applied to the structure (booming noise). The vertical axis is the modulus in dBA. The horizontal axis is the frequency expressed in rotation per minute of the engine. [Figure from [72]].

noise (acoustic pressure at a given point inside the internal acoustic cavity) induced by engine excitations applied to the structure (booming noise). The important variabilities that can be seen are due to the fluctuations inherent to the manufacturing process and are also due to some small differences in the configurations of the 20 cars.

1.4 Role Played by the Model-Parameter Uncertainties and the Modeling Errors in a Computational Model

Figure 1.3 shows the role played by the model-parameter uncertainties and the modeling errors in the computational model of the car with optional extras, for which experimental measurements have been presented in Section 1.3 (for the booming noise). The computational structural-acoustic model of the car is a finite element model with 978 733 structural degrees-of-freedom (DOFs) and 8 139 acoustic pressure DOFs in the internal acoustic cavity. For the normal structural acceleration in two given representative observation points in the structure, Figure 1.3 displays the graphs of the modulus of the measured FRFs for the 20 cars for which the excitations are induced by the engine, which are compared to the predictions given by the computational model [72]. The relatively important differences between the experimental data and the predictions are mainly due to the model uncertainties.

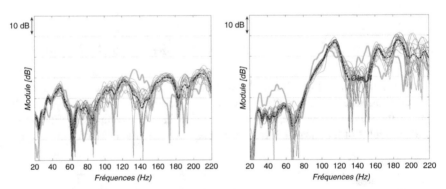

Fig. 1.3 For the normal structural acceleration in two given observation points in the structure, graph of the FRF calculated with the computational model (green thick line), graphs of the experimental FRFs (20 blue thin lines corresponding to the measurements of 20 cars of the same type with optional extras), and graph of the mean experimental FRF (dashed black line). The vertical axis is the modulus in dB. The horizontal axis is the frequency axis in Hertz. [Figure from [72]].

1.5 Major Challenges for the Computational Models

1.5.1 Which Robustness Must Be Looked for the Computational Models

For many cases, the deterministic computational models of complex real systems are generally not sufficient. Such a situation can be diagnosed as soon as their predictions are compared with experimental data. The robustness of a computational model must thus be improved in taking into account:

- the uncertainties in the computational model, for which a stochastic modeling can always be constructed by using the available mathematical properties of the uncertain physical quantities,
- the variabilities of the real system (if data are available),
- the experimental errors (if experimental data are available), which are:

  ▷ induced by the measurement noise that is generally small with respect to the second source of errors, which is described hereinafter.
  ▷ induced by the lack of knowledge of the measured experimental configurations that differ from the designed system (conceptual system) and that are not perfectly known, and which constitutes the main source of experimental errors for complex systems.

1.5.2 Why the Probability Theory and the Mathematical Statistics Are Efficient

The probability theory is a powerful mathematical tool, which allows:

- for constructing, in finite dimension or in infinite dimension, the *prior* and the *posterior stochastic models* of uncertainties (vectors, matrices, tensors, functions, fields, etc.).
- for analyzing the *propagation* of uncertainties in the computational models by using the applied mathematics.
- for *identifying* the prior and the posterior stochastic models of uncertainties, in finite or infinite dimension, using available *data* and the powerful theory of the mathematical statistics for solving statistical inverse problems. Data are either partial and limited experimental data, or numerical data calculated with a high dimensional computational model.
- the model to learn from data.

1.5.3 What Types of Stochastic Analyses Can Be Done

It is necessary to carry out computational analyses in a robust framework with re-spect to the uncertainties, with respect to the variabilities (if data are available), and with respect to the experimental errors (if experimental data are available), in order to perform with an uncertain computational model:

- *robust predictions,*
- *robust updating* (if data are available),
- *design optimization,*
- *performance optimization.*

Even if there are no available experimental data, and at stronger reasons, it is nec-essary (and one can always to do it) to take into account the uncertainties in the computational models, in order to analyze the robustness of the predictions with respect to the uncertainties and therefore, to perform a robust optimization and a robust design.

1.5.4 Uncertainty Quantification and Model Validation Must Be Carried Out

For a given deterministic computational model (mean or nominal computational model):

- a stochastic model of uncertainties must be constructed (for all the scales that are modeled), yielding the stochastic computational model.
- the propagation of uncertainties must be done by using the stochastic computa-tional model and the adapted stochastic solvers.
- if data are available, uncertainties must be quantified by using the stochastic com-putational model, the data, and the tools devoted to the statistical inverse prob-lems.

It should be noted that a probabilistic model of the uncertainties is not developed for compensating big errors introduced in the construction of the nominal computa-tional model, which must have the capability to represent the physical phenomena for which it has been constructed.

1.5.5 What Are the Major Challenges

Four main challenges can be identified.

1. The first challenge is related to the effective construction of stochastic models of uncertainties, which must include all the available information in order to

enrich the model, even if data are not available. In particular, the probability distributions, the algebraic representations, and the associated random generators must be constructed for the

- random vectors in high stochastic dimension,
- random matrices in any stochastic dimension,
- tensor-valued random fields in high stochastic dimension.

▷ *This step is fundamental but is relatively difficult. The course will present the classical tools and some advanced tools for giving a clear constructive answer to this question.*

2. The second challenge is related to the stochastic modeling of model uncertainties induced by the modeling errors when data are not available.

   ▷ *This step is an extremely challenging problem that has only received partial answers. The course will give constructive approaches in structural dynamics and coupled systems.*

3. If partial and limited data are available, the third challenge is related to the identification and the updating of the stochastic models of uncertainties by solving statistical inverse problems.

   ▷ *This step is a very challenging problem if the stochastic model of uncertainties is in high stochastic dimension. The course will give efficient tools for solving this type of problem.*

4. The fourth challenge is related to the algorithmic strategy for the robust updating, the robust optimization, and the robust design.

   ▷ *This step is a very challenging problem with respect to the computational resources and consequently, can require the introduction of reduced-order models and clever interpolation methods. The course will introduce a discussion on the base of some applications.*

1.6 Fundamental Methodologies

An introduction to mathematical statistics for uncertainty quantification devoted to uncertain parameters and to the identification of their probabilistic models can also be found in the textbooks by Smith [181] and Sullivan [212]. The approaches that are proposed in the present book are partially different, complementary, and allow for taking into account, not only uncertain parameters, but also model uncertainties induced by modeling errors in the computational models and, not only, the small stochastic dimension but also the high stochastic dimension for which adapted methodologies are proposed for solving statistical inverse problems.

1.6.1 A Partial Overview of the Methodology and What Should Never Be Done

Let us consider the notation introduced in Figure 1.1. The general case of the robust analysis for the prediction U performed with a computational model is briefly summarized in Figure 1.4. The computational model depends on an uncertain parameter X. In this figure, $x \mapsto h(x)$ is a deterministic nonlinear mapping and $H(s) = h(X)$ is a random variable that depends on s that is the hyperparameter of the probability distribution of random variable X and that allows for controlling the level of uncertainties (see the explanations given after). It should be noted that the word "hyperparameter" is used instead of the word "parameter" in order to distinguish the parameter of the probability distribution of X (called the hyperparameter) from the random variable X that is the parameter of the computational model. In order to well explain what should never be done in stochastic modeling of uncertainties and in uncertainty quantification (UQ), we introduce the most simple mechanical system (static, linear, one degree of freedom), and we present that could be the analysis performed by an "unqualified person for UQ" (as all the quantities are scalar for this simple example, the boldface letters such as x, s, U, X, H, h, and p_X are rewritten as x, s, U, X, H, h, and p_X in following the convention of notation explained in the Glossary).

☐ *Deterministic computational model.* The deterministic modeling of the computational model that is considered is made up of the linear static mechanical system with one degree-of-freedom (DOF), for which the equation is written as

$$\underline{k}\,\underline{u} = f \,, \tag{1.1}$$

in which

- the nominal stiffness is $\underline{k} = 10^6$ N/m,
- the applied external force is $f = 10^3$ N,

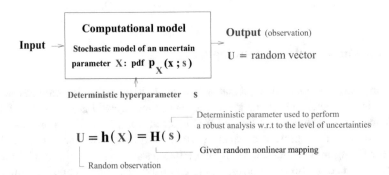

Fig. 1.4 Scheme describing the general case corresponding to a robust analysis of predictions.

  – the deterministic "solver" for computing the displacement is explicit: $\underline{u} = \underline{k}^{-1} \underline{f} = 0.001$ m.

□ *Stochastic modeling of the uncertain parameter.* We use the notions such as random variable, probability space, mean value, standard deviation, coefficient of variation, probability distribution, probability density function, Gaussian, etc. If the reader is not at all familiarized with the elements of the probability theory, Chapter 2 can be read before, but certainly, the intuition is sufficient for understanding the methodology that is presented hereinafter, even if all the notions used are not perfectly understood.

The uncertain parameter of the computational model defined by Equation (1.1) is assumed to be the nominal stiffness that is modeled by a real-valued random variable K, defined on a probability space $(\Theta, \mathcal{T}, \mathcal{P})$, and which is written as $K = \underline{k}\, X$. It is assumed that our "unqualified person for UQ" decides to choose, *a priori*, a Gaussian random variable for the stochastic modeling of X, with a mean value $m = 1$ and with a standard deviation $\sigma = s\,|m| = s$, in which $s > 0$ is defined as the coefficient of variation (such a coefficient of variation is often denoted by δ). Consequently, the probability distribution of random variable X is thus defined by the Gaussian probability density function (pdf), $x \mapsto p_X(x\,;s)$ on $\mathbb{R}$, such that

$$p_X(x\,;s) = \frac{1}{\sqrt{2\pi}\,\sigma} \exp\left\{ -\frac{(x-m)^2}{2\sigma^2} \right\} \quad , \quad \sigma = s\,|m| = s\,, \qquad (1.2)$$

for which the graph is displayed in Figure 1.5 for $s = 0.25$. For all real numbers a and b such that $a < b$, the probability that X takes its values in the interval $[a\,,b]$ is given by

$$\mathcal{P}\{X \in [a\,,b]\} = \int_a^b p_X(x\,;s)\,dx\,. \qquad (1.3)$$

The mean value $m = E\{X\}$ (E is the mathematical expectation that is a linear operator on the space of the real-valued random variables) and the second-order moment $E\{X^2\}$ of random variable X are then defined by

Fig. 1.5 Graph $x \mapsto p_X(x\,;s)$ of the Gaussian pdf of random variable X for $m = 1$ and $s = 0.25$.

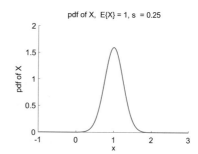

$$E\{X\} = \int_{\mathbb{R}} x \, p_X(x \,;s) \, dx \quad , \quad E\{X^2\} = \int_{\mathbb{R}} x^2 \, p_X(x \,;s) \, dx < +\infty. \quad (1.4)$$

The standard deviation is the square root of the variance $\sigma^2 = E\{(X - m)^2\} = E\{X^2\} - m^2$. Consequently,

- the mean value of K is $E\{K\} = \underline{k} \, E\{X\} = \underline{k}$ that is the nominal value.
- the hyperparameter s allows for controlling the level of uncertainties of K: larger is the value of s, larger is the uncertainty level.

☐ *Stochastic computational model.* The stochastic computational model consists in replacing $\underline{k}$ by random variable K in Equation (1.1). Therefore, the displacement becomes a random variable U that is the random solution of the following stochastic model,

$$K U = f. \quad (1.5)$$

☐ *Stochastic solver.* It should be noted that the solution of stochastic equation $K U = f$ can be written as $U = K^{-1} f$, which shows that the real-valued random variable U is a nonlinear mapping of the random parameter K. Our "unqualified person for UQ" thus decides to choose the Monte Carlo numerical simulation method (see Section 6.4), which is a nonintrusive method with respect to the commercial software. For s fixed, the steps of the stochastic solver are thus defined as follows

1. *Generation of independent realizations of random variable K.*
   - The generation of $\nu = 1,000$ independent realizations $X(\theta_1), \ldots, X(\theta_\nu)$ is performed with $\theta_1, \ldots, \theta_\nu$ in Θ. As X is a Gaussian random variable with mean value m and with standard deviation σ, the random generator could be written (using, for instance, MATLAB) as $X(\theta_\ell) = m + \sigma \times$ randn, in which randn returns a random scalar drawn from the normalized Gaussian probability distribution (Gaussian real-valued random variable with zero mean and with a unit variance, see Section 2.4.2).
   - Generation of ν independent realizations $K(\theta_1), \ldots, K(\theta_\nu)$ such that $K(\theta_\ell) = \underline{k} \, X(\theta_\ell)$.

2. *Deterministic solver.* The computation of ν independent realizations $U(\theta_1), \ldots, U(\theta_\nu)$ is carried out for the random response U such that $U(\theta_\ell) = K(\theta_\ell)^{-1} f$.

3. *Statistical postprocessing by using the mathematical statistics.* The estimates of the probabilistic quantities of interest are performed, for instance, the estimate of the second-order moment of U, which is defined by $E\{U^2\} \simeq m_2^{(\nu)}$ with $m_2^{(\nu)} = \nu^{-1} \sum_{\ell=1}^{\nu} U(\theta_\ell)^2$.

☐ *Results obtained from the stochastic solver.* Let us assume that a sensitivity analysis of the random response U is performed with respect to the level of uncertainties represented by the values $0.15, 0.20, 0.25$, and 0.30 of hyperparameter s. The pdf of random variable U is estimated by using the nonparametric statistics

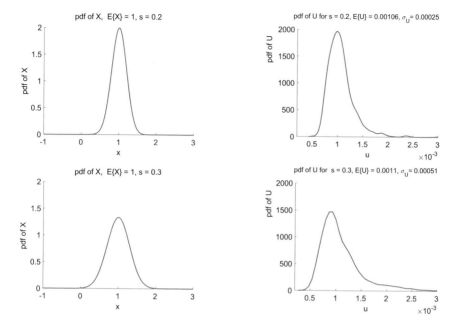

Fig. 1.6 For the two values 0.20 and 0.30 of hyperparameter s, the left figures display the pdf of random variable X, and the right figures display the pdf of random variable U, which are estimated by using the Gaussian kernel density estimation method with the $\nu = 1,000$ independent realizations.

(see Section 7.2) and is plotted in Figure 1.6 for $s = 0.20$ and for $s = 0.30$. The estimate $m_2^{(\nu)}$ of the second-order moment, $E\{U^2\}$, is plotted in Figure 1.7 for $s = 0.15, 0.20, 0.25$, and 0.30. In view of the results presented in Figure 1.7, our "unqualified person for UQ" could be relatively happy with such results, because:

- qualitatively, $E\{U^2\}$ increases with s, which is consistent.
- quantitatively, the value of $E\{U^2\}$ is of the same order that $\underline{u}^2 = 10^{-6}$.

Fig. 1.7 For $\nu = 1,000$, values of the estimate $m_2^{(\nu)}$ of $E\{U^2\}$ for $s = 0.15$, $0.20, 0.25$, and 0.30.

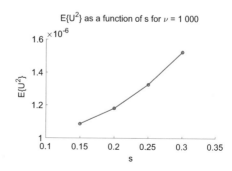

However, these results are completely erroneous and the results obtained have no meaning.

☐ *Why our "unqualified person for UQ" has not detected that the stochastic modeling of K is erroneous.* The number of realizations has been fixed to a relatively small value ($\nu = 1,000$) and a convergence analysis has not been performed with respect to ν (what must absolutely be done when a sampling method, such as the Monte Carlo numerical simulation method, is used). Such a limitation of the number of realizations is generally induced by the limitation of the available computer resources (case of the use of a high dimensional computational model for which the numerical cost for computing one realization of the response can be very high). Nevertheless, this difficulty can be circumvented, thanks to the central limit theorem that allows for estimating the error during the computation without knowing the exact solution (see Section 2.8.2). If our "unqualified person for UQ" had performed a convergence analysis with respect to ν, then he would have obtained the results displayed in Figure 1.8 and would have seen that the stochastic modeling of K was erroneous because $E\{U^2\} \simeq m_2^{(\nu)}$ is not convergent when ν increases, which is in contradiction with the physics of a passive system that is stable.

☐ *Can we directly see that the difficulties come from an erroneous stochastic modeling of parameter K.* Such a conclusion can immediately be obtained by examining the expression for $E\{U^2\}$ that is such that

$$E\{U^2\} = \frac{u^2}{\sqrt{2\pi}\,\sigma} \int_{-\infty}^{+\infty} \frac{1}{x^2} \exp\left\{-\frac{(x-m)^2}{2\sigma^2}\right\} = +\infty . \qquad (1.6)$$

In Equation (1.6), the divergence of the integral is due to the singularity x^{-2}, which is not integrable in $x = 0$.

Fig. 1.8 Graph $\nu \mapsto m_2^{(\nu)} \simeq E\{U^2\}$ for analyzing the convergence with respect to ν, for $s = 0.15$ (red line with none marker), 0.20 (green line with "cross" marker), 0.25 (black line with "plus" marker), and 0.30 (blue line with "circle" marker).

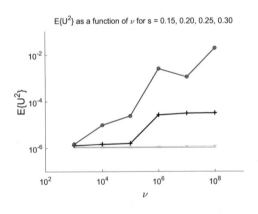

E{U²} as a function of ν for s = 0.15, 0.20, 0.25, 0.30

□ *What are the reasons for which the stochastic modeling of K is erroneous.* There are several reasons for which the stochastic modeling of K is erroneous. The analysis of these errors is important because it allows to highlight the sources of available information that will be used for construction the prior stochastic model with Information Theory (see Section 5.3).

– For constructing a stochastic model of K, the first source of information is related to its algebraic and statistical properties, which are detailed as follows:
  ▷ The random parameter K must be with positive values, which means that the support $\mathcal{S}$ of the pdf p_X must be a subset of $\mathbb{R}^+$ (this implies that, for all x not in $\mathcal{S}$, then $p_X(x) = 0$). This information has not been used (a Gaussian random variable is not positive; the support is $\mathcal{S} = \mathbb{R}$).
  ▷ The available statistical properties of K must be used. For instance, the mean value of K could be assumed given and be equal to the nominal value $\underline{k}$, which means that $E\{K\} = \underline{k}$ is given.

– The second source of information is related to the mathematical properties of the stochastic solution U of the computational model. The construction of the pdf p_X of X must be done in order that the stochastic solution U is a second-order random variable, $E\{U^2\} < +\infty$. This information has not been used and Equation (1.6) shows that $E\{U^2\} = +\infty$.

1.6.2 Summarizing the Main Steps of the Methodology for UQ

In order to help the reader to better understand what would be the methodology for constructing a probabilistic model of uncertain parameters, for studying the propagation of uncertainties in the computational model, and for identifying the probabilistic model with data, we have briefly summarized such a methodology in Figure 1.9 for which the notations of Figure 1.4 are reused (we consider the vectorial case for which the boldface letters are used). This brief summary gives a partial view about the general methods that will be presented in Chapter 5 to concerning the probabilistic methods and the statistics for performing uncertainty quantification in computational models (including model-parameters uncertainties and model uncertainties induced by modeling errors). Nevertheless, Figure 1.9 gives a first view and a support for understanding why some tools from the probability theory (Chapters 2 to 7) must be learnt. The example and the notations presented in Section 1.6.1 are reused (but for the vectorial case for which boldface letters are used). The methodology described in Figure 1.9 is therefore done in terms of the probability distribution of the uncertain parameter $\mathbf{X}$ with value in $\mathbb{R}^n$ is modeled by an $\mathbb{R}^n$-valued random variable. The probability distribution of $\mathbf{X}$ is assumed to be represented by a pdf $\mathbf{x} \mapsto p_\mathbf{X}(\mathbf{x}\,;\mathbf{s})$ on $\mathbb{R}^n$ (that is not Gaussian) and where $\mathbf{s}$ is a vector-valued hyperparameter. Here, the direct approach is thus used contrarily to the indirect approach such as the one that would correspond, for instance, to a polynomial chaos expansion of $\mathbf{X}$ (see Section 5.2).

We now explain the information given in Figure 1.9. The observation of the computational model is an $\mathbb{R}^m$-valued random variable $\mathbf{U} = \mathbf{h}(\mathbf{X})$ that is transformed from $\mathbf{X}$ by a given deterministic mapping $\mathbf{h}$ from $\mathbb{R}^n$ into $\mathbb{R}^m$ ($\mathbf{h}$ is a nonlinear mapping such that $\mathbf{U}$ be effectively a random variable).

▷ The first step consists in constructing a prior pdf $p_{\mathbf{X}}^{\text{prior}}(\mathbf{x};\mathbf{s})$ of $\mathbf{X}$ depending on an unknown vector-valued hyperparameter $\mathbf{s}$ as explained in Chapters 5 and 10 (central column entitled "Construction/Analysis" in the figure).

▷ If no data are available (left column entitled "No data" in the figure), then only a sensitivity analysis
of the stochastic solution $\mathbf{U}$ can be performed with respect to the level of uncertainties defined by the value of hyperparameter $\mathbf{s}$ (for instance, involving the coefficients of variation of some components of random vector $\mathbf{X}$):

  − a training procedure is applied for sampling values of $\mathbf{s}$ in its admissible set.
  − for each fixed value of $\mathbf{s}$, the propagation of uncertainties is analyzed by computing the stochastic solution $\mathbf{U}$ using an adapted stochastic solver (see Chapter 6).
  − the estimates of the *quantities of interest* (QoI) related to $\mathbf{U}$ are then done by using the mathematical statistics and allow for performing a sensitivity analysis with respect to $\mathbf{s}$ (for instance, with respect to the level of uncertainties).

▷ If data (coming from experiments or coming from numerical simulations) are available (right column entitled "Data" in the figure), then

  − an optimal value $\mathbf{s}^{\text{opt}}$ of hyperparameter $\mathbf{s}$ can be estimated by solving a statistical inverse problem with the data (see Chapter 7). Using the optimal prior pdf

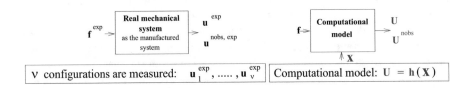

No data	Construction/ Analysis	Data
Use of the prior model	Construction of a prior model: $p_{\mathbf{X}}^{\text{prior}}(\mathbf{x};\mathbf{s})$	Use of the prior model
Sensitivity analysis with respect to hyperparameter $\mathbf{s}$		Construction of an optimal prior model: $p_{\mathbf{X}}^{\text{prior}}(\mathbf{x};\mathbf{s}^{\text{opt}})$
		Construction of a posterior model: $p_{\mathbf{X}}^{\text{post}}(\mathbf{x})$

Fig. 1.9 Summarizing the main steps of the methodology for UQ.

$p_{\mathbf{X}}^{\text{prior}}(\mathbf{x}\,;\mathbf{s}^{\text{opt}})$ for $\mathbf{X}$, the propagation of uncertainties is analyzed by computing the stochastic solution $\mathbf{U}$ and by estimating the QoI related to $\mathbf{U}$. Therefore, the uncertainty quantification has been done.

- a posterior pdf $p_{\mathbf{X}}^{\text{post}}(\mathbf{x})$ of $\mathbf{X}$ can also be carried out as explained in Chapter 7) by using the Bayes method.

Chapter 2
Elements of Probability Theory

Abstract *This chapter deals with fundamental elements of the probability theory, which must be understood for understanding the other chapters of the course. It presents a short overview on the probability distributions, on the second-order vector-valued random variables, and on some important mathematical results concerning the convergence of sequences of random variables and the central limit theorem.*

2.1 Principles of Probability Theory and Vector-Valued Random Variables

2.1.1 Principles of Probability Theory

The *stochastic modeling* of a variable $\mathbf{x}$ with values in $\mathbb{R}^n$ consists in:

1. introducing a set Θ such that each element of Θ is a combination of causes affecting the state of $\mathbf{x}$.
2. endowing set Θ with a σ-algebra $\mathcal{T}$ whose elements are called *events*.

© Springer International Publishing AG 2017

C. Soize, *Uncertainty Quantification*, Interdisciplinary Applied Mathematics 47,
DOI 10.1007/978-3-319-54339-0_2

3. endowing the measurable space $(\Theta, \mathcal{T})$ with a *probability* $\mathcal{P}$ which is a bounded positive measure with mass 1, $\mathcal{P}\{\Theta\} = 1$.

The triplet $(\Theta, \mathcal{T}, \mathcal{P})$ is called a *probability space*.

2.1.2 Conditional Probability and Independent Events

□ *Conditional probability.* Let A_1 and A_2 be in $\mathcal{T}$ such that $\mathcal{P}\{A_2\} > 0$. Then, $\mathcal{P}\{A_1|A_2\}$ that is defined by the equation

$$\mathcal{P}\{A_1 \cap A_2\} = \mathcal{P}\{A_1|A_2\} \times \mathcal{P}\{A_2\}, \tag{2.1}$$

is called the *conditional probability* of A_1 *given* A_2.

□ *Independence of two events.* Two events A_1 and A_2 in $\mathcal{T}$ are *independent* if

$$\mathcal{P}\{A_1 \cap A_2\} = \mathcal{P}\{A_1\} \times \mathcal{P}\{A_2\}. \tag{2.2}$$

If $\mathcal{P}\{A_2\} \neq 0$ and if A_1 is independent of A_2, then

$$\mathcal{P}\{A_1|A_2\} = \mathcal{P}\{A_1\}. \tag{2.3}$$

2.1.3 Random Variable with Values in $\mathbb{R}^n$ and Probability Distribution

□ *Random variable.* A random variable $\mathbf{X} = (X_1, \ldots, X_n)$, defined on the probability space $(\Theta, \mathcal{T}, \mathcal{P})$ with values in $\mathbb{R}^n$ that is endowed with the Borel σ-algebra $\mathcal{B}_n$, is a measurable mapping $\theta \mapsto \mathbf{X}(\theta)$ from Θ into $\mathbb{R}^n$, which means (definition of the measurability) that

$$\forall B \in \mathcal{B}_n \quad , \quad \mathbf{X}^{-1}(B) \in \mathcal{T}, \tag{2.4}$$

in which $\mathbf{X}^{-1}(B) = \{\theta \in \Theta | \mathbf{X}(\theta) \in B\}$ is the subset of $\mathcal{T}$, which is simply denoted by $\{\mathbf{X} \in B\}$ (using a classic abuse of notation).

□ *Realization or sample of the random variable.* For θ fixed in Θ, $\mathbf{X}(\theta)$ that is a deterministic vector in $\mathbb{R}^n$ is called a *realization* or a *sample* of the random variable $\mathbf{X}$. It should be noted that the random variable $\mathbf{X}$ is not a vector in $\mathbb{R}^n$ but is a measurable mapping $\mathbf{X} = \{\theta \mapsto \mathbf{X}(\theta)\}$ from Θ into $\mathbb{R}^n$.

□ *Probability distribution.* The probability distribution $P_{\mathbf{X}}(d\mathbf{x})$ on $\mathbb{R}^n$ of $\mathbf{X}$ is the probability measure (that is to say, a bounded positive measure of mass 1), $B \mapsto P_{\mathbf{X}}(B) = \mathcal{P}\{\mathbf{X}^{-1}(B)\}$, from $\mathcal{B}_n$ into $[0, 1]$ such that

$$0 \leq \int_B P_\mathbf{X}(d\mathbf{x}) = P_\mathbf{X}(B) = \mathcal{P}\{\mathbf{X} \in B\} \leq 1, \tag{2.5}$$

$$\int_{\mathbb{R}^n} P_\mathbf{X}(d\mathbf{x}) = P_\mathbf{X}(\mathbb{R}^n) = \mathcal{P}\{\mathbf{X} \in \mathbb{R}^n\} = 1. \tag{2.6}$$

□ *Marginal probability distribution.* The marginal probability distribution $P_{X_j}(dx_j)$ of X_j with values in $\mathbb{R}$ is such that

$$\forall B_j \in \mathcal{B}_1 \quad , \quad P_{X_j}(B_j) = \mathcal{P}(X_1 \in \mathbb{R}, \dots, X_j \in B_j, \dots, X_n \in \mathbb{R})$$

$$= \int_{x_1 \in \mathbb{R}} \cdots \int_{x_j \in B_j} \cdots \int_{x_n \in \mathbb{R}} P_\mathbf{X}(dx_1, \dots, dx_j, \dots, dx_n). \tag{2.7}$$

□ *Cumulative distribution function.* The cumulative distribution function $\mathbf{x} \mapsto F_\mathbf{X}(\mathbf{x})$ on $\mathbb{R}^n$ with values in $[0,1]$ is defined by

$$F_\mathbf{X}(\mathbf{x}) = \mathcal{P}\{\mathbf{X} \leq \mathbf{x}\} = P_\mathbf{X}\{B_\mathbf{x}\} = \int_{\mathbf{y} \in B_\mathbf{x}} P_\mathbf{X}(d\mathbf{y}), \tag{2.8}$$

with $B_\mathbf{x} =\,]-\infty, x_1] \times \dots \times\,]-\infty, x_n] \in \mathcal{B}_n$, and thus,

$$\mathcal{P}\{\mathbf{X} \leq \mathbf{x}\} = P_\mathbf{X}\{B_\mathbf{x}\} = \int_{-\infty}^{x_1} \cdots \int_{-\infty}^{x_n} dF_\mathbf{X}(\mathbf{y}), \tag{2.9}$$

□ *Probability density function.* The probability density function with respect to $d\mathbf{x}$ (if it exists) is such that
$$P_\mathbf{X}(d\mathbf{x}) = p_\mathbf{X}(\mathbf{x})\, d\mathbf{x}, \tag{2.10}$$
where $\mathbf{x} \mapsto p_\mathbf{X}(\mathbf{x})$ is a function from $\mathbb{R}^n$ into $[0, +\infty[$ and its integrability condition is written as

$$P_\mathbf{X}(\mathbb{R}^n) = \int_{\mathbb{R}^n} P_\mathbf{X}(d\mathbf{x}) = \int_{\mathbb{R}^n} p_\mathbf{X}(\mathbf{x})\, d\mathbf{x} = 1. \tag{2.11}$$

If the cumulative distribution function $F_\mathbf{X}$ is differentiable on $\mathbb{R}^n$, then

$$p_\mathbf{X}(\mathbf{x}) = \frac{\partial^n}{\partial x_1 \dots \partial x_n} F_\mathbf{X}(\mathbf{x}). \tag{2.12}$$

□ *Case of independence of $X_1, \dots, X_n$.* A necessary and sufficient condition for that the real random variables $X_1, \dots, X_n$ be independent is that

$$P_\mathbf{X}(d\mathbf{x}) = P_{X_1}(dx_1) \otimes \dots \otimes P_{X_n}(dx_n), \tag{2.13}$$

which implies that

$$F_{\mathbf{X}}(\mathbf{x}) = F_{X_1}(x_1) \times \ldots \times F_{X_n}(x_n) , \tag{2.14}$$

and, if a probability density function exists, which implies that

$$p_{\mathbf{X}}(\mathbf{x}) = p_{X_1}(x_1) \times \ldots \times p_{X_n}(x_n) . \tag{2.15}$$

□ *Support of the probability distribution.* If $\mathcal{S}_n \subset \mathbb{R}^n$ is the *support* of the probability distribution $P_{\mathbf{X}}(d\mathbf{x})$, this implies that

1. if $B \in \mathcal{B}_n$ is such that $B \cap \mathcal{S}_n = \{\emptyset\}$, then $P_{\mathbf{X}}(B) = 0$.
2. if $\mathbf{x} \in \mathbb{R}^n$ is such that $B_{\mathbf{x}} \cap \mathcal{S}_n = \{\emptyset\}$, then $F_{\mathbf{X}}(\mathbf{x}) = 0$.
3. if there is a density, then $p_{\mathbf{X}}(\mathbf{x}) = 0$ for all $\mathbf{x} \notin \mathcal{S}_n$, and the pdf can be rewritten as

$$p_{\mathbf{X}}(\mathbf{x}) = \mathbb{1}_{\mathcal{S}_n}(\mathbf{x}) f(\mathbf{x}) , \tag{2.16}$$

in which $\mathbf{x} \mapsto \mathbb{1}_{\mathcal{S}_n}(\mathbf{x})$ is the indicator function of the set $\mathcal{S}_n$ and where $\mathbf{x} \mapsto f(\mathbf{x})$ is an integrable function from $\mathcal{S}_n$ into $[0, +\infty[$ such that

$$\int_{\mathcal{S}_n} f(\mathbf{x}) \, d\mathbf{x} = 1 . \tag{2.17}$$

2.1.4 Mathematical Expectation and Integration of $\mathbb{R}^n$-Valued Random Variables

Let $\mathbf{X} = (X_1, \ldots, X_n)$ be a random variable defined on the probability space $(\Theta, \mathcal{T}, \mathcal{P})$ with values in $\mathbb{R}^n$.

□ *Definition of a q-order random variable.* Let $1 \leq q < +\infty$ be an integer. The random variable $\mathbf{X}$ is a q-order random variable if

$$\int_{\Theta} \|\mathbf{X}(\theta)\|^q \, d\mathcal{P}(\theta) = \int_{\mathbb{R}^n} \|\mathbf{x}\|^q \, P_{\mathbf{X}}(d\mathbf{x}) < +\infty . \tag{2.18}$$

□ *Mathematical expectation and fundamental equation.* Let $\mathbf{x} \mapsto \mathbf{h}(\mathbf{x})$ be a mapping from $\mathbb{R}^n$ into $\mathbb{R}^m$ such that $\mathbf{Y} = \mathbf{h}(\mathbf{X})$ is an $\mathbb{R}^m$-valued random variable. Then the mathematical expectation E of $\mathbf{Y}$ is defined by

$$\begin{aligned} E\{\mathbf{Y}\} &= \int_{\mathbb{R}^m} \mathbf{y} \, P_{\mathbf{Y}}(d\mathbf{y}) \\ &= E\{\mathbf{h}(\mathbf{X})\} \\ &= \int_{\mathbb{R}^n} \mathbf{h}(\mathbf{x}) \, P_{\mathbf{X}}(d\mathbf{x}) . \end{aligned} \tag{2.19}$$

The first equation corresponds to the definition of the mathematical expectation $E\{\mathbf{Y}\}$ that is expressed using the probability distribution $P_{\mathbf{Y}}(d\mathbf{y})$ of $\mathbf{Y}$. The second and the third equation correspond to a fundamental method that allows for expressing the mathematical expectation $E\{\mathbf{Y}\}$ of random variable $\mathbf{Y}$ that is transformed by $\mathbf{h}$ from random variable $\mathbf{X}$, by using its probability distribution $P_{\mathbf{X}}(d\mathbf{x})$. It should be noted that such an equation avoids the calculation of the probability distribution $P_{\mathbf{Y}}(d\mathbf{y})$ of $\mathbf{Y}$ for calculating $E\{\mathbf{Y}\}$.

Taking $m = 1$ and $h(\mathbf{x}) = \|\mathbf{x}\|^q$, it can be seen that $\mathbf{X}$ is a *q-order random variable* if $E\{\|\mathbf{X}\|^q\} < +\infty$.

□ *Vector space* $L^0(\Theta, \mathbb{R}^n)$. This is the set of all the $\mathbb{R}^n$-valued random variables defined on $(\Theta, \mathcal{T}, \mathcal{P})$ (set of equivalence classes of the almost surely equal random variables).

□ *Mathematical expectation is a linear operator.* Let $\mathbf{X}$ and $\mathbf{Y}$ be two random variables in $L^0(\Theta, \mathbb{R}^n)$. Then, for all λ and μ in $\mathbb{R}$,

$$E\{\lambda \mathbf{X} + \mu \mathbf{Y}\} = \lambda E\{\mathbf{X}\} + \mu E\{\mathbf{Y}\}. \tag{2.20}$$

□ *Vector space* $L^q(\Theta, \mathbb{R}^n)$. For $q \geq 1$, the vector space $L^q(\Theta, \mathbb{R}^n) \subset L^0(\Theta, \mathbb{R}^n)$ of all the q-order $\mathbb{R}^n$-valued random variables is a Banach space (complete normed vector space) with the norm,

$$\|\mathbf{X}\|_{\Theta,q} = (E\{\|\mathbf{X}\|^q\})^{1/q}. \tag{2.21}$$

□ *Vector space* $L^2(\Theta, \mathbb{R}^n)$. The vector space $L^2(\Theta, \mathbb{R}^n) \subset L^0(\Theta, \mathbb{R}^n)$ of all the second-order random variables with values in $\mathbb{R}^n$ is a Hilbert space for the inner product and the associated norm,

$$<\mathbf{X}, \mathbf{Y}>_\Theta = E\{<\mathbf{X}, \mathbf{Y}>\} = \int_{\mathbb{R}^n} \int_{\mathbb{R}^n} <\mathbf{x}, \mathbf{y}> P_{\mathbf{XY}}(d\mathbf{x}, d\mathbf{y}),$$
$$\|\mathbf{X}\|_\Theta = (E\{\|\mathbf{X}\|^2\})^{1/2}, \tag{2.22}$$

in which $P_{\mathbf{XY}}(d\mathbf{x}, d\mathbf{y})$ is the joint probability distribution of $\mathbf{X}$ and $\mathbf{Y}$.

For simplifying the writing, $\|\mathbf{X}\|_{\Theta,2}$ will be simply denoted by $\|\mathbf{X}\|_\Theta$.

2.1.5 Characteristic Function of an $\mathbb{R}^n$-Valued Random Variable

The characteristic function of the $\mathbb{R}^n$-valued random variable $\mathbf{X}$ is the continuous function $\mathbf{u} \mapsto \Phi_{\mathbf{X}}(\mathbf{u})$ on $\mathbb{R}^n$ with values in $\mathbb{C}$, defined by

$$\Phi_{\mathbf{X}}(\mathbf{u}) = E\{\exp(i < \mathbf{u}, \mathbf{X} >)\}$$

$$= \int_{\mathbb{R}^n} e^{i<\mathbf{u},\mathbf{x}>} P_{\mathbf{X}}(d\mathbf{x}). \tag{2.23}$$

The characteristic function is a mean for describing the probability distribution $P_{\mathbf{X}}(d\mathbf{x})$ of $\mathbb{R}^n$-valued random variable $\mathbf{X}$. If $\Phi_{\mathbf{X}}$ is integrable or square integrable on $\mathbb{R}^n$, then $P_{\mathbf{X}}(d\mathbf{x}) = p_{\mathbf{X}}(\mathbf{x}) \, d\mathbf{x}$ and

$$p_{\mathbf{X}}(\mathbf{x}) = \frac{1}{(2\pi)^n} \int_{\mathbb{R}^n} e^{-i<\mathbf{u},\mathbf{x}>} \Phi_{\mathbf{X}}(\mathbf{u}) \, d\mathbf{u}. \tag{2.24}$$

2.1.6 Moments of an $\mathbb{R}^n$-Valued Random Variable

Let $\boldsymbol{\alpha} = (\alpha_1, \ldots, \alpha_n) \in \mathbb{N}^n$ be any multi-index of order n with α_k a positive or null integer for each $k \in \{1, \ldots, n\}$. The $\boldsymbol{\alpha}$-order moment of the random variable $\mathbf{X}$ is defined by

$$m_{\boldsymbol{\alpha}} = E\{X_1^{\alpha_1} \times \ldots \times X_n^{\alpha_n}\} = \int_{\mathbb{R}^n} x_1^{\alpha_1} \times \ldots \times x_n^{\alpha_n} P_{\mathbf{X}}(d\mathbf{x}). \tag{2.25}$$

If $\mathbf{X}$ is a q-order random variable, then $|m_{\boldsymbol{\alpha}}| < +\infty$ for any multi-index $\boldsymbol{\alpha}$ such that $|\boldsymbol{\alpha}| = \alpha_1 + \ldots + \alpha_n \leq q$, and

$$\left\{ \frac{\partial^{\alpha_1}}{\partial u_1^{\alpha_1}} \times \ldots \times \frac{\partial^{\alpha_n}}{\partial u_n^{\alpha_n}} \Phi_{\mathbf{X}}(\mathbf{u}) \right\}_{\mathbf{u}=0} = m_{\boldsymbol{\alpha}} \, i^{|\boldsymbol{\alpha}|}. \tag{2.26}$$

2.1.7 Summary: How the Probability Distribution of a Random Vector X Can Be Described

The probability distribution of a random vector $\mathbf{X}$ can be described:

- either by the probability distribution $P_{\mathbf{X}}(d\mathbf{x})$, which is a bounded positive measure on $\mathbb{R}^n$ with mass 1.
- either by the cumulative distribution function $F_{\mathbf{X}}(\mathbf{x})$, which is a function from $\mathbb{R}^n$ into $[0, 1]$.
- either, if it exists, by the *probability density function* (pdf) $p_{\mathbf{X}}(\mathbf{x})$, which is a function from $\mathbb{R}^n$ into $[0, +\infty[$ with the normalization condition $\int_{\mathbb{R}^n} p_{\mathbf{X}}(\mathbf{x}) \, d\mathbf{x} = 1$.
- or by the characteristic function $\mathbf{u} \mapsto \Phi_{\mathbf{X}}(\mathbf{u}) = E\{\exp(i < \mathbf{u}, \mathbf{X} >)\}$.

Example. Let us consider the following probability distribution: $P_{\mathbf{X}}(d\mathbf{x}) = \delta_{x_1^0}(x_1) \otimes \delta_{x_2^0}(x_2) \otimes \{\mathbb{1}_{\mathbb{R}^+}(x_3) \, f(x_3) \, dx_3\} \otimes \{g(x_4, \ldots, x_n) \, dx_4 \ldots dx_n\}$. It can be seen that such a probability distribution has no density with respect to $d\mathbf{x}$, that the coordinates $X_1 = x_1^0$ and $X_2 = x_2^0$ are deterministic, that the pdf of X_3 is $\mathbb{1}_{\mathbb{R}^+}(x_3)\, f(x_3)$ (its support is $\mathbb{R}^+$), that the pdf of $\mathbf{Y} = (X_4, \ldots, X_n)$ is $g(x_4, \ldots, x_n)$), and finally, that the random variables X_3 and $\mathbf{Y}$ are independent.

2.2 Second-Order Vector-Valued Random Variables

Let $\mathbf{X}$ and $\mathbf{Y}$ be in $L^2(\Theta, \mathbb{R}^n)$-space, which means that $\mathbf{X}$ and $\mathbf{Y}$ are two second-order random variables with values in $\mathbb{R}^n$. Consequently, we have

$$\|\mathbf{X}\|_\Theta^2 = E\{\|\mathbf{X}\|^2\} < +\infty \quad , \quad \|\mathbf{Y}\|_\Theta^2 = E\{\|\mathbf{Y}\|^2\} < +\infty . \tag{2.27}$$

2.2.1 Mean Vector and Centered Random Variable

The mean vector of $\mathbf{X}$ is defined by

$$\mathbf{m}_{\mathbf{X}} = E\{\mathbf{X}\} \in \mathbb{R}^n , \tag{2.28}$$

and its components are such that

$$\{\mathbf{m}_{\mathbf{X}}\}_j = E\{X_j\} = m_{X_j} = \int_{\mathbb{R}^n} x_j P_{\mathbf{X}}(d\mathbf{x}) = \int_{\mathbb{R}} x_j P_{X_j}(dx_j) . \tag{2.29}$$

If $\mathbf{X}$ is not centered ($\mathbf{m}_{\mathbf{X}} \neq \mathbf{0}$), then

$$\mathbf{Y} = \mathbf{X} - \mathbf{m}_{\mathbf{X}} , \tag{2.30}$$

is a *centered* random variable because $\mathbf{m}_{\mathbf{Y}} = \mathbf{0}$. Therefore, a non-centered second-order random vector can always be centered by using Equation (2.30).

2.2.2 Correlation Matrix

The correlation matrix of $\mathbf{X}$ is defined by

$$[R_{\mathbf{X}}] = E\{\mathbf{X}\,\mathbf{X}^T\} \in \mathbb{M}_n^{+0}(\mathbb{R}) . \tag{2.31}$$

The entries of the semidefinite-positive matrix $[R_{\mathbf{X}}]$ are such that

$$\begin{aligned}
[R_{\mathbf{X}}]_{jk} &= E\{X_j\, X_k\} \\
&= \int_{\mathbb{R}^n} x_j\, x_k P_{\mathbf{X}}(d\mathbf{x}) = \int_{\mathbb{R}} \int_{\mathbb{R}} x_j\, x_k P_{X_j X_k}(dx_j, dx_k) .
\end{aligned} \tag{2.32}$$

The trace of the correlation matrix is finite and is such that

$$\text{tr}[R_{\mathbf{X}}] = E\{<\mathbf{X}, \mathbf{X}>\} = ||\mathbf{X}||_\Theta^2 < +\infty .\tag{2.33}$$

2.2.3 Covariance Matrix

□ The *covariance matrix* of $\mathbf{X}$ is defined by

$$[C_{\mathbf{X}}] = E\{(\mathbf{X} - \mathbf{m}_{\mathbf{X}})(\mathbf{X} - \mathbf{m}_{\mathbf{X}})^T\} = [R_{\mathbf{X}}] - \mathbf{m}_{\mathbf{X}}\mathbf{m}_{\mathbf{X}}^T \in \mathbb{M}_n^{+0}(\mathbb{R}) .\tag{2.34}$$

□ The *second-order quantities of the real-valued random variable* X_j are defined as follows:

  – Mean value: $m_{X_j} = E\{X_j\}$.
  – Second-order moment: $E\{X_j^2\} = [R_{\mathbf{X}}]_{jj}$.
  – Variance: $\sigma_{X_j}^2 = E\{(X_j - m_{X_j})^2\} = [C_{\mathbf{X}}]_{jj} = E\{X_j^2\} - m_{X_j}^2$.
  – Standard deviation: $\sigma_{X_j} = (\sigma_{X_j}^2)^{1/2}$.
  – Coefficient of variation: $\delta_X = \sigma_{X_j}/|m_{X_j}|$ for $m_{X_j} \neq 0$.

□ The *correlation coefficient of the real-valued random variables* X_j and X_k is defined by $r_{X_j X_k} = [C_{\mathbf{X}}]_{jk}/(\sigma_{X_j}\sigma_{X_k})^{-1}$ and is such that $-1 \leq r_{X_j X_k} \leq 1$.

□ The *orthogonality and the noncorrelation of the real-valued random variables* X_j and X_k are defined as follows.

  – Orthogonality: $<X_j, X_k>_\Theta = E\{X_j X_k\} = 0$.
  – Noncorrelation: $r_{X_j X_k} = 0$.

2.2.4 Cross-Correlation Matrix

□ The *cross-correlation matrix* of $\mathbf{X}$ and $\mathbf{Y}$ is defined by

$$[R_{\mathbf{XY}}] = E\{\mathbf{X}\mathbf{Y}^T\} \in \mathbb{M}_n(\mathbb{R}) \quad , \quad \text{tr}[R_{\mathbf{XY}}] = <\mathbf{X}, \mathbf{Y}>_\Theta .\tag{2.35}$$

□ *Orthogonality.* The random variables $\mathbf{X}$ and $\mathbf{Y}$ are orthogonal if

$$<\mathbf{X}, \mathbf{Y}>_\Theta = \text{tr}[R_{\mathbf{XY}}] = 0 .\tag{2.36}$$

2.2.5 Cross-Covariance Matrix

- The *cross-covariance matrix* of $\mathbf{X}$ and $\mathbf{Y}$ is defined by

$$[C_{\mathbf{XY}}] = E\{(\mathbf{X} - \mathbf{m}_{\mathbf{X}})(\mathbf{Y} - \mathbf{m}_{\mathbf{Y}})^T\} = [R_{\mathbf{XY}}] - \mathbf{m}_{\mathbf{X}}\mathbf{m}_{\mathbf{Y}}^T \in \mathbb{M}_n(\mathbb{R}). \quad (2.37)$$

$$\mathrm{tr}[C_{\mathbf{XY}}] = <\mathbf{X},\mathbf{Y}>_\Theta - <\mathbf{m}_{\mathbf{X}},\mathbf{m}_{\mathbf{Y}}> . \quad (2.38)$$

- *Noncorrelation* of $\mathbf{X}$ and $\mathbf{Y}$: $\mathrm{tr}[C_{\mathbf{XY}}] = 0$.

2.2.6 Summary: What Are the Second-Order Quantities That Describe Second-Order Random Vectors

If $\mathbf{X}$ and $\mathbf{Y}$ belong to $L^2(\Theta, \mathbb{R}^n)$, then the second-order quantities are the

- mean vector, $\mathbf{m}_{\mathbf{X}} \in \mathbb{R}^n$.
- correlation matrix, $[R_{\mathbf{X}}] \in \mathbb{M}_n^{+0}(\mathbb{R})$.
- covariance matrix, $[C_{\mathbf{X}}] \in \mathbb{M}_n^{+0}(\mathbb{R})$.
- cross-correlation matrix, $[R_{\mathbf{XY}}] \in \mathbb{M}_n(\mathbb{R})$.
- cross-covariance matrix, $[C_{\mathbf{XY}}] \in \mathbb{M}_n(\mathbb{R})$.

2.3 Markov and Tchebychev Inequalities

2.3.1 Markov Inequality

Let Y be an $\mathbb{R}^+$-valued random variable defined on $(\Theta, \mathcal{T}, \mathcal{P})$, with a mean value $E\{Y\} > 0$. Then, for all positive number $\varepsilon > 0$,

$$\mathcal{P}\{Y \geq \varepsilon\} \leq \frac{E\{Y\}}{\varepsilon}. \quad (2.39)$$

- *Application 1.* If $\mathbf{X}$ is an $\mathbb{R}^n$-valued random variable, defined on $(\Theta, \mathcal{T}, \mathcal{P})$, and if $\mathbf{x} \mapsto h(\mathbf{x})$ is a mapping from $\mathbb{R}^n$ into $\mathbb{R}$, taking $Y = |h(\mathbf{X})|$ yields

$$\mathcal{P}\{|h(\mathbf{X})| \geq \varepsilon\} \leq \frac{1}{\varepsilon}\int_{\mathbb{R}^n} |h(\mathbf{x})| \, P_{\mathbf{X}}(d\mathbf{x}). \quad (2.40)$$

- *Application 2.* Let Z be a random variable defined on $(\Theta, \mathcal{T}, \mathcal{P})$, with values in $\mathbb{R}$, and let $y \mapsto f(y)$ be a function defined on $\mathbb{R}^+$, monotone and not decreasing. Then for all $\varepsilon > 0$,

$$\mathcal{P}\{|Z| \geq \varepsilon\} \leq \frac{E\{f(|Z|)\}}{f(\varepsilon)}. \quad (2.41)$$

2.3.2 Tchebychev Inequality

The Tchebychev inequality is deduced from Equation (2.41).

□ Let X be a second-order $\mathbb{R}$-valued random variable defined on $(\Theta, \mathcal{T}, \mathcal{P})$, with the mean value m_X and with the variance σ_X^2. Then, for all $\varepsilon > 0$,

$$\mathcal{P}\{|X - m_X| \geq \varepsilon\} \leq \frac{\sigma_X^2}{\varepsilon^2}, \qquad (2.42)$$

or, if $m_X \neq 0$, for all $k > 0$, $\mathcal{P}\{|X - m_X| \geq k\,|m_X|\} \leq \frac{\delta_X^2}{k^2}$.

□ Let $\mathbf{X}$ be a second-order $\mathbb{R}^n$-valued random variable defined on $(\Theta, \mathcal{T}, \mathcal{P})$, with the mean value $\mathbf{m_X}$ and with the covariance matrix $[C_{\mathbf{X}}]$. Then, for all $\varepsilon > 0$,

$$\mathcal{P}\{\|\mathbf{X} - \mathbf{m_X}\| \geq \varepsilon\} \leq \frac{\mathrm{tr}[C_{\mathbf{X}}]}{\varepsilon^2}. \qquad (2.43)$$

Equation (2.43) shows the following important result in which the notion of the convergence in probability is defined in Section 2.7.2.

▷ For $n \geq 1$ and for any probability distribution of the second-order random variable $\mathbf{X}$, if $\sum_{j=1}^{n} \sigma_{X_j}^2 \to 0$, then $\mathbf{X} \to \mathbf{m_X}$ (deterministic vector) in probability.

2.4 Examples of Probability Distributions

2.4.1 Poisson Distribution on $\mathbb{R}$ with Parameter $\lambda \in \mathbb{R}^+$

The Poisson distribution on $\mathbb{R}$ with parameter $\lambda \in \mathbb{R}^+$ is the probability distribution of a *discrete* random variable X taking its value in $\mathbb{N}$ with a nonzero probability, which is written as

$$P_X(dx) = \sum_{k=0}^{+\infty} (k!)^{-1} \lambda^k e^{-\lambda} \delta_k(x). \qquad (2.44)$$

The graph of the Poisson distribution $k \mapsto \mathrm{Prob}\{X = k\}$ is displayed in Figure 2.1 for $\lambda = 4$. The characteristic function is such that $\Phi_X(u) = \exp\{\lambda(e^{iu} - 1)\}$ and the mean value is $m_X = \lambda$. The second-order moment is $E\{X^2\} = \lambda(\lambda + 1)$ and the variance is $\sigma_X^2 = \lambda$. For B_k such that $B_k \cap \mathbb{N} = \{k\}$, we have $P_X(B_k) = \int_{B_k} P_X(dx) = \frac{\lambda^k}{k!} e^{-\lambda}$.

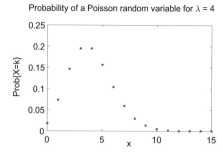

Fig. 2.1 Graph of the Poisson distribution $k \mapsto \mathrm{Prob}\{X = k\}$ for $\lambda = 4$.

2.4.2 Gaussian (or Normal) Distribution on $\mathbb{R}^n$

☐ *Definition.* The probability distribution $P_{\mathbf{X}}(d\mathbf{x})$ of a Gaussian second-order $\mathbb{R}^n$-valued random variable $\mathbf{X}$, with a mean vector $\mathbf{m_X} \in \mathbb{R}^n$ and with a covariance matrix $[C_{\mathbf{X}}] \in \mathbb{M}_n^{+0}(\mathbb{R})$, is defined by the characteristic function on $\mathbb{R}^n$,

$$\Phi_{\mathbf{X}}(\mathbf{u}) = \exp\left\{i < \mathbf{m_X}, \mathbf{u} > -\frac{1}{2} < [C_{\mathbf{X}}]\mathbf{u}, \mathbf{u} >\right\}. \qquad (2.45)$$

☐ *Case of a positive-definite covariance matrix.* If $[C_{\mathbf{X}}] \in \mathbb{M}_n^+(\mathbb{R})$, then $P_{\mathbf{X}}(d\mathbf{x}) = p_{\mathbf{X}}(\mathbf{x})\,d\mathbf{x}$ is defined by the pdf,

$$p_{\mathbf{X}}(\mathbf{x}) = (2\pi)^{-n/2}(\det[C_{\mathbf{X}}])^{-1/2}e^{-\frac{1}{2}<[C_{\mathbf{X}}]^{-1}(\mathbf{x}-\mathbf{m_X}),(\mathbf{x}-\mathbf{m_X})>}. \qquad (2.46)$$

☐ The *normalized Gaussian probability distribution* (also called the *canonical Gaussian measure*) corresponds to $\mathbf{m_X} = \mathbf{0}$ and $[C_{\mathbf{X}}] = [I_n]$, and is written as $P_{\mathbf{X}}(d\mathbf{x}) = p_{\mathbf{X}}(\mathbf{x})\,d\mathbf{x}$ in which the *normalized Gaussian pdf* is defined by

$$p_{\mathbf{X}}(\mathbf{x}) = (2\pi)^{-n/2}\exp\left\{-\frac{1}{2}\|\mathbf{x}\|^2\right\}. \qquad (2.47)$$

For $n = 1$, the normalized Gaussian distribution is defined by the normalized Gaussian pdf that is written as $p_X(x) = (2\pi)^{-1/2}\exp\{-\frac{1}{2}x^2\}$.

☐ *Illustration.* The graph of the pdf of a Gaussian $\mathbb{R}^2$-valued random variable $\mathbf{X}$ for which

$$\mathbf{m_X} = (2, 1) \quad \text{and} \quad [C_{\mathbf{X}}] = \begin{bmatrix} 0.4 & 0.2 \\ 0.2 & 1.5 \end{bmatrix}, \qquad (2.48)$$

is displayed in Figure 2.2.

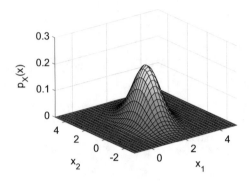

Fig. 2.2 Graph of the pdf of a Gaussian $\mathbb{R}^2$-valued random variable **X**.

2.5 Linear and Nonlinear Transformations of Random Variables

2.5.1 A Method for a Nonlinear Bijective Mapping

Let $\mathbf{x} \mapsto \mathbf{y} = \mathbf{h}(\mathbf{x}) = (h_1(\mathbf{x}), \dots, h_n(\mathbf{x}))$ be a bijective mapping (one-to-one mapping) from Ω_n onto Ω'_n (two any open subsets of $\mathbb{R}^n$), such that the inverse mapping $\mathbf{y} \mapsto \mathbf{x} = \mathbf{h}^{-1}(\mathbf{y})$ is continuously differentiable from Ω'_n into Ω_n, and such that, $\forall \mathbf{y} \in \Omega'_n$, the Jacobian matrix $[J(\mathbf{y})]_{jk} = \frac{\partial}{\partial y_k} h_j^{-1}(\mathbf{y})$ is invertible.

Let $\mathbf{X} = (X_1, \dots, X_n)$ be an $\mathbb{R}^n$-valued random vector whose pdf $p_\mathbf{X}(\mathbf{x})$ is a continuous function on Ω_n. Then, the probability distribution of the $\mathbb{R}^n$-valued random variable $\mathbf{Y} = \mathbf{h}(\mathbf{X})$ admits a density on Ω'_n, which is written as

$$p_\mathbf{Y}(\mathbf{y}) = p_\mathbf{X}(\mathbf{h}^{-1}(\mathbf{y})) \, |\det[J(\mathbf{y})]| \,. \tag{2.49}$$

□ *Example 1. Orthogonal transformation.* The pdf of $\mathbf{Y} = \mathbf{h}(\mathbf{X}) = [H]\mathbf{X}$, in which $[H][H]^T = [H]^T[H] = [I_n]$, is written as

$$p_\mathbf{Y}(\mathbf{y}) = p_\mathbf{X}([H]^T\mathbf{y}) \,. \tag{2.50}$$

□ *Example 2. Sum of two random variables.* Let $\mathbf{X} = (X_1, X_2)$ be an $\mathbb{R}^2$-valued random variable whose pdf is $p_{X_1 X_2}(x_1, x_2)$. Then the pdf $p_{Y_1}(y_1)$ of the real-valued random variable $Y_1 = X_1 + X_2$ is written as

$$p_{Y_1}(y_1) = \int_{\mathbb{R}} p_{X_1 X_2}(y_1 - y_2, y_2) \, dy_2 \,. \tag{2.51}$$

If X_1 and X_2 are independent, then $p_{Y_1} = p_{X_1} * p_{X_2}$ (convolution product).

2.5.2 *Method of the Characteristic Function*

The characteristic function, $\mathbf{v} \mapsto \Phi_{\mathbf{Y}}(\mathbf{v})$ on $\mathbb{R}^m$, of the $\mathbb{R}^m$-valued random variable

$$\mathbf{Y} = \mathbf{h}(\mathbf{X}),\tag{2.52}$$

in which $\mathbf{x} \mapsto \mathbf{y} = \mathbf{h}(\mathbf{x})$ is a given measurable mapping from $\mathbb{R}^n$ into $\mathbb{R}^m$, is

$$\Phi_{\mathbf{Y}}(\mathbf{v}) = E\{\exp(i <\mathbf{v}, \mathbf{h}(\mathbf{X}) >)\} = \int_{\mathbb{R}^n} e^{i<\mathbf{v},\mathbf{h}(\mathbf{x})>}\, P_{\mathbf{X}}(d\mathbf{x}),\tag{2.53}$$

where $P_{\mathbf{X}}(d\mathbf{x})$ is the probability distribution of the $\mathbb{R}^n$-valued random variable $\mathbf{X}$.

☐ *Example 1. Case of an affine transformation from $\mathbb{R}^n$ into $\mathbb{R}^m$.* The characteristic function, $\mathbf{v} \mapsto \Phi_{\mathbf{Y}}(\mathbf{v})$ on $\mathbb{R}^m$, of the $\mathbb{R}^m$-valued random variable

$$\mathbf{Y} = [A]\mathbf{X} + \mathbf{b},\tag{2.54}$$

in which $\mathbf{b} \in \mathbb{R}^m$ and $[A] \in \mathbb{M}_{m,n}(\mathbb{R})$, is

$$\Phi_{\mathbf{Y}}(\mathbf{v}) = e^{i<\mathbf{v},\mathbf{b}>}\Phi_{\mathbf{X}}([A]^T\mathbf{v}),\tag{2.55}$$

where $\mathbf{u} \mapsto \Phi_{\mathbf{X}}(\mathbf{u})$ is the characteristic function of $\mathbf{X}$ on $\mathbb{R}^n$.

☐ *Example 2. Fundamental example of an affine transformation of an $\mathbb{R}^n$-valued Gaussian random variable.* Let $\mathbf{X}$ be a Gaussian second-order $\mathbb{R}^n$-valued random variable with the mean vector $\mathbf{m}_{\mathbf{X}} \in \mathbb{R}^n$ and with the covariance matrix $[C_{\mathbf{X}}] \in \mathbb{M}_n^{+0}(\mathbb{R})$. The characteristic function, $\mathbf{v} \mapsto \Phi_{\mathbf{Y}}(\mathbf{v})$ on $\mathbb{R}^m$, of the $\mathbb{R}^m$-valued random variable

$$\mathbf{Y} = [A]\mathbf{X} + \mathbf{b},\tag{2.56}$$

in which $\mathbf{b} \in \mathbb{R}^m$ and $[A] \in \mathbb{M}_{m,n}(\mathbb{R})$, is

$$\Phi_{\mathbf{Y}}(\mathbf{v}) = \exp\left\{i <\mathbf{m}_{\mathbf{Y}}, \mathbf{v}> - \frac{1}{2} <[C_{\mathbf{Y}}]\mathbf{v}, \mathbf{v}>\right\},\tag{2.57}$$

where

$$\mathbf{m}_{\mathbf{Y}} = [A]\mathbf{m}_{\mathbf{X}} + \mathbf{b} \in \mathbb{R}^m \quad, \quad [C_{\mathbf{Y}}] = [A][C_{\mathbf{X}}][A]^T \in \mathbb{M}_m^{+0}(\mathbb{R}).\tag{2.58}$$

We have then proved the following theorem in finite dimension, but which also holds in infinite dimension.

Theorem 2.1. *Any vector-valued affine transformation of a vector-valued Gaussian random variable is a vector-valued Gaussian random variable.*

2.5.3 Summary: What Is the Efficiency of These Tools for UQ

☐ Two *fundamental methods* (that cannot be ignored) have been presented:

- (i) a method for a nonlinear bijective mapping (which is based on the use of a hypothesis that is very restrictive for the applications).
- (ii) the method of the characteristic function (which is, *a priori*, the most general method).

☐ These two methods are important for understanding and obtaining fundamental theoretical results for the linear or the affine transformations, but, in general, for nonlinear transformations, these methods are not constructive tools for UQ problems. Why?

- For method (i), in general, the nonlinear transformations are not bijective. If the nonlinear transformation is bijective, the computation of $\mathbf{h}^{-1}$ and its derivatives cannot be done (in particular, in high dimension).
- Method (ii) requires an integration on $\mathbb{R}^n$ for which there is, in general, no explicit solution (such an approach cannot be used, except for applications in dimension 1 or in dimension 2).

☐ The statistical computational methods adapted to UQ problems will be presented in Chapter 4 using the theoretical tools presented in Chapter 3.

2.6 Second-Order Calculations

Let $\mathbf{X}$ be a second-order $\mathbb{R}^n$-valued random variable, with the mean vector $\mathbf{m_X}$, with correlation matrix $[R_\mathbf{X}]$, and with the covariance matrix $[C_\mathbf{X}]$. Let $\mathbf{h}$ be a measurable mapping from $\mathbb{R}^n$ into $\mathbb{R}^m$. Then, the second-order quantities of the $\mathbb{R}^m$-valued random variable

$$\mathbf{Y} = [A]\mathbf{X} + \mathbf{b}, \qquad (2.59)$$

in which $\mathbf{b} \in \mathbb{R}^m$ and $[A] \in \mathbb{M}_{m,n}(\mathbb{R})$, are

$$\mathbf{m_Y} = E\{\mathbf{Y}\} = [A]\mathbf{m_X} + \mathbf{b}, \qquad (2.60)$$

$$[R_\mathbf{Y}] = E\{\mathbf{YY}^T\} = [A][R_\mathbf{X}][A]^T + [A]\mathbf{m_X}\mathbf{b}^T + \mathbf{b}\mathbf{m_X}^T[A]^T + \mathbf{b}\mathbf{b}^T, \quad (2.61)$$

$$[C_\mathbf{Y}] = [R_\mathbf{Y}] - \mathbf{m_Y}\mathbf{m_Y}^T = [A][C_\mathbf{X}][A]^T, \qquad (2.62)$$

▷ *For such an affine transformation, the second-order quantities can be calculated without knowing the probability distributions of the random variables. This last statement is not true for a nonlinear transformation.*

2.7 Convergence of Sequences of Random Variables

Let $\mathbf{X}$ and $\{\mathbf{X}_m\}_{m \in \mathbb{N}}$ be $\mathbb{R}^n$-valued random variables, defined on $(\Theta, \mathcal{T}, \mathcal{P})$.

2.7.1 Mean-Square Convergence or Convergence in $L^2(\Theta, \mathbb{R}^n)$

☐ A sequence $\{\mathbf{X}_m\}_{m \in \mathbb{N}}$ of random variables in $L^2(\Theta, \mathbb{R}^n)$ *converges in mean square* to a second-order random variable $\mathbf{X}$ belonging to $L^2(\Theta, \mathbb{R}^n)$ if

$$\lim_{m \to +\infty} ||\mathbf{X}_m - \mathbf{X}||_\Theta = 0, \tag{2.63}$$

Since $L^2(\Theta, \mathbb{R}^n)$ is a Hilbert space, it is complete. Therefore, any Cauchy sequence in $L^2(\Theta, \mathbb{R}^n)$ is convergent in $L^2(\Theta, \mathbb{R}^n)$ (in norm). It can easily be proved by the following criterion.

☐ **Criterion.** *The sequence $\{\mathbf{X}_m\}_{m \in \mathbb{N}}$ of random variables in $L^2(\Theta, \mathbb{R}^n)$ converges in mean square to a random variable $\mathbf{X}$ belonging to $L^2(\Theta, \mathbb{R}^n)$ with $||\mathbf{X}||_\Theta^2 = x$, if and only if,*

$$\lim_{m, m' \to +\infty} <\mathbf{X}_m, \mathbf{X}_{m'}>_\Theta = x. \tag{2.64}$$

2.7.2 Convergence in Probability or Stochastic Convergence

A sequence $\{\mathbf{X}_m\}_{m \in \mathbb{N}}$ of random vectors in $L^0(\Theta, \mathbb{R}^n)$ *converges in probability* to an $\mathbb{R}^n$-valued random variable $\mathbf{X}$ if, for every $\varepsilon > 0$,

$$\lim_{m \to +\infty} \mathcal{P}\{||\mathbf{X}_m - \mathbf{X}|| \geq \varepsilon\} = 0. \tag{2.65}$$

The convergence in probability is also called the *stochastic convergence* or the *convergence in measure*.

2.7.3 Almost-Sure Convergence

A sequence $\{\mathbf{X}_m\}_{m \in \mathbb{N}}$ of random vectors in $L^0(\Theta, \mathbb{R}^n)$ *converges almost surely* (a.s.) to an $\mathbb{R}^n$-valued random variable $\mathbf{X}$ if,

$$\mathcal{P}\{\lim_{m \to +\infty} \mathbf{X}_m \neq \mathbf{X}\} = 0, \tag{2.66}$$

which means that the subset $A_0 = \{\theta \in \Theta \mid \lim_{m \to +\infty} \mathbf{X}_m(\theta) \neq \mathbf{X}(\theta)\}$ is
$\mathcal{P}$-negligible, *i.e.*, is such that $\mathcal{P}(A_0) = 0$.

2.7.4 Convergence in Probability Distribution

A sequence $\{\mathbf{X}_m\}_{m \in \mathbb{N}}$ of random variables in $L^0(\Theta, \mathbb{R}^n)$ *converges in probability distribution* to an $\mathbb{R}^n$-valued random variable $\mathbf{X}$ with probability distribution $P_{\mathbf{X}}(d\mathbf{x})$ if, the sequence $\{P_{\mathbf{X}_m}(d\mathbf{x})\}_{m \in \mathbb{N}}$ of probability distributions converges weakly to $P_{\mathbf{X}}(d\mathbf{x})$, *i.e.* if, for all real continuous functions f on $\mathbb{R}^n$ such that $f(\mathbf{x}) \to 0$ as $\|\mathbf{x}\| \to +\infty$, we have

$$\lim_{m \to +\infty} \int_{\mathbb{R}^n} f(\mathbf{x}) P_{\mathbf{X}_m}(d\mathbf{x}) = \int_{\mathbb{R}^n} f(\mathbf{x}) P_{\mathbf{X}}(d\mathbf{x}). \qquad (2.67)$$

2.7.5 Summary: What Are the Relationships Between the Four Types of Convergence

The relationships between the four types of convergence are summarized hereinafter:

Almost-sure
convergence ↘
                        Convergence in   Convergence in
                        probability   ⟶   probability
Mean-square ↗   distribution
convergence

▷ *The understanding of the modes of convergence is important for analyzing the convergence of representations of the stochastic models introduced in Uncertainty Quantification, and for analyzing the convergence of the stochastic solvers.*

2.8 Central Limit Theorem and Computation of Integrals in High Dimensions by the Monte Carlo Method

2.8.1 Central Limit Theorem

The central limit theorem is of the type of the *law of large numbers* and gives the behavior of a sum of *independent* random variables. Two theorems will be given:

- a first one for independent random variables with the *same* probability distribution.
- a second one for independent random variables with probability distributions that can be *different*.

▷ *The central limit theorem is a fundamental tool for the mathematical statistics, which is useful for studying convergence properties of sequences of random estimators, and consequently, is a fundamental tool in computational statistics for Uncertainty Quantification.*

Theorem 2.2. *Let $\{X_\ell\}_{\ell \geq 1}$ be a sequence of independent random variables in $L^2(\Theta, \mathbb{R}^n)$ with the same probability distribution, with the mean vector $\boldsymbol{m} = E\{X_\ell\}$ and with the covariance matrix $[C] = E\{(X_\ell - \boldsymbol{m})(X_\ell - \boldsymbol{m})^T\}$. Let Y_ν be the $\mathbb{R}^n$-valued random variable defined by*

$$Y_\nu = \frac{1}{\sqrt{\nu}} \sum_{\ell=1}^{\nu} (X_\ell - \boldsymbol{m}). \tag{2.68}$$

Then, for $\nu \to +\infty$, the sequence $\{Y_\nu\}_\nu$ of random variables converges in probability distribution to a Gaussian second-order centered $\mathbb{R}^n$-valued random variable with a covariance matrix $[C]$.

▷ *Use in dimension $n = 1$.* Theorem 2.2 implies that the sequence $\{Z_\nu\}_{\nu \geq 1}$,

$$Z_\nu = \frac{1}{\sqrt{\nu}} \left(\frac{X_1 + \ldots + X_\nu - \nu m}{\sigma} \right), \tag{2.69}$$

in which $m = E\{X_\ell\}$ and $\sigma^2 = E\{(X_\ell - m)^2\}$, converges in probability distribution to a Gaussian, second-order and centered, real-valued random variable Z with a unit variance.

Theorem 2.3. *Let $\{X_\ell\}_{\ell \geq 1}$ be a sequence of independent real-valued random variables in $L^2(\Theta, \mathbb{R})$, with a mean vector $m_\ell = E\{X_\ell\}$ and with a variance $\sigma_\ell^2 = E\{(X_\ell - m_\ell)^2\}$, and for which the cumulative distribution function $F_\ell(x) = P\{X_\ell \leq x\}$ can be different. Let $s_\nu^2 = \sum_{\ell=1}^{\nu} \sigma_\ell^2$. If the random variables $(X_\ell - m_\ell)/s_\nu$ are uniformly small with a large probability (Lindenberg's condition), i.e., if for all $\varepsilon > 0$,*

$$\lim_{\nu \to +\infty} \left\{ \frac{1}{s_\nu^2} \sum_{\ell=1}^{\nu} \int_{|x - m_\ell| > \varepsilon s_\nu} (x - m_\ell)^2 \, dF_\ell(x) \right\} = 0, \tag{2.70}$$

then the sequence of random variables

$$Y_\nu = \frac{1}{s_\nu} \sum_{\ell=1}^{\nu} (X_\ell - m_\ell), \tag{2.71}$$

converges in probability distribution to a normalized Gaussian real-valued random variable (centered and with a unit variance).

2.8.2 Computation of Integrals in High Dimension by Using the Monte Carlo Method

□ *Probabilistic interpretation of a deterministic multidimensional integral.* The problem consists in computing

$$\mathfrak{j} = \int_{\Omega} h(\mathbf{x}) \, d\mathbf{x} \quad , \quad \Omega \subset \mathbb{R}^n \, , \tag{2.72}$$

in which $\mathbf{x} \mapsto h(\mathbf{x})$ is a square-integrable function from $\mathbb{R}^n$ into $\mathbb{R}$, with a compact support $\Omega = [a_1, b_1] \times \ldots \times [a_n, b_n] \subset \mathbb{R}^n$ (thus $h(\mathbf{x}) = 0$ for all $\mathbf{x} \notin \Omega$, and since Ω is compact, $L^2(\Omega, \mathbb{R}) \subset L^1(\Omega, \mathbb{R})$ and consequently, h is integrable). Integral $\mathfrak{j}$ can be rewritten as (probabilistic interpretation),

$$\mathfrak{j} = |\Omega| \, E\{h(\mathbf{X})\} = |\Omega| \int_{\mathbb{R}^n} h(\mathbf{x}) \, P_{\mathbf{X}}(d\mathbf{x}) \, , \tag{2.73}$$

in which $|\Omega| = (b_1 - a_1) \times \ldots \times (b_n - a_n)$ and where $\mathbf{X}$ is a uniformly distributed $\mathbb{R}^n$-valued random variable,

$$P_{\mathbf{X}}(d\mathbf{x}) = \frac{1}{|\Omega|} \, \mathbb{1}_{\Omega}(\mathbf{x}) \, d\mathbf{x} \, . \tag{2.74}$$

□ *Introducing a statistical estimator of $\mathfrak{j}$ (Monte Carlo method).* Let $\{J_{\nu}\}_{\nu \geq 1}$ be the sequence of random estimators

$$J_{\nu} = |\Omega| \frac{1}{\nu} \sum_{\ell=1}^{\nu} h(\mathbf{X}^{(\ell)}) \, , \tag{2.75}$$

in which $\mathbf{X}^{(1)}, \ldots, \mathbf{X}^{(\nu)}$ are ν independent copies of $\mathbf{X}$ (it means that $\mathbf{X}^{(1)}, \ldots, \mathbf{X}^{(\nu)}$ are independent random variables and that all these random variables have the same probability distribution that is the probability distribution of $\mathbf{X}$). The mean vector $m_{J_{\nu}} = E\{J_{\nu}\}$ and the variance $\sigma_{J_{\nu}}^2 = E\{(J_{\nu} - m_{J_{\nu}}))^2\}$ of J_{ν} are such that

$$m_{J_{\nu}} = \mathfrak{j} \quad , \quad \sigma_{J_{\nu}} = \frac{|\Omega|}{\sqrt{\nu}} \, \sigma_{h(\mathbf{X})} \, , \tag{2.76}$$

with $\sigma_{h(\mathbf{X})}^2 = E\{(h(\mathbf{X}) - E\{h(\mathbf{X})\})^2\}$. Since $h \in L^2(\Omega, \mathbb{R})$, then $\sigma_{h(\mathbf{X})}^2$ is finite and thus $\sigma_{J_{\nu}}^2 \to 0$ when $\nu \to +\infty$. The random variables $h(\mathbf{X}^{(1)}), \ldots, h(\mathbf{X}^{(\nu)})$

are independent, and consequently, Theorem 2.2 can be used for analyzing the convergence of the sequence $\{J_\nu\}_{\nu \geq 1}$.

□ *Convergence analysis using Theorem 2.2.* The sequence $\{J_\nu\}_\nu$ converges in probability to $\hat{\jmath}$, *i.e.*, for all $\varepsilon > 0$,

$$\lim_{\nu \to +\infty} \mathcal{P}\{|J_\nu - \hat{\jmath}| \geq \varepsilon\} = 0. \tag{2.77}$$

and for all $\eta > 0$,

$$\lim_{\nu \to +\infty} \mathcal{P}\left\{|J_\nu - \hat{\jmath}| \leq \frac{\eta\,|\Omega|\,\sigma_{h(\mathbf{X})}}{\sqrt{\nu}}\right\} = F_G(\eta), \tag{2.78}$$

in which $F_G(\eta) = (2\pi)^{-1/2} \int_{-\infty}^\eta e^{-g^2/2}\,dg$ is the cumulative distribution function of the Gaussian normalized real-valued random variable G.

□ *Construction of the confidence interval associated with the probability P_c.* Let η_c be such that $F_G(\eta_c) = P_c$ (for instance, for $\eta_c = 3$, one has $P_c = 0.955$). For any given $\varepsilon_c > 0$, for $\nu \geq (|\Omega|\,\sigma_{h(\mathbf{X})}\,\eta_c/\varepsilon_c)^2$, one has

$$\mathcal{P}\{|J_\nu - \hat{\jmath}| \leq \varepsilon_c\} \geq P_c. \tag{2.79}$$

During the computation, and without knowing the exact value of $\hat{\jmath}$, for each value of ν, the level of accuracy can be controlled in calculating the confidence interval by using the following estimation $\widehat{\sigma}_{h(\mathbf{X})}^{(\nu)}$ of $\sigma_{h(\mathbf{X})}$

$$\widehat{\sigma}_{h(\mathbf{X})}^{(\nu)} = \frac{1}{\nu}\sum_{\ell=1}^\nu h(\mathbf{x}^{(\ell)})^2 - \left(\frac{1}{\nu}\sum_{\ell=1}^\nu h(\mathbf{x}^{(\ell)})\right)^2, \tag{2.80}$$

in which $\mathbf{x}^{(1)}, \ldots, \mathbf{x}^{(\nu)}$ are the ν independent realizations of $\mathbf{X}$, which are used for computing the estimation $\hat{\jmath}_\nu = |\Omega|\frac{1}{\nu}\sum_{\ell=1}^\nu h(\mathbf{x}^{(\ell)})$ of J_ν.

2.9 Notions of Stochastic Processes

2.9.1 Definition of a Continuous-Parameter Stochastic Process

A continuous-parameter stochastic process, or in short a *stochastic process*, defined on a probability space $(\Theta, \mathcal{T}, \mathcal{P})$, indexed by any finite or infinite subset T of $\mathbb{R}$, with values in $\mathbb{R}^n$, is a mapping $t \mapsto \mathbf{X}(t)$ from T into $L^0(\Theta, \mathbb{R}^n)$.

For all t fixed in T, $\mathbf{X}(t)$ is an $\mathbb{R}^n$-valued random variable on $(\Theta, \mathcal{T}, \mathcal{P})$. Thus, a stochastic process is an uncountable family $\{\mathbf{X}(t), t \in T\}$ of random variables.

For all θ fixed in Θ, the mapping $t \mapsto \mathbf{X}(t, \theta)$ from T into $\mathbb{R}^n$ is defined as a *trajectory* or a *sample path* of stochastic process $\mathbf{X}$.

2.9.2 System of Marginal Distributions and System of Marginal Characteristic Functions

Let $\mathcal{J}$ be the set of all the finite, nonempty, and nonordered subsets of T. For all $j = \{t_1, \ldots, t_m\}$ in $\mathcal{J}$, the mapping

$$\theta \mapsto \mathbb{X}^j(\theta) = (\mathbf{X}(t_1, \theta), \ldots, \mathbf{X}(t_m, \theta)), \tag{2.81}$$

is a random variable with values in $\mathbb{R}^n \times \ldots \times \mathbb{R}^n = (\mathbb{R}^n)^m$. Let $\mathbb{x}^j = (\mathbf{x}^1, \ldots, \mathbf{x}^m)$ be in $(\mathbb{R}^n)^m$. The probability distribution of $\mathbb{X}^j$ on $(\mathbb{R}^n)^m$ is

$$P_{\mathbb{X}^j}(j \, ; d\mathbb{x}^j) = P_{\mathbf{X}(t_1)\ldots\mathbf{X}(t_m)}(t_1, d\mathbf{x}^1 \, ; \ldots ; t_m, d\mathbf{x}^m). \tag{2.82}$$

When j runs through $\mathcal{J}$, the collection $\{P_{\mathbb{X}^j}(j \, ; d\mathbb{x}^j)\}_{j \in \mathcal{J}}$ of probability distributions is called the *system of marginal distributions* of stochastic process $\mathbf{X}$.

For all j in $\mathcal{J}$, let $\mathbb{u}^j = (\mathbf{u}^1, \ldots, \mathbf{u}^m)$ be in $(\mathbb{R}^n)^m$. The system of marginal characteristic functions $\{\Phi_{\mathbb{X}^j}(j \, ; \mathbb{u}^j)\}_{j \in \mathcal{J}}$ of stochastic process $\mathbf{X}$, which is associated with its system of marginal distributions, is such that

$$\Phi_{\mathbb{X}^j}(j \, ; \mathbb{u}^j) = E\{\exp(i \sum_{k=1}^{m} < \mathbf{u}^k, \mathbf{X}(t_k) >)\}. \tag{2.83}$$

Definition 2.1. The probability distribution of stochastic process $\mathbf{X}$ is defined by its system of marginal distributions, or equivalently, by its system of marginal characteristic functions.

2.9.3 Stationary Stochastic Process

□ *Stationary stochastic process on* $\mathbb{R}$. The stochastic process $\mathbf{X}$ indexed by $T = \mathbb{R}$ is stationary on $\mathbb{R}$ (for the shift operator $t \mapsto t + u$ with u in $\mathbb{R}$) if for all $j = \{t_1, \ldots, t_m\}$ in $\mathcal{J}$ and for all u in $\mathbb{R}$,

$$P_{\mathbf{X}(t_1)\ldots\mathbf{X}(t_m)}(t_1, d\mathbf{x}^1 \, ; \ldots ; t_m, d\mathbf{x}^m) =$$
$$P_{\mathbf{X}(t_1+u)\ldots\mathbf{X}(t_m+u)}(t_1 + u, d\mathbf{x}^1 \, ; \ldots ; t_m + u, d\mathbf{x}^m). \tag{2.84}$$

In particular, for all t and t' in $\mathbb{R}$, we have

$$P_{\mathbf{X}(t)}(t, d\mathbf{x}) = P_{\mathbf{X}(t)}(d\mathbf{x}), \qquad (2.85)$$

$$P_{\mathbf{X}(t)\mathbf{X}(t')}(t, d\mathbf{x}; t', d\mathbf{x}') = P_{\mathbf{X}(t)\mathbf{X}(t')}(t - t'; d\mathbf{x}, d\mathbf{x}'), \qquad (2.86)$$

☐ *Stationary stochastic process on* $\mathbb{R}^+$. If $T = \mathbb{R}^+$, the stationarity is defined for the right shift operator on $\mathbb{R}^+$, which is defined by the positive shifts $t \mapsto t + u$ with $u \geq 0$.

2.9.4 Fundamental Examples of Stochastic Processes

☐ *Definition of a stochastic process with independent increments.* A stochastic process $\mathbf{X}$, indexed by $T = [t_I, t_S[\subset \mathbb{R}$, with values in $\mathbb{R}^n$ is said to be a stochastic process with independent increments if, for all finite ordered subsets $t_I < t_1 < t_2 < \ldots < t_m < t_S$ of T, the increments

$$\mathbf{X}(t_I) \quad, \quad \mathbf{X}(t_1) - \mathbf{X}(t_I) \quad, \quad \ldots \quad, \quad \mathbf{X}(t_m) - \mathbf{X}(t_{m-1}), \qquad (2.87)$$

are mutually independent random variables.

☐ *Definition of an* $\mathbb{R}^n$-*valued Gaussian stochastic process.* A stochastic process $\{\mathbf{X}(t), t \in T\}$ indexed by T with values in $\mathbb{R}^n$ is *Gaussian* if for all $j = \{t_1, \ldots, t_m\}$ in $\mathcal{J}$, the random variable $\mathbb{X}^j = (\mathbf{X}(t_1), \ldots, \mathbf{X}(t_m))$ is Gaussian. Consequently, the system of marginal characteristic functions $\{\Phi_{\mathbb{X}^j}(j; \mathbf{u}^j)\}_{j \in \mathcal{J}}$, defined by Equation (2.83), is written (see Equation (2.45)), for all $\mathbf{u}^j = (\mathbf{u}^1, \ldots, \mathbf{u}^m)$ in $(\mathbb{R}^n)^m$, as

$$\Phi_{\mathbb{X}^j}(j; \mathbf{u}^j) = \exp\left\{i < \mathbf{m}_{\mathbb{X}^j}, \mathbf{u}^j > -\frac{1}{2} < [C_{\mathbb{X}^j}]\mathbf{u}^j, \mathbf{u}^j >\right\}. \qquad (2.88)$$

2.9.5 Continuity of Stochastic Processes

☐ *Types of continuity for a stochastic process.* Let $\{\mathbf{X}(t), t \in T\}$ be a stochastic process defined on $(\Theta, \mathcal{T}, \mathcal{P})$, indexed by T, with values in $\mathbb{R}^n$. The types of continuity of stochastic process $\mathbf{X}$ are related to the types of convergence that have been defined in Section 2.7 for a sequence of random variables. Consequently, the types of continuity for $\mathbf{X}$ are:

— the mean-square continuity.
— the continuity in probability or the stochastically continuity.
— the almost-sure continuity.

☐ *Example corresponding to the mean-square continuity.* Stochastic process $\mathbf{X}$ is mean-square continuous on T if the mapping $t \mapsto \mathbf{X}(t)$ is continuous from T into $L^2(\Theta, \mathbb{R}^n)$,

$$\forall t \in T \quad , \quad \lim_{t' \to t} ||\mathbf{X}(t') - \mathbf{X}(t)||_\Theta = 0. \tag{2.89}$$

☐ *Stochastic process having almost-sure continuous trajectories.* The trajectories (or the sample paths) of stochastic process $\{\mathbf{X}(t), t \in T\}$ are *almost surely continuous* if

$$\mathcal{P}(A_1) = 1 \quad , \quad A_1 = \{\theta \in \Theta \,|\, t \mapsto \mathbf{X}(t, \theta) \in C^0(T, \mathbb{R}^n)\}. \tag{2.90}$$

Kolmogorov's lemma. *If for every compact subset κ of T, one can find three real positive constants $\alpha > 0$, $\beta > 0$, and $c_\kappa > 0$ such that*

$$\forall t \in \kappa, \forall t' \in \kappa \quad , \quad E\{||\mathbf{X}(t) - \mathbf{X}(t')||^\alpha\} \leq c_\kappa ||t - t'||^{1+\beta}, \tag{2.91}$$

then X has almost-sure continuous trajectories.

2.9.6 Second-Order Stochastic Processes with Values in $\mathbb{R}^n$

Let $\{\mathbf{X}(t) \in T\}$ be a stochastic process, defined on the probability space $(\Theta, \mathcal{T}, \mathcal{P})$, indexed by T, with values in $\mathbb{R}^n$.

☐ *Definition of a second-order stochastic process.* Stochastic process $\mathbf{X}$ is second order if $t \mapsto \mathbf{X}(t)$ is a mapping from T into $L^2(\Theta, \mathbb{R}^n)$,

$$||\mathbf{X}(t)||_\Theta = (E\{||\mathbf{X}(t)||^2\})^{1/2} < +\infty \quad , \quad \forall t \in T, \tag{2.92}$$

i.e. if for all fixed t in T, $\mathbf{X}(t)$ is a second-order $\mathbb{R}^n$-valued random variable.

☐ *Mean function.* The mean function of stochastic process $\mathbf{X}$ is the mapping $t \mapsto \mathbf{m}_{\mathbf{X}}(t)$ from T into $\mathbb{R}^n$ such that for all t in T,

$$\mathbf{m}_{\mathbf{X}}(t) = E\{\mathbf{X}(t)\} = \int_{\mathbb{R}^n} \mathbf{x}\, P_{\mathbf{X}(t)}(t, d\mathbf{x}). \tag{2.93}$$

The stochastic process *is centered* if, for all t in T, $\mathbf{m}_{\mathbf{X}}(t) = \mathbf{0}$.

☐ *Autocorrelation function.* The autocorrelation function of stochastic process $\mathbf{X}$ is the mapping $(t, t') \mapsto [R_{\mathbf{X}}(t, t')]$ from $T \times T$ into $\mathbb{M}_n(\mathbb{R})$ such that for all t and t' in T,

$$[R_{\mathbf{X}}(t,t')] = E\{\mathbf{X}(t)\,\mathbf{X}(t')^T\} = \int_{\mathbb{R}^n}\int_{\mathbb{R}^n} \mathbf{x}\,\mathbf{y}^T\, P_{\mathbf{X}(t)\mathbf{X}(t')}(t,d\mathbf{x}\,;t',d\mathbf{y})\,. \quad (2.94)$$

For all $t \in T$, we have

$$\mathrm{tr}[R_{\mathbf{X}}(t,t)] = E\{\|\mathbf{X}(t)\|^2\} < +\infty\,. \quad (2.95)$$

□ *Covariance function.* The covariance function of stochastic process $\mathbf{X}$ is the mapping $(t,t') \mapsto [C_{\mathbf{X}}(t,t')]$ from $T \times T$ into $\mathbb{M}_n(\mathbb{R})$ which is defined as the autocorrelation function of the centered stochastic process $\mathbf{Y}(t) = \mathbf{X}(t) - \mathbf{m}_{\mathbf{X}}(t)$. For all t and t' in T,

$$\begin{aligned}
[C_{\mathbf{X}}(t,t')] &= E\{\left(\mathbf{X}(t) - \mathbf{m}_{\mathbf{X}}(t)\right)\left(\mathbf{X}(t') - \mathbf{m}_{\mathbf{X}}(t')\right)^T\} \\
&= [R_{\mathbf{X}}(t,t')] - \mathbf{m}_{\mathbf{X}}(t)\mathbf{m}_{\mathbf{X}}(t')^T\,. \quad (2.96)
\end{aligned}$$

□ *Mean-square stationarity.* If stochastic process $\mathbf{X}$ is indexed by $T = \mathbb{R}$, is stationary, and is of second-order, then, for all t in $\mathbb{R}$,

$$\mathbf{m}_{\mathbf{X}}(t) = \mathbf{m}_1\,, \quad (2.97)$$

in which $\mathbf{m}_1$ is a constant vector in $\mathbb{R}^n$ and, for all t and u in $\mathbb{R}$,

$$[R_{\mathbf{X}}(t+u,t)] = E\{\mathbf{X}(t+u)\,\mathbf{X}(t)^T\} = [R_{\mathbf{X}}(u)]\,, \quad (2.98)$$

which means that the autocorrelation function depends only on u and not on t.

If a second-order $\mathbb{R}^n$-valued stochastic process $\mathbf{X}$ indexed by $T = \mathbb{R}$ verifies Equations (2.97) and (2.98), then the stochastic process is mean-square stationary. In general, a mean-square stationary stochastic process is not stationary (exception: a Gaussian stochastic process).

□ *Matrix-valued spectral density function for a mean-square stationary and mean-square continuous $\mathbb{R}^n$-valued stochastic process.* If $\{\mathbf{X}(t), t \in \mathbb{R}\}$ is a second-order $\mathbb{R}^n$-valued stochastic process indexed by $T = \mathbb{R}$, which is centered, mean-square stationary, and mean-square continuous (thus function $u \mapsto [R_{\mathbf{X}}(u)]$ is continuous and bounded), and if $[R_{\mathbf{X}}(u)]$ goes to $[0]$ when $|u|$ goes to $+\infty$, then there exists a matrix-valued spectral density function $\omega \mapsto [S_{\mathbf{X}}(\omega)]$, which is integrable from $\mathbb{R}$ into $\mathbb{M}_n^{+0}(\mathbb{C})$ (hermitian positive) such that, for all u in $\mathbb{R}$,

$$[R_{\mathbf{X}}(u)] = E\{\mathbf{X}(t+u)\,\mathbf{X}(t)^T\} = \int_{\mathbb{R}} e^{i\omega u}[S_{\mathbf{X}}(\omega)]\,d\omega\,, \quad (2.99)$$

$$E\{\|\mathbf{X}(t)\|^2\} = \int_{\mathbb{R}} \mathrm{tr}[S_{\mathbf{X}}(\omega)]\,d\omega\,. \quad (2.100)$$

For all $k = 1, \ldots, n$, $\omega \mapsto [S_{\mathbf{X}}(\omega)]_{kk}$ is a function from $\mathbb{R}$ into $\mathbb{R}^+$, which is called the *power spectral density function* of the $\mathbb{R}$-valued stochastic process $\{X_k(t), t \in \mathbb{R}\}$, and is such that $E\{X_k(t)^2\} = \int_{\mathbb{R}} [S_{\mathbf{X}}(\omega)]_{kk} \, d\omega$. If, in addition, function $u \mapsto [R_{\mathbf{X}}(u)]$ belongs to $L^1(\mathbb{R}, \mathbb{M}_n(\mathbb{R}))$ or belongs to $L^2(\mathbb{R}, \mathbb{M}_n(\mathbb{R}))$, then, for all ω in $\mathbb{R}$,

$$[S_{\mathbf{X}}(\omega)] = \frac{1}{2\pi} \int_{\mathbb{R}} e^{-i\omega u} [R_{\mathbf{X}}(u)] \, du \, . \tag{2.101}$$

2.9.7 Summary and What Is the Error That Has to Be Avoided

☐ The probability distribution of a stochastic process $\{\mathbf{X}(t) \in T\}$ is not defined by the family of the probability distribution $P_{\mathbf{X}(t)}(t, d\mathbf{x})$ for $t \in T$, but is defined by the family $P_{\mathbf{X}(t_1)\ldots\mathbf{X}(t_m)}(t_1, d\mathbf{x}^1; \ldots; t_m, d\mathbf{x}^m)$ for all subsets $j = (t_1, \ldots, t_m)$ of $\mathcal{J}$.

☐ If $\{\mathbf{X}(t) \in T\}$ is a second-order $\mathbb{R}^n$-valued stochastic process, then the second-order quantities are

  – the mean function: $t \mapsto \mathbf{m}_{\mathbf{X}}(t)$ from T into $\mathbb{R}^n$.
  – the autocorrelation function: $(t, t') \mapsto [R_{\mathbf{X}}(t, t')]$ from $T \times T$ into $\mathbb{M}_n(\mathbb{R})$.
  – the covariance function: $(t, t') \mapsto [C_{\mathbf{X}}(t, t')]$ from $T \times T$ into $\mathbb{M}_n(\mathbb{R})$.

☐ If $\{\mathbf{X}(t) \in \mathbb{R}\}$ is a second-order and mean-square stationary $\mathbb{R}^n$-valued stochastic process, then $\mathbf{m}_{\mathbf{X}}(t)$ is independent of t and $[R_{\mathbf{X}}(t, t')]$ depends only on $t - t'$. Under appropriate conditions, there exists an integrable function $\omega \mapsto [S_{\mathbf{X}}(\omega)]$ from $\mathbb{R}$ into the set of all the hermitian positive $(n \times n)$ complex matrices, such that

$$[R_{\mathbf{X}}(u)] = \int_{\mathbb{R}} e^{i\omega u} [S_{\mathbf{X}}(\omega)] \, d\omega \, , \; [S_{\mathbf{X}}(\omega)] = \frac{1}{2\pi} \int_{\mathbb{R}} e^{-i\omega u} [R_{\mathbf{X}}(u)] \, du \, . \tag{2.102}$$

Chapter 3
Markov Process and Stochastic Differential Equation

Abstract *The developments that are presented in this chapter are fundamental for understanding some important tools for UQ, in particular those presented in Chapter 4, which are devoted to the Markov Chain Monte Carlo (MCMC) methods that are a class of algorithms for constructing realizations from a probability distribution.*

3.1 Markov Process

Let $\mathbf{X}(t) = (X_1(t), \ldots, X_n(t))$ be a stochastic process, defined on the probability space $(\Theta, \mathcal{T}, \mathcal{P})$, indexed by $\mathbb{R}^+ = [0, +\infty[$, with values in $\mathbb{R}^n$.

3.1.1 Notation

□ *Defining the conditional probability.* For all B in $\mathcal{B}_n$ (Borel σ-algebra of $\mathbb{R}^n$) and for all $0 \leq s < t < +\infty$, the conditional probability of the event $\{\mathbf{X}(t) \in B\}$ given $\mathbf{X}(s) = \mathbf{x}$ in $\mathbb{R}^n$ is written as

$$P(s, \mathbf{x}; t, B) = \mathcal{P}\{\mathbf{X}(t) \in B \mid \mathbf{X}(s) = \mathbf{x}\} = \int_{\mathbf{y} \in B} P(s, \mathbf{x}; t, d\mathbf{y}). \qquad (3.1)$$

© Springer International Publishing AG 2017
C. Soize, *Uncertainty Quantification*, Interdisciplinary Applied Mathematics 47,
DOI 10.1007/978-3-319-54339-0_3

The probability measure $P(s, \mathbf{x}; t, d\mathbf{y})$ is defined as the conditional probability distribution.

□ *Case of a density.* If there is a density of the conditional probability distribution with respect to $d\mathbf{y}$, we will write

$$P(s, \mathbf{x}; t, d\mathbf{y}) = \rho(s, \mathbf{x}; t, \mathbf{y}) \, d\mathbf{y}, \tag{3.2}$$

in which $\mathbf{y} \mapsto \rho(s, \mathbf{x}; t, \mathbf{y})$ is defined as the conditional probability density function.

3.1.2 Markov Property

□ *Notion of the Markov property.* Stochastic process $\mathbf{X}$ has the Markov property if it has no post action or no memory. This means that the state at time $t + dt$ depends only on the state at time t and does not depend on the states for $s < t$.

□ *Definition of the Markov property.* Stochastic process $\mathbf{X}$ has the Markov property if, for all B in $\mathcal{B}_n$, for all finite integer K, and for all

$$0 \le t_1 < t_2 < \ldots < t_K < s < t, \tag{3.3}$$

we have

$$\mathcal{P}\{\mathbf{X}(t) \in B \mid \mathbf{X}(t_1) = \mathbf{x}^1, \ldots, \mathbf{X}(t_K) = \mathbf{x}^K, \mathbf{X}(s) = \mathbf{x}\} = \mathcal{P}\{\mathbf{X}(t) \in B \mid \mathbf{X}(s) = \mathbf{x}\}$$
$$= P(s, \mathbf{x}; t, B), \tag{3.4}$$

in which $\mathbf{x}^1, \ldots, \mathbf{x}^K$ and $\mathbf{x}$ are any vectors in $\mathbb{R}^n$.

3.1.3 Chapman-Kolmogorov Equation

If stochastic process $\mathbf{X}$ verifies the Markov property, then, for all $0 \le s < u < t < +\infty$, for all B in $\mathcal{B}_n$, and for all $\mathbf{x}$ in $\mathbb{R}^n$, we have the following Chapman-Kolmogorov equation:

$$P(s, \mathbf{x}; t, B) = \int_{\mathbf{z} \in \mathbb{R}^n} P(s, \mathbf{x}; u, d\mathbf{z}) \, P(u, \mathbf{z}; t, B). \tag{3.5}$$

If there exists a density such that $P(s, \mathbf{x}; t, d\mathbf{y}) = \rho(s, \mathbf{x}; t, \mathbf{y}) \, d\mathbf{y}$, then, for all $\mathbf{y}$ in $\mathbb{R}^n$, Equation (3.5) can be rewritten as

$$\rho(s, \mathbf{x}; t, \mathbf{y}) = \int_{\mathbf{z} \in \mathbb{R}^n} \rho(s, \mathbf{x}; u, \mathbf{z}) \, \rho(u, \mathbf{z}; t, \mathbf{y}) \, d\mathbf{z}. \tag{3.6}$$

3.1.4 Transition Probability

☐ *Definition of a transition probability.* For $0 \le s < t < +\infty$, for $\mathbf{x}$ in $\mathbb{R}^n$, and for B in $\mathcal{B}_n$, the conditional probability $P(s, \mathbf{x}; t, B)$ is a transition probability if

1. $B \mapsto P(s, \mathbf{x}; t, B)$ is a probability distribution: $\int_{\mathbf{y} \in \mathbb{R}^n} P(s, \mathbf{x}; t, d\mathbf{y}) = 1$.
2. the mapping $\mathbf{x} \mapsto P(s, \mathbf{x}; t, B)$ is measurable from $\mathbb{R}^n$ into $[0, 1]$.
3. P verifies the Chapman-Kolmogorov equation.

☐ *Homogeneous transition probability and homogeneous Chapman-Kolmogorov equation.* The transition probability $P(s, \mathbf{x}; t, B)$ is homogeneous if it verifies the following property:

$$P(s, \mathbf{x}; t, B) = P(0, \mathbf{x}; t - s, B).$$

In such a case, $P(0, \mathbf{x}; t - s, B)$ is simply rewritten as $P(\mathbf{x}; t - s, B)$, and consequently, transition probability $P(s, \mathbf{x}; t, B)$ is homogeneous if

$$P(s, \mathbf{x}; t, B) = P(\mathbf{x}; t - s, B), \tag{3.7}$$

$$\mathcal{P}\{\mathbf{X}(t) \in B \mid \mathbf{X}(s) = \mathbf{x}\} = \int_{\mathbf{y} \in B} P(\mathbf{x}; t - s, d\mathbf{y}). \tag{3.8}$$

For a homogeneous transition probability, the Chapman-Kolmogorov equation defined by Equation (3.5) becomes the following homogeneous Chapman-Kolmogorov equation that is written, for all $0 \le s < u < t < +\infty$, for all B in $\mathcal{B}_n$, and for all $\mathbf{x}$ in $\mathbb{R}^n$, as

$$P(\mathbf{x}; t - s, B) = \int_{\mathbf{z} \in \mathbb{R}^n} P(\mathbf{x}; u - s, d\mathbf{z}) P(\mathbf{z}; t - u, B). \tag{3.9}$$

3.1.5 Definition of a Markov Process

Stochastic process $\mathbf{X}$ is a Markov process if

1. $\mathbf{X}(0)$ is a random variable with any probability distribution on $\mathbb{R}^n$.
2. stochastic process $\mathbf{X}$ verifies the Markov property defined by Equations (3.3) and (3.4).
3. the conditional probabilities $\{B \mapsto P(s, \mathbf{x}; t, B), 0 \le s < t < +\infty, \mathbf{x} \in \mathbb{R}^n\}$ is a system of transition probabilities (see Section 3.1.4).

3.1.6 Fundamental Consequences

For a Markov process, it is sufficient to know the system of transition probabilities to construct the system of marginal probability distributions.

For any $0 \leq t_1 < \ldots < t_K$ in $\mathbb{R}^+$ and for any Borel sets $B_1, \ldots, B_K$ in $\mathcal{B}_n$,

$$
\begin{aligned}
&\mathcal{P}\{\mathbf{X}(t_1) \in B_1, \ldots, \mathbf{X}(t_K) \in B_K\} \\
&= \int_{B_1} \cdots \int_{B_K} P_{\mathbf{X}(t_1)\ldots\mathbf{X}(t_K)}(t_1, d\mathbf{x}^1\,; \ldots \,; t_K, d\mathbf{x}^K) \qquad\qquad (3.10) \\
&= \int_{B_1} \cdots \int_{B_K} P_{\mathbf{X}(t_1)}(t_1, d\mathbf{x}^1)\, \mathrm{P}(t_1, \mathbf{x}^1\,; t_2, d\mathbf{x}^2) \ldots \mathrm{P}(t_{K-1}, \mathbf{x}^{K-1}\,; t_K, d\mathbf{x}^K)\,,
\end{aligned}
$$

in which $\mathbf{x}^k = (x_1^k, \ldots, x_n^k)$ and $d\mathbf{x}^k = dx_1^k \ldots dx_n^k$, where $P_{\mathbf{X}(t_1)}(t_1, d\mathbf{x}^1)$ is the probability distribution on $\mathbb{R}^n$ of the random variable $\mathbf{X}(t_1)$, and where $\mathrm{P}(s, \mathbf{x}\,; t, d\mathbf{y})$ is the transition probability of $\mathbf{X}$.

3.2 Stationary Markov Process, Invariant Measure, and Ergodic Average

Let $\mathbf{X}(t) = (X_1(t), \ldots, X_n(t))$ be a Markov process, defined on the probability space $(\Theta, \mathcal{T}, \mathcal{P})$, indexed by $\mathbb{R}^+ = [0, +\infty[$, with values in $\mathbb{R}^n$.

3.2.1 Stationarity

We recall (see Section 2.9.3) that stochastic process $\mathbf{X}$ is stationary on $\mathbb{R}^+$ for the right shift operator, $t \mapsto t + u$ with $u \geq 0$, if for any finite integer K, for any $0 \leq t_1 < \ldots < t_K$ in $\mathbb{R}^+$, for any $B_1, \ldots, B_K$ in $\mathcal{B}_n$, and for any $u \geq 0$, we have

$$
\mathcal{P}\{\mathbf{X}(t_1 + u) \in B_1, \ldots, \mathbf{X}(t_K + u) \in B_K\} = \mathcal{P}\{\mathbf{X}(t_1) \in B_1, \ldots, \mathbf{X}(t_K) \in B_K\}\,.
$$
$$(3.11)$$

Since stochastic process $\mathbf{X}$ is a Markov process, a necessary and sufficient condition for that $\mathbf{X}$ be stationary is that

1. the probability distribution $P_{\mathbf{X}(t)}(t, d\mathbf{x})$ of $\mathbf{X}(t)$ be independent of t.
2. the transition probabilities be homogeneous, *i.e.*, for all $0 \leq s < t < +\infty$,

$$
\mathcal{P}\{\mathbf{X}(t) \in B \mid \mathbf{X}(s) = \mathbf{x}\} = \mathrm{P}(\mathbf{x}\,; t - s, B) = \int_{\mathbf{y} \in B} \mathrm{P}(\mathbf{x}\,; t - s, d\mathbf{y})\,. \qquad (3.12)
$$

3.2.2 Invariant Measure

☐ *Definition of an invariant measure for Markov process X*. If there exists a time-independent probability distribution $P_S(d\mathbf{x})$ on $\mathbb{R}^n$, which is a solution of the integral equation defined by Equation (3.9),

$$\forall t > 0 \quad, \quad P_S(d\mathbf{y}) = \int_{\mathbf{x} \in \mathbb{R}^n} P_S(d\mathbf{x}) \, \mathrm{P}(\mathbf{x} ; t, d\mathbf{y}), \qquad (3.13)$$

then $P_S(d\mathbf{x})$ is called an invariant measure.

☐ *Invariant measure and stationarity of Markov process X*. If Markov process $\mathbf{X}$ has a system of homogeneous transition probabilities and admits an associated invariant measure $P_S(d\mathbf{x})$, and if we choose $P_{\mathbf{X}(0)}(0, d\mathbf{x}) = P_S(d\mathbf{x})$, then stochastic process $\mathbf{X}$ is stationary on $\mathbb{R}^+$, and for all $t \geq 0$, $P_{\mathbf{X}(t)}(t, d\mathbf{x}) = P_S(d\mathbf{x})$. In addition, for all $\mathbf{x}$ in $\mathbb{R}^n$ and for all B in $\mathcal{B}_n$,

$$\lim_{t \to +\infty} \mathcal{P}\{\mathbf{X}(t) \in B \mid \mathbf{X}(0) = \mathbf{x}\} = \lim_{t \to +\infty} \mathrm{P}(\mathbf{x} ; t, B) = P_S(B). \qquad (3.14)$$

3.2.3 Ergodic Average

Let $\mathbf{Z}$ be the $\mathbb{R}^n$-valued random variable for which the probability distribution $P_{\mathbf{Z}}(d\mathbf{z}) = P_S(d\mathbf{z})$ is the invariant measure of stationary Markov process $\mathbf{X}$. Let $\mathbf{f}$ be any function from $\mathbb{R}^n$ into $\mathbb{R}^m$ such that

$$E\{\|\mathbf{f}(\mathbf{Z})\|\} = \int_{\mathbb{R}^n} \|\mathbf{f}(\mathbf{z})\| \, P_S(d\mathbf{z}) < +\infty. \qquad (3.15)$$

Let $\{\mathbf{x}(t), t \in \mathbb{R}^+\}$ be any realization of stochastic process $\{\mathbf{X}(t), t \in \mathbb{R}^+\}$. The ergodic formula for computing $E\{\mathbf{f}(\mathbf{Z})\}$ is written as

$$E\{\mathbf{f}(\mathbf{Z})\} = \int_{\mathbb{R}^n} \mathbf{f}(\mathbf{z}) \, P_S(d\mathbf{z}) = \lim_{\tau \to +\infty} \frac{1}{\tau} \int_0^\tau \mathbf{f}(\mathbf{x}(t)) \, dt. \qquad (3.16)$$

For Δt sufficiently small and K sufficiently large, the following approximation can be used for computation:

$$E\{\mathbf{f}(\mathbf{Z})\} \simeq \frac{1}{K} \sum_{k=0}^{K} \mathbf{f}(\mathbf{x}(k \Delta t)). \qquad (3.17)$$

3.3 Fundamental Examples of Markov Processes

3.3.1 Process with Independent Increments

☐ *Definition of a stochastic process with independent increments.* An $\mathbb{R}^n$-valued stochastic process, $\{\mathbf{X}(t), t \in T\}$ indexed by $T = [t_I, t_S[\subset \mathbb{R}$, is said to be a process with independent increments if, for all finite ordered subsets $t_I < t_1 < t_2 < \ldots < t_K < t_S$ of T, the increments

$$\mathbf{X}(t_I) \quad , \quad \mathbf{X}(t_1) - \mathbf{X}(t_I) \quad , \quad \ldots \quad , \quad \mathbf{X}(t_K) - \mathbf{X}(t_{K-1}), \qquad (3.18)$$

are mutually independent random variables.

☐ *A stochastic process with independent increments is a Markov process.* Let $\{\mathbf{X}(t), t \in \mathbb{R}^+\}$ be an $\mathbb{R}^n$-valued stochastic process with independent increments such that $\mathbf{X}(0) = \mathbf{0}$ almost surely. Then, for all $0 \leq s < t < +\infty$, the $\mathbb{R}^n$-valued random variables $\Delta\mathbf{X}_{st} = \mathbf{X}(t) - \mathbf{X}(s)$ and $\Delta\mathbf{X}_{0s} = \mathbf{X}(s) - \mathbf{X}(0) = \mathbf{X}(s)$ are independent, and then

$$\mathbf{X}(t) = \Delta\mathbf{X}_{st} + \mathbf{X}(s) \quad , \quad 0 \leq s < t < +\infty, \qquad (3.19)$$

where $\Delta\mathbf{X}_{st}$ and $\mathbf{X}(s)$ are independent $\mathbb{R}^n$-valued random variables. Consequently, $\{\mathbf{X}(t), t \in \mathbb{R}^+\}$ verifies the Markov property with the transition probability $P(s, \mathbf{x}; t, B) = \mathcal{P}\{\mathbf{X}(t) \in B \,|\, \mathbf{X}(s) = \mathbf{x}\}$ that can be rewritten as

$$P(s, \mathbf{x}; t, B) = \mathcal{P}\{\Delta\mathbf{X}_{st} + \mathbf{x} \in B\}, \ 0 \leq s < t < +\infty, \ \mathbf{x} \in \mathbb{R}^n. \qquad (3.20)$$

It can then be concluded that $\{\mathbf{X}(t), t \in \mathbb{R}^+\}$ is a Markov process.

3.3.2 Poisson Process with Mean Function $\lambda(t)$

☐ *Definition of a Poisson process with mean function $\lambda(t)$.* Let $\{X(t), t \in \mathbb{R}^+\}$ be a stochastic process with values in $\mathbb{N}$. Let $t \mapsto \lambda(t)$ be an increasing measurable positive function on $\mathbb{R}^+$ such that $\lambda(0) = 0$. Then, $\{X(t), t \in \mathbb{R}^+\}$ is a Poisson process with mean function $t \mapsto \lambda(t) = E\{X(t)\}$ if

1. $\{X(t), t \in \mathbb{R}^+\}$ is a process with independent increments.
2. $X(0) = 0$ a.s.
3. for all $0 \leq s < t$, the increment $\Delta X_{st} = X(t) - X(s)$ is a Poisson real-valued random variable with mean value $\lambda(t) - \lambda(s)$ (see Section 2.4.1),

$$\forall k \in \mathbb{N} \quad , \quad \mathcal{P}\{\Delta X_{st} = k\} = (k!)^{-1}(\lambda(t) - \lambda(s))^k \, e^{-(\lambda(t) - \lambda(s))}. \quad (3.21)$$

Taking into account the conditions $X(0) = 0$ and $\lambda(0) = 0$, Equation (3.21) yields

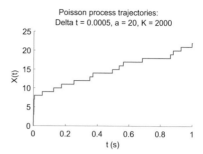

Fig. 3.1 Graph of a trajectory of a non-usual Poisson process.

$$\forall k \in \mathbb{N} \quad , \quad \mathcal{P}\{X(t) = k\} = (k!)^{-1}\,\lambda(t)^k\,e^{-\lambda(t)}. \tag{3.22}$$

The usual Poisson process is obtained for $\lambda(t) = at$ with $a > 0$.

☐ *The trajectories of the Poisson process are not continuous.* A graph of a trajectory of the non-usual Poisson process ($\lambda(t) \neq at$) for which the mean function is $\lambda(t) = a\sqrt{t}$ with $a = 20$, and for which the time sampling is the set $\{t_k = k\,\Delta t, k = 0, \ldots, K\}$ with $\Delta t = 0.0005$ and $K = 2000$, is displayed in Figure 3.1.

☐ *The Poisson process is a Markov process.* The stochastic process $\{X(t), t \in \mathbb{R}^+\}$ verifies the Markov property, with the inhomogeneous transition probability, which is such that, for all $0 \leq s < t < +\infty$ and for all integers $j \leq k$,

$$P(s, j\,; t, k) = \mathcal{P}\{X(t) = k \mid X(s) = j\} = \mathcal{P}\{\Delta X_{st} + j = k\}. \tag{3.23}$$

Therefore, $\{X(t), t \in \mathbb{R}^+\}$ is a nonstationary Markov process.

For the usual Poisson process defined by $\lambda(t) = at$, we have $\lambda(t) - \lambda(s) = a(t - s)$. Consequently, the system of transition probabilities is homogeneous and $\{X(t), t \in \mathbb{R}^+\}$ is a stationary Markov process.

3.3.3 Vector-Valued Normalized Wiener Process

☐ *Definition of an $\mathbb{R}^n$-valued normalized Wiener process.* The $\mathbb{R}^n$-valued stochastic process $\{\mathbf{W}(t) = (W_1(t), \ldots, W_n(t)), t \in \mathbb{R}^+\}$ is said to be a normalized Wiener process if

1. the $\mathbb{R}$-valued stochastic processes $W_1, \ldots, W_n$ are mutually independent.
2. $\mathbf{W}(0) = \mathbf{0}$ a.s.
3. $\{\mathbf{W}(t), t \in \mathbb{R}^+\}$ is a process with independent increments.
4. for all $0 \leq s < t < +\infty$, increment $\Delta\mathbf{W}_{st} = \mathbf{W}(t) - \mathbf{W}(s)$ is an $\mathbb{R}^n$-valued second-order random variable which is Gaussian, centered, and with a covariance matrix $[C_{\Delta\mathbf{W}_{st}}] \in \mathbb{M}_n^+(\mathbb{R})$, which is written as

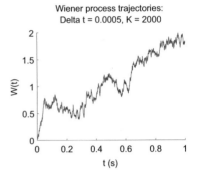

Fig. 3.2 Graph of a trajectory
of the real-valued normalized
Wiener process.

$$[C_{\Delta \mathbf{W}_{st}}] = E\{\Delta \mathbf{W}_{st} \ \Delta \mathbf{W}_{st}{}^T\} = (t - s)[I_n] \,. \qquad (3.24)$$

☐ *Properties of the $\mathbb{R}^n$-valued normalized Wiener process.* The $\mathbb{R}^n$-valued normalized Wiener process $\{\mathbf{W}(t), t \in \mathbb{R}^+\}$ is Gaussian, second-order, and centered

$$\mathbf{m_W}(t) = E\{\mathbf{W}(t)\} = \mathbf{0} \quad , \quad E\{\|\mathbf{W}(t)\|^2\} < +\infty \quad , \quad \forall t \in \mathbb{R}^+ \,. \qquad (3.25)$$

This process is not stationary and its covariance function, defined on $\mathbb{R}^+ \times \mathbb{R}^+$, with values in $\mathbb{M}_n^{+0}(\mathbb{R})$, is written as

$$[C_{\mathbf{W}}(t, s)] = E\{\mathbf{W}(t) \ \mathbf{W}(s)^T\} = \min(t, s) \ [I_n] \,. \qquad (3.26)$$

For $s = 0$ or $t = 0$, we have $[C_{\mathbf{W}}(t, s)] = [0]$ and consequently, $[C_{\mathbf{W}}(t, s)]$ does not belong to $\mathbb{M}_n^+(\mathbb{R})$, but for $0 < s < t$, matrix $[C_{\mathbf{W}}(t, s)]$ belongs to $\mathbb{M}_n^+(\mathbb{R})$.

☐ *The trajectories of the $\mathbb{R}^n$-valued normalized Wiener process are continuous.* A graph of a trajectory of the real-valued normalized Wiener process, for which the time sampling is defined by the set $\{t_k = k \, \Delta t, k = 0, \ldots, K\}$ with $\Delta t = 0.0005$ and $K = 2000$, is displayed in Figure 3.2.

☐ *Continuity but nondifferentiability of the $\mathbb{R}^n$-valued normalized Wiener process.* The trajectories of $\{\mathbf{W}(t), t \in \mathbb{R}^+\}$ are continuous but are not differentiable. The second-order stochastic process $\{\mathbf{W}(t), t \in \mathbb{R}^+\}$ is not mean-square differentiable on $\mathbb{R}^+$. The generalized derivative $D_t \mathbf{W}$ with respect to t of $\mathbf{W}$ (see [118]) is

— an $\mathbb{R}^n$-valued generalized stochastic process indexed by $\mathbb{R}^+$,
— whose extension on $\mathbb{R}$, always denoted as $D_t \mathbf{W}$, is the *normalized Gaussian white noise* $\mathbf{N}_\infty$, and we write $D_t \mathbf{W} = \mathbf{N}_\infty$.

☐ *Definition of the $\mathbb{R}^n$-valued normalized Gaussian white noise.* The normalized Gaussian white noise $\mathbf{N}_\infty$ is a generalized stationary and centered stochastic process, but is not a classical second-order stochastic process [118]. Its matrix-valued spectral density function (see the analogy with Equations (2.99) and

(2.101)) is written as

$$[S_{\mathbf{N}_\infty}(\omega)] = \frac{1}{2\pi}[I_n] \quad , \quad \forall\, \omega \in \mathbb{R}\,, \tag{3.27}$$

$$[R_{\mathbf{N}_\infty}(\tau)] = \int_{\mathbb{R}} e^{i\omega\tau}[S_{\mathbf{N}_\infty}(\omega)]\, d\omega = \delta_0(\tau)[I_n] \quad , \quad \forall\, \tau \in \mathbb{R}\,, \tag{3.28}$$

in which $\delta_0(\tau)$ is the Dirac generalized function on $\mathbb{R}$ at point 0. It should be noted that $E\{\|\mathbf{N}_\infty(t)\|^2\} = \int_{\mathbb{R}} \mathrm{tr}\{[S_{\mathbf{N}_\infty}(\omega)]\}\, d\omega = +\infty$ and consequently, the normalized Gaussian white noise is effectively not a classical second-order stochastic process for which its matrix-valued autocorrelation function is a generalized function.

☐ *The $\mathbb{R}^n$-valued normalized Wiener process is a Markov process.* The stochastic process $\{\mathbf{W}(t), t \in \mathbb{R}^+\}$ is a Markov process whose system of transition probabilities is homogeneous and is written, for $0 \leq s < t < +\infty$ and for $\mathbf{x}$ in $\mathbb{R}^n$, as

$$P(s, \mathbf{x}; t, d\mathbf{y}) = P(\mathbf{x}; t - s, d\mathbf{y}) = \rho(\mathbf{x}; t - s, \mathbf{y})\, d\mathbf{y}\,, \tag{3.29}$$

$$\rho(\mathbf{x}; t - s, \mathbf{y}) = \{2\pi(t - s)\}^{-n/2} \exp\left\{-\frac{1}{2}\frac{\|\mathbf{y} - \mathbf{x}\|^2}{t - s}\right\}. \tag{3.30}$$

For all fixed $t > 0$, the probability distribution $P_{\mathbf{W}(t)}(t, d\mathbf{w})$ of the random vector $\mathbf{W}(t)$ is written as $P_{\mathbf{W}(t)}(t, d\mathbf{w}) = p_{\mathbf{W}(t)}(t, \mathbf{w})\, d\mathbf{w}$ with

$$p_{\mathbf{W}(t)}(t, \mathbf{w}) = (2\pi t)^{-n/2} \exp\left\{-\frac{1}{2}\frac{\|\mathbf{w}\|^2}{t}\right\}, \tag{3.31}$$

and thus depends on t. Consequently, the Markov process $\{\mathbf{W}(t), t \in \mathbb{R}^+\}$ is not stationary.

3.4 Itô Stochastic Integral

3.4.1 Definition of an Itô Stochastic Integral and Why It Is Not a Classical Integral

An *Itô stochastic integral* is the $\mathbb{R}^n$-valued stochastic process $\{\mathbf{Z}(t), t \in \mathbb{R}^+\}$ that is defined by

$$\mathbf{Z}(t) = \int_0^t [\mathbf{A}(s)]\, d\mathbf{W}(s)\,, \tag{3.32}$$

in which $\{[\mathbf{A}(s)], s \geq 0\}$ is a nonanticipative $\mathbb{M}_{n,m}(\mathbb{R})$-valued stochastic process (see Section 3.4.2) and where $\{\mathbf{W}(s), s \in \mathbb{R}^+\}$ is the $\mathbb{R}^m$-valued normalized Wiener process. The integral defined by Equation (3.32) cannot be interpreted neither as a mean-square integral (because $\mathbf{W}$ is nowhere differentiable) nor as a Riemann-Stieltjes integral for each trajectory (sample path), because the continuous trajectories of $\mathbf{W}$ are not of bounded variation on any bounded time interval. Nevertheless, if $[\mathbf{A}(s)]$ is a constant matrix $[a]$, we want that

$$\mathbf{Z} = \int_{t_0}^{t_1} [a] \, d\mathbf{W}(s) = [a] \, (\mathbf{W}(t_1) - \mathbf{W}(t_0)) \quad \text{for} \quad 0 \leq t_0 \leq t_1 < +\infty . \quad (3.33)$$

3.4.2 Definition of Nonanticipative Stochastic Processes with Respect to the $\mathbb{R}^m$-Valued Normalized Wiener Process

A stochastic process $\{\mathbf{B}(t), t \in T\}$ (or $\{[\mathbf{A}(t)], t \in T\}$), defined on $(\Theta, \mathcal{T}, \mathcal{P})$, indexed by $T = [0, t_1[\subset \mathbb{R}^+$, with values in $\mathbb{R}^n$ (or in $\mathbb{M}_{n,m}(\mathbb{R})$), is nonanticipative with respect to the $\mathbb{R}^m$-valued normalized Wiener process $\{\mathbf{W}(t), t \in \mathbb{R}^+\}$, if, for all fixed t in T, the family of random variables

$$\{\mathbf{B}(s), 0 \leq s \leq t\} \quad , \quad (\text{or} \quad \{[\mathbf{A}(s)], 0 \leq s \leq t\}), \quad (3.34)$$

are independent of the family of random variables

$$\{\Delta \mathbf{W}_{t\tau} = \mathbf{W}(\tau) - \mathbf{W}(t) \, , \, \tau > t\} . \quad (3.35)$$

3.4.3 Definition of the Space $M_W^2(n, m)$

The space $M_W^2(n, m)$ is the set of all the second-order stochastic processes $\{[\mathbf{A}(t)], t \in T\}$, defined on $(\Theta, \mathcal{T}, \mathcal{P})$, indexed by $T = [0, t_1[\subset \mathbb{R}^+$, with values in $\mathbb{M}_{n,m}(\mathbb{R})$, which are

1. nonanticipative with respect to $\mathbb{R}^m$-valued Wiener process $\mathbf{W}$.
2. such that $E\{\int_T \|[\mathbf{A}(s)]\|_F^2 \, ds\} < +\infty$.

3.4.4 Approximation of a Process in $M_W^2(n, m)$ by a Step Process

□ *Definition of a step process.* For every t fixed in T, an $\mathbb{M}_{n,m}(\mathbb{R})$-valued stochastic process $\{[\mathbf{A}_K(s)], s \in [0, t]\}$, defined on $(\Theta, \mathcal{T}, \mathcal{P})$, is a step process if there exists a partition $0 = t_1 < t_2 < \ldots < t_K < t_{K+1} = t$ of $[0, t]$ such that

$$[\mathbf{A}_K(s)] = [\mathbf{A}_K(t_k)] \quad \text{if} \quad t_k \leq s < t_{k+1} \quad , \quad \forall k \in \{1, \ldots, K\} . \quad (3.36)$$

□ *Approximation of a process in $M_W^2(n, m)$ by a step process.* Let $\{[\mathbf{A}(t)], t \in T\}$ be a stochastic process in $M_W^2(n, m)$. For every t fixed in T, there exists a sequence of step processes $\{[\mathbf{A}_K]\}_K$ in $M_\mathbf{W}^2(n, m)$ such that

$$\lim_{K \to +\infty} E\left\{\int_0^t \|\mathbf{A}(s) - \mathbf{A}_K(s)\|_F^2 \, ds\right\} = 0. \tag{3.37}$$

3.4.5 Definition of the Itô Stochastic Integral for a Stochastic Process in $M_\mathbf{W}^2(n, m)$

Let $\{[\mathbf{A}(t)], t \in T\}$ be a stochastic process in $M_\mathbf{W}^2(n, m)$, $\{\mathbf{W}(s), s \in \mathbb{R}^+\}$ be the $\mathbb{R}^m$-valued normalized Wiener process, t fixed in T, $\{[\mathbf{A}_K]\}_K$ be a sequence of step processes in $M_\mathbf{W}^2(n, m)$, indexed by $[0, t]$, which approximates $[\mathbf{A}]$ on $[0, t]$. Then, the Itô stochastic integral on $[0, t]$ of $[\mathbf{A}]$ is defined by

$$\mathbf{Z}(t) = \int_0^t [\mathbf{A}(s)] \, d\mathbf{W}(s) = \text{l.i.m}_{K \to +\infty} \sum_{k=1}^K [\mathbf{A}_K(t_k)] \Big(\mathbf{W}(t_{k+1}) - \mathbf{W}(t_k)\Big), \tag{3.38}$$

in which l.i.m denotes the limit in mean square of the sequence of the $\mathbb{R}^n$-valued second-order random variables, and then $E\{\|\mathbf{Z}(t)\|^2\} < +\infty$ for all t. Since mean-square convergence implies convergence in probability, Equation (3.38) holds for the convergence in probability.

3.4.6 Properties of the Itô Stochastic Integral of a Stochastic Process in $M_\mathbf{W}^2(n, m)$

We have the following important properties for the Itô stochastic integral of a stochastic process in $M_\mathbf{W}^2(n, m)$.

▫ The mapping defined by $[\mathbf{A}] \mapsto \int_0^t [\mathbf{A}(s)] \, d\mathbf{W}(s)$ is linear from $M_\mathbf{W}^2(n, m)$ into $L^2(\Theta, \mathbb{R}^n)$ and

$$E\left\{\int_0^t [\mathbf{A}(s)] \, d\mathbf{W}(s)\right\} = \mathbf{0}, \tag{3.39}$$

$$E\{\|\mathbf{Z}(t)\|^2\} = E\left\{\|\int_0^t [\mathbf{A}(s)] \, d\mathbf{W}(s)\|^2\right\} = E\left\{\int_0^t \|\mathbf{A}(s)\|_F^2 \, ds\right\}. \tag{3.40}$$

▫ For each fixed t, the $\mathbb{R}^n$-valued second-order stochastic process $t \mapsto \mathbf{Z}(t) = \int_0^t [\mathbf{A}(s)] \, d\mathbf{W}(s)$ indexed by T has almost-surely continuous trajectories.

▫ If stochastic process $[\mathbf{A}] \in M_\mathbf{W}^2(n, m)$ has almost-surely continuous trajectories on T, then it is not necessary to introduce a sequence of step processes $\{[\mathbf{A}_K]\}_K$. The Itô stochastic integral on $[0, t]$ of $[\mathbf{A}]$ is then defined by

$$\int_0^t [\mathbf{A}(s)]\, d\mathbf{W}(s) = \text{l.i.m}_{K \to +\infty} \sum_{k=1}^{K} [\mathbf{A}(t_k)] \left(\mathbf{W}(t_{k+1}) - \mathbf{W}(t_k) \right). \quad (3.41)$$

3.5 Itô Stochastic Differential Equation and Fokker-Planck Equation

3.5.1 Definition of an Itô Stochastic Differential Equation (ISDE)

The Itô stochastic differential equation (ISDE) is defined by

$$d\mathbf{X}(t) = \mathbf{b}(\mathbf{X}(t), t)\, dt + [a(\mathbf{X}(t), t)]\, d\mathbf{W}(t) \quad , \quad t \in]0, t_1], \quad (3.42)$$

with the initial condition

$$\mathbf{X}(0) = \mathbf{X}_0 \quad \text{a.s.}, \quad (3.43)$$

in which

$\{\mathbf{X}(t), t \in T\}$ is an $\mathbb{R}^n$-valued stochastic process indexed by $T = [0, t_1]$,
$\{\mathbf{W}(s), s \in \mathbb{R}^+\}$ is the $\mathbb{R}^m$-valued normalized Wiener process,
$\mathbf{X}_0$ is an $\mathbb{R}^n$-valued random variable that is independent of $\{\mathbf{W}(s), s \in \mathbb{R}^+\}$,
$(\mathbf{x}, t) \mapsto \mathbf{b}(\mathbf{x}, t)$ is a function from $\mathbb{R}^n \times T$ into $\mathbb{R}^n$,
$(\mathbf{x}, t) \mapsto [a(\mathbf{x}, t)]$ is a function from $\mathbb{R}^n \times T$ into $\mathbb{M}_{n,m}(\mathbb{R})$,
$[\mathbf{A}(t)] = [a(\mathbf{X}(t), t)]$ belongs to $M_{\mathbf{W}}^2(n, m)$.

The ISDE defined by Equation (3.42), with the initial condition defined by Equation (3.43), can be rewritten, for all t in $[0, t_1]$, as

$$\mathbf{X}(t) = \mathbf{X}_0 + \int_0^t \mathbf{b}(\mathbf{X}(s), s)\, ds + \int_0^t [a(\mathbf{X}(s), s)]\, d\mathbf{W}(s), \quad (3.44)$$

in which the second integral in the right-hand side of Equation (3.44) is an Itô stochastic integral.

3.5.2 Existence of a Solution of the ISDE as a Diffusion Process

Under suitable hypotheses [110, 115, 183] for $\mathbf{b}(\mathbf{x}, t)$ and $[a(\mathbf{x}, t)]$, the ISDE admits a unique solution $\{\mathbf{X}(t), t \in T\}$ in $M_{\mathbf{W}}^2(n, 1)$ such as

1. $\{\mathbf{X}(t), t \in T\}$ has almost-surely continuous trajectories.
2. $\{\mathbf{X}(t), t \in T\}$ is a Markov process with the system of transition probabilities $P(s, \mathbf{x}; t, B) = \mathcal{P}\{\mathbf{X}(t) \in B \mid \mathbf{X}(s) = \mathbf{x}\}$ for all $0 \le s < t < t_1$.

3. If, in addition, $\mathbf{b}(\mathbf{x}, t)$ and $[a(\mathbf{x}, t)]$ are continuous functions in t on T, then $\{\mathbf{X}(t), t \in T\}$ is a *diffusion process* with a *drift vector* $\mathbf{b}(\mathbf{x}, t) \in \mathbb{R}^n$ and a *diffusion matrix* $[\sigma(\mathbf{x}, t)] \in \mathbb{M}_n^{+0}(\mathbb{R})$ that are defined as follows. For all $0 \le t < t + h$ and $\varepsilon > 0$,

$$\mathbf{b}(\mathbf{x}, t) = \lim_{h \downarrow 0} \frac{1}{h} \int_{\|\mathbf{y} - \mathbf{x}\| < \varepsilon} (\mathbf{y} - \mathbf{x}) \, P(t, \mathbf{x}; t + h, d\mathbf{y}), \qquad (3.45)$$

$$[\sigma(\mathbf{x}, t)] = \lim_{h \downarrow 0} \frac{1}{h} \int_{\|\mathbf{y} - \mathbf{x}\| < \varepsilon} (\mathbf{y} - \mathbf{x}) \, (\mathbf{y} - \mathbf{x})^T \, P(t, \mathbf{x}; t + h, d\mathbf{y}), \qquad (3.46)$$

in which $h \downarrow 0$ means $h \to 0$ with $h > 0$. For all $\mathbf{x}$ and t, the diffusion matrix $[\sigma(\mathbf{x}, t)]$ is written as

$$[\sigma(\mathbf{x}, t)] = [a(\mathbf{x}, t)] \, [a(\mathbf{x}, t)]^T, \qquad (3.47)$$

which is a symmetric and positive matrix, but which is not necessarily a positive-definite matrix.

3.5.3 Fokker-Planck (FKP) Equation for the Diffusion Process

If, for all $0 \le s < t < t_1$, the transition probability $P(s, \mathbf{x}; t, d\mathbf{y})$ admits a density

$$P(s, \mathbf{x}; t, d\mathbf{y}) = \rho(s, \mathbf{x}; t, \mathbf{y}) \, d\mathbf{y}, \qquad (3.48)$$

then, for all $\mathbf{x}$ and $\mathbf{y}$ in $\mathbb{R}^n$, the density $\rho(s, \mathbf{x}; t, \mathbf{y})$ verifies the FKP equation,

$$\frac{\partial}{\partial t} \rho(s, \mathbf{x}; t, \mathbf{y}) + \sum_{j=1}^n \frac{\partial}{\partial y_j} (b_j(\mathbf{y}, t) \, \rho(s, \mathbf{x}; t, \mathbf{y}))$$

$$- \frac{1}{2} \sum_{j,k=1}^n \frac{\partial^2}{\partial y_j \partial y_k} (\sigma_{jk}(\mathbf{y}, t) \, \rho(s, \mathbf{x}; t, \mathbf{y})) = 0 \quad , \quad t \in]s, t_1[, \quad (3.49)$$

with, for $t = s$, the initial condition that is written as

$$\lim_{t \downarrow s} \rho(s, \mathbf{x}; t, \mathbf{y}) \, d\mathbf{y} = \delta_0(\mathbf{y} - \mathbf{x}), \qquad (3.50)$$

in which $\delta_0(\mathbf{y})$ is the Dirac measure at the origin of $\mathbb{R}^n$ and where $t \downarrow s$ means $t \to s$ with $t > s$.

3.5.4 Itô Formula for Stochastic Differentials

Let $\{\mathbf{X}(t), t \in T\}$ be an $\mathbb{R}^n$-valued stochastic process admitting the Itô stochastic differential,

$$d\mathbf{X}(t) = \mathbf{b}(\mathbf{X}(t), t)\, dt + [a(\mathbf{X}(t), t)]\, d\mathbf{W}(t)\,. \tag{3.51}$$

Let $(\mathbf{x}, t) \mapsto u(\mathbf{x}, t)$ be a continuous mapping from $\mathbb{R}^n \times T$ into $\mathbb{R}$ with continuous partial derivatives $\partial u / \partial t$, $\{\nabla_{\mathbf{x}} u\}_j = \partial u / \partial x_j$, and $[\partial_{\mathbf{x}}^2 u]_{jk} = \partial^2 u / \partial x_j \partial x_k$. Then the stochastic process $U(t) = u(\mathbf{X}(t), t)$ admits an Itô stochastic differential given by the *Itô formula* that is written as

$$dU(t) = \frac{\partial u}{\partial t}(\mathbf{X}(t), t)\, dt + \; <\nabla_{\mathbf{x}} u(\mathbf{X}(t), t)\,, d\mathbf{X}(t)>$$
$$+ \frac{1}{2} \mathrm{tr}\Big\{ [\partial_{\mathbf{x}}^2 u(\mathbf{X}(t), t)]\, [\mathbf{A}(t)]\, [\mathbf{A}(t)]^T \Big\}\, dt\,, \tag{3.52}$$

in which $[\mathbf{A}(t)] = [a(\mathbf{X}(t), t)]$. It should be noted (see Equation (3.52)) that the Itô stochastic differential involves a complementary terms in the expression of the differential dU with respect to the usual differential calculus.

Example. For integer $q \geq 2$, $d(W(t)^q) = q\, W(t)^{q-1}\, dW(t) + \frac{1}{2} q(q - 1) W(t)^{q-2}\, dt$.

3.6 Itô Stochastic Differential Equation Admitting an Invariant Measure

3.6.1 ISDE with Time-Independent Coefficients

We consider the ISDE with initial condition defined by Equations (3.42) to (3.44) for which the drift vector and the diffusion matrix are independent of t, and for which $t \geq 0$. Consequently, the ISDE is rewritten as

$$d\mathbf{X}(t) = \mathbf{b}(\mathbf{X}(t))\, dt + [a(\mathbf{X}(t))]\, d\mathbf{W}(t) \quad , \quad t > 0\,, \tag{3.53}$$

with the initial condition

$$\mathbf{X}(0) = \mathbf{X}_0 \quad \text{a.s.}\,, \tag{3.54}$$

in which

    $\{\mathbf{X}(t), t \geq 0\}$ is an $\mathbb{R}^n$-valued stochastic process indexed by $\mathbb{R}^+$,
    $\{\mathbf{W}(s), s \geq 0\}$ is the $\mathbb{R}^m$-valued normalized Wiener process,
    $\mathbf{X}_0$ is an $\mathbb{R}^n$-valued random variable that is independent of $\{\mathbf{W}(s), s \geq 0\}$,
    $\mathbf{x} \mapsto \mathbf{b}(\mathbf{x})$ is a function from $\mathbb{R}^n$ into $\mathbb{R}^n$,
    $\mathbf{x} \mapsto [a(\mathbf{x})]$ is a function from $\mathbb{R}^n$ into $\mathbb{M}_{n,m}(\mathbb{R})$,
    $[\mathbf{A}(t)] = [a(\mathbf{X}(t))]$ belongs to $\mathbb{M}_{\mathbf{W}}^2(n, m)$.

The ISDE defined by Equation (3.53), with the initial condition defined by Equation (3.54), can be rewritten, for all $t \geq 0$, as

$$\mathbf{X}(t) = \mathbf{X}_0 + \int_0^t \mathbf{b}(\mathbf{X}(s))\, ds + \int_0^t [a(\mathbf{X}(s))]\, d\mathbf{W}(s) , \qquad (3.55)$$

in which, in the right-hand side of Equation (3.55), the second integral is an Itô stochastic integral.

3.6.2 Existence and Uniqueness of a Solution

Under suitable hypotheses [115, 183] for $\mathbf{b}(\mathbf{x})$ and $[a(\mathbf{x})]$, the ISDE admits a unique solution $\{\mathbf{X}(t), t \geq 0\}$, which

1. is defined almost surely for all $t \geq 0$ (no explosion of the solution).
2. has almost-surely continuous trajectories on $\mathbb{R}^+$.
3. is a Markov process with homogeneous transition probabilities,

$$\mathrm{P}(\mathbf{x}; t - s, B) = \mathcal{P}\{\mathbf{X}(t) \in B \,|\, \mathbf{X}(s) = \mathbf{x}\} \quad , \quad 0 \leq s < t < +\infty . \qquad (3.56)$$

4. is a diffusion process with the drift vector $\mathbf{b}(\mathbf{x})$ and with the diffusion matrix $[\sigma(\mathbf{x})] = [a(\mathbf{x})]\,[a(\mathbf{x})]^T$.

3.6.3 Invariant Measure and Steady-State FKP Equation

If there exists an invariant measure $P_S(d\mathbf{x})$, *i.e.* if there exists a probability distribution on $\mathbb{R}^n$ independent of time t, which verifies Equation (3.9),

$$P_S(d\mathbf{y}) = \int_{\mathbf{x} \in \mathbb{R}^n} P_S(d\mathbf{x})\, \mathrm{P}(\mathbf{x}; t, d\mathbf{y}) \quad , \quad \forall t > 0 , \qquad (3.57)$$

and if $P_S(d\mathbf{y}) = \rho(\mathbf{y})\, d\mathbf{y}$, then the density $\rho(\mathbf{y})$ verifies the following *steady-state FKP*, which is written, for all $\mathbf{y} \in \mathbb{R}^n$, as

$$\sum_{j=1}^n \frac{\partial}{\partial y_j} (b_j(\mathbf{y})\, \rho(\mathbf{y})) - \frac{1}{2} \sum_{j,k=1}^n \frac{\partial^2}{\partial y_j \partial y_k} (\sigma_{jk}(\mathbf{y})\, \rho(\mathbf{y})) = 0 , \qquad (3.58)$$

with the normalization condition,

$$\int_{\mathbb{R}^n} \rho(\mathbf{y})\, d\mathbf{y} = 1 . \qquad (3.59)$$

Some sufficient conditions for obtaining the existence of an invariant measure can be found in [115, 183].

3.6.4 Stationary Solution

If $\{\mathbf{X}(t), t \geq 0\}$ is stationary, it is rewritten as $\{\mathbf{X}_S(t), t \geq 0\}$. If there exists an invariant measure $P_S(d\mathbf{x})$, and if the probability distribution of the initial condition $\mathbf{X}_S(0)$ is such that

$$P_{\mathbf{X}_S(0)}(d\mathbf{x}) = P_S(d\mathbf{x}), \tag{3.60}$$

then the unique solution $\{\mathbf{X}_S(t), t \in \mathbb{R}^+\}$ is stationary on $\mathbb{R}^+$ for the right shift operator $t \mapsto t + u$, $u \geq 0$, and

$$\forall t \geq 0 \quad , \quad P_{\mathbf{X}_S(t)}(d\mathbf{x}) = P_S(d\mathbf{x}), \tag{3.61}$$

in which $P_{\mathbf{X}_S(t)}(d\mathbf{x})$ is the probability distribution (independent of t) of the random vector $\mathbf{X}_S(t)$.

3.6.5 Asymptotic Stationary Solution

If there exists an invariant measure $P_S(d\mathbf{x})$, and if the probability distribution of the initial condition $\mathbf{X}(0)$ is such that

$$P_{\mathbf{X}(0)}(0, d\mathbf{x}) \neq P_S(d\mathbf{x}), \tag{3.62}$$

then the unique solution $\{\mathbf{X}(t), t \geq 0\}$ is not stationary. However, for $t_0 \to +\infty$, the nonstationary stochastic process $\{\mathbf{X}(t), t \geq t_0\}$ is stochastically equivalent (or is isonomic) to the stationary process $\{\mathbf{X}_S(t), t \geq 0\}$, *i.e.* admits the same system of marginal distributions (see Section 2.9.2), and we have

$$\lim_{t \to +\infty} P_{\mathbf{X}(t)}(t, d\mathbf{x}) = P_S(d\mathbf{x}), \tag{3.63}$$

in which $P_{\mathbf{X}(t)}(t, d\mathbf{x})$ is the probability distribution of the random vector $\mathbf{X}(t)$.

3.6.6 Ergodic Average: Formulas for Computing Statistics by MCMC Methods

In this section, we present some important results concerning the Markov Chain Monte Carlo (MCMC) method for computing statistics. Let $\{\mathbf{X}_S(t), t \geq 0\}$ be the stationary solution for which a given realization is denoted by $\{\mathbf{x}_S(t), t \geq 0\}$. Let

$\mathbf{Z}$ be the $\mathbb{R}^n$-valued random variable whose probability distribution is the invariant measure P_S, i.e. $P_{\mathbf{Z}}(d\mathbf{z}) = P_S(d\mathbf{z})$. Let $\{\mathbf{X}(t), t \geq 0\}$ be the asymptotic stationary solution for which a given realization is denoted by $\{\mathbf{x}(t), t \geq 0\}$. Let $\mathbf{f}$ be a function from $\mathbb{R}^n$ into $\mathbb{R}^m$ such that $E\{\|\mathbf{f}(\mathbf{Z})\|\} = \int_{\mathbb{R}^n} \|\mathbf{f}(\mathbf{z})\| P_S(d\mathbf{z}) < +\infty$.

□ *First formula using the stationary solution.* Such a formula can be used if a realization $\mathbf{X}_S(0) \sim P_S(d\mathbf{z})$ can be computed and thus is available,

$$E\{\mathbf{f}(\mathbf{Z})\} = \lim_{\tau \to +\infty} \frac{1}{\tau} \int_0^\tau \mathbf{f}(\mathbf{x}_S(t))\, dt \simeq \frac{1}{K} \sum_{k=0}^{K} \mathbf{f}(\mathbf{x}_S(k\Delta t)), \qquad (3.64)$$

for Δt sufficiently small and for K sufficiently large.

□ *Second formula using the asymptotic stationary solution.* Such a formula can be used even if a realization of $\mathbf{X}_S(0) \sim P_S(d\mathbf{z})$ cannot be computed and thus is not available,

$$E\{\mathbf{f}(\mathbf{Z})\} = \lim_{\tau \to +\infty} \frac{1}{\tau} \int_{t_0}^{t_0+\tau} \mathbf{f}(\mathbf{x}(t))\, dt \simeq \frac{1}{K} \sum_{k=k_0}^{k_0+K} \mathbf{f}(\mathbf{x}(k\Delta t)), \qquad (3.65)$$

in which $t_0 = k_0\Delta t$ with $t_0 > 0$ such that, for $t \geq t_0$, $\{\mathbf{X}(t), t \geq t_0\}$ is stochastically equivalent (isonomic) to $\{\mathbf{X}_S(t), t \geq 0\}$.

3.7 Markov Chain

A Markov chain is the discretized version of a Markov process, which is adapted to computation and to the digital signal processing.

3.7.1 Connection with the Markov Processes

An $\mathbb{R}^n$-valued Markov chain is an ordered set $\{\mathbf{X}^j, j \in \mathbb{N}\}$ of random vectors $\mathbf{X}^0, \mathbf{X}^1, \mathbf{X}^2, \ldots$ with values in $\mathbb{R}^n$, which can be viewed as the time sampling of a $\mathbb{R}^n$-valued Markov process $\{\mathbf{X}(t), t \in \mathbb{R}^+\}$ such as $\mathbf{X}^j = \mathbf{X}(t_j)$ for $j = 0, 1, 2, \ldots$, in which $t_0 = 0 < t_1 < t_2 <, \ldots$ are the sampling times. Consequently,

- all the previous results can directly be used for defining and for analyzing the properties of a Markov chain.

- in order to limit the presentation, we restrict the developments to the case of the time-homogeneous Markov chain.

3.7.2 Definition of a Time-Homogeneous Markov Chain

A time-homogeneous $\mathbb{R}^n$-valued Markov chain is an ordered set $\{\mathbf{X}^k, k \in \mathbb{N}\}$ of random vectors $\mathbf{X}^0, \mathbf{X}^1, \mathbf{X}^2, \ldots$, defined on $(\Theta, \mathcal{T}, \mathcal{P})$, with values in $\mathbb{R}^n$, which satisfies the Markov property,

$$\mathcal{P}\{\mathbf{X}^{k+1} \in B^{k+1} \,|\, \mathbf{X}^0 = \mathbf{x}^0, \ldots, \mathbf{X}^k = \mathbf{x}^k\} = \mathcal{P}\{\mathbf{X}^{k+1} \in B^{k+1} \,|\, \mathbf{X}^k = \mathbf{x}^k\}, \quad (3.66)$$

for which the transition probability (or transition kernel) is homogeneous and is written, for $\mathbf{x}^j$ in $\mathbb{R}^n$, for B^k in $\mathcal{B}_n$, and for $0 \leq j < k$, as

$$\mathcal{P}\{\mathbf{X}^k \in B^k \,|\, \mathbf{X}^j = \mathbf{x}^j\} = \mathrm{P}(\mathbf{x}^j ; k - j, B^k) = \int_{\mathbf{y} \in B^k} \mathrm{P}(\mathbf{x}^j ; k - j, d\mathbf{y}). \quad (3.67)$$

If the homogeneous transition kernel $\mathrm{P}(\mathbf{x} ; k, d\mathbf{y}), k > 0$, admits a density, then

$$\mathrm{P}(\mathbf{x}^j ; k - j, d\mathbf{y}) = \rho(\mathbf{x}^j ; k - j, \mathbf{y}) \, d\mathbf{y}. \quad (3.68)$$

3.7.3 Properties of the Homogeneous Transition Kernel and Homogeneous Chapman-Kolmogorov Equation

For $\mathbf{x}^j$ in $\mathbb{R}^n$, for B^k in $\mathcal{B}_n$, and for $0 \leq j < k$, the homogeneous transition kernel $\mathrm{P}(\mathbf{x} ; k, d\mathbf{y})$ verifies the following properties.

1. The mapping $B^k \mapsto \mathrm{P}(\mathbf{x}^j ; k - j, B^k)$ is a probability distribution, which means that $\int_{\mathbf{y} \in \mathbb{R}^n} \mathrm{P}(\mathbf{x}^j ; k - j, d\mathbf{y}) = 1$.

2. The mapping $\mathbf{x}^j \mapsto \mathrm{P}(\mathbf{x}^j ; k - j, B^k)$ is measurable from $\mathbb{R}^n$ into $[0, 1]$.

3. The homogeneous transition kernel verifies the homogeneous Chapman-Kolmogorov equation: for all $j \geq 0$, for all $k \geq 2$, for all B^{j+k} in $\mathcal{B}_n$, and for all $\mathbf{x}^j$ in $\mathbb{R}^n$,

$$\mathrm{P}(\mathbf{x}^j ; k, B^{j+k}) = \int_{\mathbf{x}^{j+k-1} \in \mathbb{R}^n} \mathrm{P}(\mathbf{x}^j ; k-1, d\mathbf{x}^{j+k-1}) \mathrm{P}(\mathbf{x}^{j+k-1} ; 1, B^{j+k}). \quad (3.69)$$

4. The conditional probability $\mathrm{P}(\mathbf{x}^j ; 1, B^{j+1})$ is rewritten as $\mathrm{P}(\mathbf{x}^j, B^{j+1})$. The probability distribution $P_S(d\mathbf{x}^j)$ on $\mathbb{R}^n$ is an invariant measure for the homogeneous transition kernel

$$B^{j+1} \mapsto \mathrm{P}(\mathbf{x}^j, B^{j+1}), \quad (3.70)$$

if $P_S(B^{j+1})$ satisfies the following equation:

$$P_S(B^{j+1}) = \int_{\mathbf{x}^j \in \mathbb{R}^n} P_S(d\mathbf{x}^j) \, \mathrm{P}(\mathbf{x}^j, B^{j+1}). \quad (3.71)$$

3.7.4 Asymptotic Stationarity for Time-Homogeneous Markov Chains

Let $\{\mathbf{X}^k, k \in \mathbb{N}\}$ be a time-homogeneous Markov chain that admits an invariant measure $B \mapsto P_S(B)$ on $\mathbb{R}^n$, and for which the homogeneous transition kernel

$$P(\mathbf{x}^0 ; k, B) = \mathcal{P}\{\mathbf{X}^k \in B \mid \mathbf{X}^0 = \mathbf{x}^0\} \ , \ k \geq 1 \ , \ \mathbf{x}^0 \in \mathbb{R}^n \ , \ B \in \mathcal{B}_n \ , \quad (3.72)$$

is *not periodic* and is *irreducible*, *i.e.*, for all $\mathbf{x}^0$ in $\mathbb{R}^n$, and for all B in $\mathcal{B}_n$ such that $P_S(B) > 0$, there exists $k \geq 1$ such that $P(\mathbf{x}^0 ; k, B) > 0$.

□ *Stationary case.* If $P_{\mathbf{X}^0}(d\mathbf{x}^0) = P_S(d\mathbf{x}^0)$, then $\{\mathbf{X}^k, k \in \mathbb{N}\}$ is stationary and,

$$P_{\mathbf{X}^k}(d\mathbf{x}^k) = P_S(d\mathbf{x}^k) \quad , \quad \forall k \geq 0 , \quad (3.73)$$

which means that $P_{\mathbf{X}^k}(d\mathbf{x}^k)$ is independent of k.

□ *Asymptotically stationary case.* If $P_{\mathbf{X}^0}(d\mathbf{x}^0) \neq P_S(d\mathbf{x}^0)$, then $\{\mathbf{X}^k, k \in \mathbb{N}\}$ is not stationary, but is asymptotically stationary. For all $\mathbf{x}^0$ in $\mathbb{R}^n$ and for all B in $\mathcal{B}_n$,

$$\lim_{k \to +\infty} \mathcal{P}\{\mathbf{X}^k \in B \mid \mathbf{X}^0 = \mathbf{x}^0\} = \lim_{k \to +\infty} P(\mathbf{x}^0 ; k, B) = P_S(B) . \quad (3.74)$$

3.7.5 Ergodic Average Using the Asymptotic Stationarity of a Time-Homogeneous Markov Chain: Formulas for Computing Statistics by MCMC Methods

Let $\{\mathbf{X}^k, k \in \mathbb{N}\}$ be a time-homogeneous Markov chain that admits an invariant measure $B \mapsto P_S(d\mathbf{z})$ on $\mathbb{R}^n$, for which the homogeneous transition kernel is irreducible and not periodic. Let us assume that $P_{\mathbf{X}^0}(d\mathbf{x}^0) \neq P_S(d\mathbf{x}^0)$. Let $\{\mathbf{x}^k, k \in \mathbb{N}\}$ be a realization of $\{\mathbf{X}^k, k \in \mathbb{N}\}$. Let $\mathbf{Z}$ be an $\mathbb{R}^n$-valued random variable whose probability distribution is $P_{\mathbf{Z}}(d\mathbf{z}) = P_S(d\mathbf{z})$ and let $\mathbf{f}$ be a nonlinear mapping from $\mathbb{R}^n$ into $\mathbb{R}^m$ such that

$$E\{\|\mathbf{f}(\mathbf{Z})\|\} = \int_{\mathbb{R}^n} \|\mathbf{f}(\mathbf{z})\| \, P_S(d\mathbf{z}) < +\infty . \quad (3.75)$$

Then, for all $k_0 \geq 0$, we have

$$E\{\mathbf{f}(\mathbf{Z})\} = \lim_{K \to +\infty} \frac{1}{K} \sum_{k=k_0}^{k_0+K} \mathbf{f}(\mathbf{x}^k) , \quad (3.76)$$

and, for $k_0 > 0$ and for K sufficiently large,

$$E\{\mathbf{f}(\mathbf{Z})\} \simeq \frac{1}{K} \sum_{k=k_0}^{k_0+K} \mathbf{f}(\mathbf{x}^k) . \quad (3.77)$$

Chapter 4
MCMC Methods for Generating Realizations and for Estimating the Mathematical Expectation of Nonlinear Mappings of Random Vectors

4.1. Definition of the Problem to Be Solved
4.2. Metropolis-Hastings Method
4.3. Algorithm Based on an ISDE for the High Dimensions

Abstract *The Markov Chain Monte Carlo (MCMC) method constitutes a fundamental tool for UQ and in computational statistics, which allows for generating realizations of a random vector for which the pdf is given, and then for computing the mathematical expectation of nonlinear mappings of this random vector. The theoretical bases of MCMC methods have been presented in Chapter 3.*

4.1 Definition of the Problem to Be Solved

4.1.1 Computation of Integrals in Any Dimension, in Particular in High Dimension

Let $\mathbf{Z}$ be an $\mathbb{R}^n$-valued random variable with the probability distribution

$$P_{\mathbf{Z}}(d\mathbf{z}) = p_{\mathbf{Z}}(\mathbf{z}) \, d\mathbf{z} \,, \tag{4.1}$$

which is represented by a pdf $\mathbf{z} \mapsto p_{\mathbf{Z}}(\mathbf{z})$ on $\mathbb{R}^n$, and let $\mathbf{h}$ be a nonlinear mapping from $\mathbb{R}^n$ into $\mathbb{R}^m$ such that $\mathbf{h}(\mathbf{Z})$ be an $\mathbb{R}^m$-valued random variable verifying $E\{\|\mathbf{h}(\mathbf{Z})\|\} < +\infty$. The problem to be solved is the computation of an estimation of

$$E\{\mathbf{h}(\mathbf{Z})\} = \int_{\mathbb{R}^n} \mathbf{h}(\mathbf{z}) \, p_{\mathbf{Z}}(\mathbf{z}) \, d\mathbf{z} \,, \tag{4.2}$$

with n any positive integer that can be large (for instance, 10, 100 1 000, etc.).

© Springer International Publishing AG 2017
C. Soize, *Uncertainty Quantification*, Interdisciplinary Applied Mathematics 47,
DOI 10.1007/978-3-319-54339-0_4

4.1.2 Is the Usual Deterministic Approach Can Be Used

With the usual numerical quadrature methods, a set of points $\mathbf{z}^1, \ldots, \mathbf{z}^K$ in $\mathbb{R}^n$ and the corresponding weights $w_1, \ldots, w_K$ are defined for computing the following approximation,

$$\int_{\mathbb{R}^n} \mathbf{h}(\mathbf{z})\, p_{\mathbf{z}}(\mathbf{z})\, d\mathbf{z} \quad \simeq \quad \sum_{k=1}^{K} w_k\, \mathbf{h}(\mathbf{z}^k). \tag{4.3}$$

The quadrature methods cannot be extended for the high dimensions, because the product rule generates quadrature grids of dimension $K = m^n$ if there are m terms for each coordinate of $\mathbb{R}^n$. Such approaches collapse due to the curse of dimensionality and the statistical methods must be used.

4.1.3 The Computational Statistics Yield Efficient Approaches for the High Dimensions

Two statistical methods are available, the Monte Carlo method and the Markov Chain Monte Carlo (MCMC) methods.

□ *Monte Carlo method.* In Section 2.8.2, the Monte Carlo method has been presented for computing an integral in high dimension (big value of n),

$$\int_{\mathbb{R}^n} \mathbf{h}(\mathbf{z})\, p_{\mathbf{z}}(\mathbf{z})\, d\mathbf{z} \quad \simeq \quad \frac{1}{K} \sum_{k=1}^{K} \mathbf{h}(\mathbf{z}^k), \tag{4.4}$$

in which $\mathbf{z}^1, \ldots, \mathbf{z}^K$ are K independent realizations of the random variable $\mathbf{Z}$ for which its probability distribution is $P_{\mathbf{Z}}(d\mathbf{z}) = p_{\mathbf{z}}(\mathbf{z})\, d\mathbf{z}$. Such a method thus requires a generator of independent realizations.

▷ *We have seen that the rate of convergence is independent of dimension n, even if the convergence is only of order $K^{-1/2}$, but in counter part there is no curse of dimensionality.*

□ *MCMC methods.* The MCMC methods are based on the theoretical results for the ergodic average presented in Sections 3.6.6 and 3.7.5. This class of methods consists in calculating one realization, $\{\mathbf{x}^k, k = 0, \ldots, k_0 + K\}$, of the first $k_0 + K + 1$ terms of an asymptotic stationary time-homogeneous Markov chain, $\{\mathbf{X}^k, k \in \mathbb{N}\}$, starting from $\mathbf{X}^0 = \mathbf{x}^0 \in \mathbb{R}^n$, which admits the probability $P_{\mathbf{z}}(d\mathbf{z}) = p_{\mathbf{z}}(\mathbf{z})\, d\mathbf{z}$ as an invariant measure $P_S(d\mathbf{z})$. For all $k_0 \geq 0$, the ergodic average yields

$$\int_{\mathbb{R}^n} \mathbf{h}(\mathbf{z})\, p_{\mathbf{z}}(\mathbf{z})\, d\mathbf{z} \quad \simeq \quad \frac{1}{K} \sum_{k=k_0}^{k_0+K} \mathbf{h}(\mathbf{x}^k), \tag{4.5}$$

in which,

- $\{\mathbf{x}^k, k = 0, \ldots, k_0 + K\}$ is the set of the vectors that correspond to one realization of dependent random vectors $\{\mathbf{X}^k, k = 0, \ldots, k_0 + K\}$.
- the integer k_0 can be chosen for that $\{\mathbf{x}^k, k = k_0, \ldots, k_0 + K\}$ be one realization of the asymptotic stationary part of the Markov chain.

4.1.4 Algorithms in the Class of the MCMC Methods

□ *Metropolis-Hastings algorithm* (see Section 4.2). The Metropolis-Hastings algorithm is a very general method, but can present some difficulties in the high dimensions, in particular if there are attraction regions that do not correspond to the invariant measure $P_{\mathbf{Z}}(d\mathbf{z}) = p_{\mathbf{Z}}(\mathbf{z}) \, d\mathbf{z}$.

□ *Gibbs sampler* [89]. The Gibbs sampling is a slightly different algorithm of the Metropolis-Hastings method but is better adapted to the multidimensional case. In counter part, this algorithm requires the construction of a random generator for univariate or multivariate conditional probability distributions corresponding to marginal distributions of $P_{\mathbf{Z}}(d\mathbf{z})$, what can induce some difficulties for the high dimensions.

□ *Langevin Metropolis-Hastings Algorithm* [89]. This is a variant of the Metropolis-Hastings algorithm for which the random generator of the proposal distribution is constructed by using a time-discretized Langevin stochastic differential equation, which is an ISDE corresponding to a first-order nonlinear dynamical system driven by a stochastic Gaussian white noise.

□ *Algorithm based on an ISDE associated with a stochastic Hamiltonian dynamical system adapted to the high dimensions* [192, 206] (see Section 4.3). This new algorithm has been developed for the high dimensions and is based on the time discretization of an ISDE corresponding to a second-order nonlinear dynamical system (dissipative Hamiltonian system) driven by a stochastic Gaussian white noise.

4.2 Metropolis-Hastings Method

The Metropolis-Hastings algorithm [136, 137, 105] is a popular MCMC algorithm, based on the theory of the asymptotic stationary time-homogeneous Markov chains for which the invariant measure is a given probability distribution. The theory has been presented in Section 3.7. The use of such an algorithm requires an expertise for constructing a good proposal distribution or the candidate-generating kernel, related to the transition kernel of the Markov chain.

This algorithm depends on a step-size parameter σ:

- if σ is small, the rejection rate is small, but the convergence rate is slow.
- if σ is large, the convergence rate is fast, but the rejection rate is high.

4.2.1 Metropolis-Hastings Algorithm

Let $\mathbf{Z}$ be an $\mathbb{R}^n$-valued random variable whose probability distribution is $P_{\mathbf{Z}}(d\mathbf{z}) = p_{\mathbf{z}}(\mathbf{z})\, d\mathbf{z}$. The Metropolis-Hastings algorithm allows for constructing an asymptotic stationary time-homogeneous Markov chain, $\{\mathbf{X}^k, k \in \mathbb{N}\}$, starting from $\mathbf{X}^0 = \mathbf{x}^0 \in \mathbb{R}^n$, admitting the invariant measure $P_S(d\mathbf{z}) = p_{\mathbf{z}}(\mathbf{z})\, d\mathbf{z}$, and for which $\{\mathbf{x}^k, k \in \mathbb{N}\}$ is a given realization.

The homogeneous transition kernel $\mathrm{P}(\mathbf{x}^k, B) = \mathcal{P}\{\mathbf{X}^{k+1} \in B \mid \mathbf{X}^k = \mathbf{x}^k\}$ is constructed by using a *proposal distribution* $\rho(\mathbf{x}^k, \mathbf{y})$, which is the conditional pdf of a $\mathbb{R}^n$-valued random variable $\mathbf{Y}$ given $\mathbf{X}^k = \mathbf{x}^k$ (and thus $\int_{\mathbb{R}^n} \rho(\mathbf{x}^k, \mathbf{y})\, d\mathbf{y} = 1$).

If $\mathbf{h}$ is a nonlinear mapping from $\mathbb{R}^n$ into $\mathbb{R}^m$ such that $E\{\|\mathbf{h}(\mathbf{Z})\|\} < +\infty$, then, for any $k_0 \geq 0$, and for K sufficiently large, the use of the ergodic average yields

$$E\{\mathbf{h}(\mathbf{Z})\} = \int_{\mathbb{R}^n} \mathbf{h}(\mathbf{z})\, p_{\mathbf{z}}(\mathbf{z})\, d\mathbf{z} \quad \simeq \quad \frac{1}{K} \sum_{k=k_0}^{k_0+K} \mathbf{h}(\mathbf{x}^k). \tag{4.6}$$

Algorithm. The following algorithm allows for generating the realizations $\mathbf{x}^k$ of $\mathbf{X}^k$ for $k = 0, 1 \ldots, K$.

— *Start of the algorithm* —

1. Chose the initial value $\mathbf{x}^0$ in $\mathbb{R}^n$ and set $k = 1$.
2. Calculate a realization $\mathbf{y}$ of $\mathbf{Y}$ with the probability distribution $\rho(\mathbf{x}^k, \mathbf{y})\, d\mathbf{y}$.
3. Calculate the acceptance ratio:

$$a(\mathbf{x}^k, \mathbf{y}) = \min\left\{1, \frac{p_{\mathbf{z}}(\mathbf{y})\, \rho(\mathbf{y}, \mathbf{x}^k)}{p_{\mathbf{z}}(\mathbf{x}^k)\, \rho(\mathbf{x}^k, \mathbf{y})}\right\}.$$

4. Calculate an independent realization u of a uniform random variable U on $[0, 1]$.
5. If $a(\mathbf{x}^k, \mathbf{y}) \geq u$
     set $\mathbf{x}^{k+1} = \mathbf{y}$
     k = k+1
     go to step 2
   else
     reject $\mathbf{y}$
     go to step 2
   end

— *End of the algorithm* —

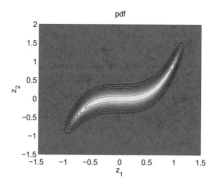

Fig. 4.1 Graph of the pdf $p_{\mathbf{Z}}$ on $\mathbb{R}^2$ for which the MCMC method is used.

4.2.2 Example and Analysis of the Metropolis-Hastings Algorithm

☐ *Giving the probability density function of $\mathbf{Z}$ as the data.* The pdf $p_{\mathbf{Z}}$ on $\mathbb{R}^2$ for which the MCMC method is used is defined by

$$p_{\mathbf{Z}}(\mathbf{z}) = c_0 \exp\{-15(z_1^3 - z_2)^2 - (z_2 - 0.3)^4\} \quad , \quad \mathbf{z} = (z_1, z_2) \in \mathbb{R}^2 \,. \quad (4.7)$$

The graph of the pdf $p_{\mathbf{Z}}$ on $\mathbb{R}^2$ is displayed in Figure 4.1.

☐ *Constructing a proposal distribution $\rho(\mathbf{x}^j, \mathbf{y})$.* As we have explained, a proposal distribution must be constructed for implementing the Metropolis-Hastings algorithm.

1. The proposal distribution $\rho(\mathbf{x}^k, \mathbf{y})$ (conditional pdf) must be constructed by the "expertise" (there is no general methodology for constructing it, contrary to the approach presented in Section 4.3).

2. For instance, the proposal distribution can be constructed using the time discretization of the following ISDE (see Section 3.5.1),

$$d\mathbf{Y}(t) = d\mathbf{W}(t) \quad , \quad t > 0 \,, \quad (4.8)$$

with the initial condition $\mathbf{Y}(0) = \mathbf{x}^0$, in which $\mathbf{x}^0$ is chosen in $\mathbb{R}^2$, and where $\{\mathbf{W}(t), t \in \mathbb{R}^+\}$ is the $\mathbb{R}^2$-valued normalized Wiener process (see Section 3.3.3). Consequently, $\{\mathbf{Y}(t), t \in \mathbb{R}^+\}$ is a Markov process with homogeneous transition probability, for which the time sampling, $\{\mathbf{Y}(j \, \Delta t), j \in \mathbb{N}\}$, yields a time-homogeneous Markov chain: $\{\mathbf{Y}^j, j \in \mathbb{N}\}$ with $\mathbf{Y}^0 = \mathbf{x}^0$ a.s.

3. The time discretization of $d\mathbf{Y}(t) = d\mathbf{W}(t)$ yields

$$\mathbf{Y}^{k+1} - \mathbf{Y}^k = \Delta \mathbf{W}_{t_k, t_{k+1}} = \mathbf{W}(t_{k+1}) - \mathbf{W}(t_k) = \sigma \, \mathbf{N}^{k+1} \,, \quad (4.9)$$

in which $\sigma = \sqrt{\Delta t}$ is a parameter controlling the step size and where $\mathbf{N}^1, \mathbf{N}^2, \ldots$ is a sequence of independent $\mathbb{R}^2$-valued normalized Gaussian vectors.

4. The *random generator* for drawing $\mathbf{y}$ of random vector $\mathbf{Y}$ from the conditional probability density function $\rho(\mathbf{x}^k, \mathbf{y})$ (the proposal distribution), given $\mathbf{Y}^k = \mathbf{x}^k$, is thus written as

$$\mathbf{Y} = \mathbf{x}^k + \sigma\,\mathbf{N}^{k+1}, \qquad (4.10)$$

and corresponds to the Gaussian pdf for $\mathbf{Y}$ with mean value $\mathbf{x}^k$ and with covariance matrix $\sigma^2\,[I_2]$,

$$\rho(\mathbf{x}^k, \mathbf{y}) = c_g \times \exp\left\{\frac{1}{2\sigma^2}\|\mathbf{y} - \mathbf{x}^k\|^2\right\}, \qquad (4.11)$$

which is symmetric because $\rho(\mathbf{x}^k, \mathbf{y}) = \rho(\mathbf{y}, \mathbf{x}^k)$.

☐ *Numerical simulation with the Metropolis-Hastings algorithm.* For the example that is presented, the initial condition is $\mathbf{x}^0 = (1.4, -0.75)$ and the number of draws, ν_{draw}, is fixed to the value $10\,000$. The numerical simulations depend on the step-size parameter σ. Figure 4.2 displays the simulations for 4 different values of σ that are $0.02, 0.05, 0.1$, and 0.25.

☐ *Number of realizations and rejection rate as a function of the step size parameter σ.* For each fixed value of σ, Table 4.1 gives the number K of realizations

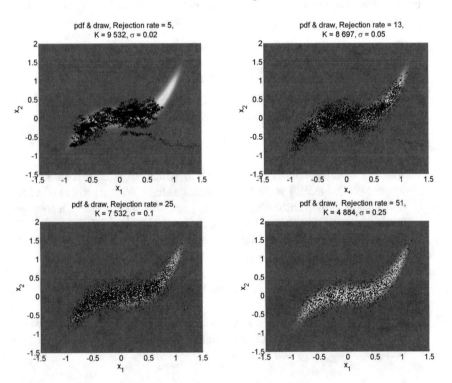

Fig. 4.2 Numerical simulation obtained with the Metropolis-Hastings algorithm for $\sigma = 0.02$ (up left), 0.05 (up right), 0.1 (down left), and 0.25 (down right).

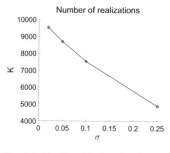 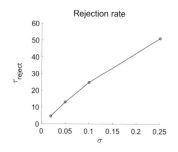

Fig. 4.3 Number K of realizations (left figure) and rejection rate τ_{reject} in percent (right figure) as a function of the step-size parameter σ.

(the draws that are not rejected) and the rejection rate (in percent) such that $\tau_{\text{reject}} = 100 \times (1 - K/\nu_{\text{draw}})$. Figure 4.3 displays the number K of realizations and the rejection rate τ_{reject} as a function of the step-size parameter σ.

□ *Ergodic average for computing the second-order moment m_2 of the random vector.* For the step-size parameter σ fixed to the value 0.02, the convergence of the second-order moment of the random vector $\mathbf{Z}$ is analyzed as a function of the number K of realizations. As an illustration, for $K = 944\,863$, Figure 4.4 displays the sample path $\{x_1^k, k = 1, \ldots, K\}$ of the random sequence $\{X_1^k, k = 1, \ldots, K\}$ and the sample path $\{x_2^k, k = 1, \ldots, K\}$ of the random sequence $\{X_2^k, k = 1, \ldots, K\}$. The second-order moment $E\{\|\mathbf{Z}\|^2\}$ of random vector $\mathbf{Z}$ is estimated by $m_2(K)$ that is computed with the ergodic average as a function of k,

Table 4.1 Values of τ_{reject} in percent and K as a function of σ.

σ	$\tau_{\text{reject}}(\%)$	K
0.02	5	9 532
0.05	13	8 697
0.1	25	7 532
0.25	51	4 884

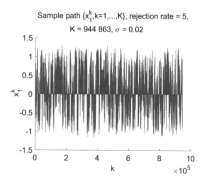 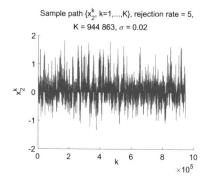

Fig. 4.4 Sample path $\{x_1^k, k = 1, \ldots, K\}$ (left figure) and sample path of $\{x_2^k, k = 1, \ldots, K\}$ (right figure).

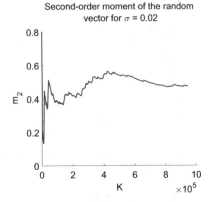

Fig. 4.5 Graph $K \mapsto$
$m_2(K)$ of the convergence of
the second-order moment.

$$m_2(K) = \frac{1}{K} \sum_{k=1}^{K} \|\mathbf{x}^k\|^2 . \qquad (4.12)$$

Figure 4.5 displays the graph $K \mapsto m_2(K)$ and shows that the convergence is
obtained for $K = 900{,}000$.

4.3 Algorithm Based on an ISDE for the High Dimensions

This algorithm, introduced for the high dimensions, is based on the use of an ISDE
associated with a stochastic dissipative Hamiltonian dynamical system, driven by
a stochastic Gaussian white noise [192, 100, 206, 207], and for which a damping
parameter allows for controlling the obtention of the stationary solution. The developments are based on the use of the theoretical results that have been presented in
Section 3.6.

4.3.1 Summarizing the Methodology for Constructing
the Algorithm

It is assumed first that the support of the probability distribution $P_{\mathbf{Z}}(d\mathbf{z}) = p_{\mathbf{z}}(\mathbf{z}) \, d\mathbf{z}$
is the entire space $\mathbb{R}^n$. The ISDE is constructed in order that the probability distri-
bution $P_{\mathbf{Z}}(d\mathbf{z})$ of the $\mathbb{R}^n$-valued random variable $\mathbf{Z}$ is a unique invariant measure.
A time-discretization is introduced and allows for constructing a random generator
for $\mathbf{Z}$ and for computing $E\{\mathbf{h}(\mathbf{Z})\} = \int_{\mathbb{R}^n} \mathbf{h}(\mathbf{z}) \, p_{\mathbf{z}}(\mathbf{z}) \, d\mathbf{z}$.

Secondly, it is assumed that the support of the probability distribution $P_{\mathbf{Z}}(d\mathbf{z}) = p_{\mathbf{z}}(\mathbf{z})\, d\mathbf{z}$ is a bounded subset $\mathcal{S}_n \subset \mathbb{R}^n$. A first extension (by rejection) is introduced for computing $E\{\mathbf{h}(\mathbf{Z})\} = \int_{\mathcal{S}_n} \mathbf{h}(\mathbf{z})\, p_{\mathbf{z}}(\mathbf{z})\, d\mathbf{z}$. A second extension (by regularization) can then be introduced for computing $E\{\mathbf{h}(\mathbf{Z})\} = \int_{\mathcal{S}_n} \mathbf{h}(\mathbf{z})\, p_{\mathbf{z}}(\mathbf{z})\, d\mathbf{z}$.

4.3.2 First Case: The Support Is the Entire Space $\mathbb{R}^n$

The support of the pdf $\mathbf{z} \mapsto p_{\mathbf{z}}(\mathbf{z})$ on $\mathbb{R}^n$ is assumed to be the entire space $\mathbb{R}^n$. There is a potential function $\mathbf{z} \mapsto \Phi(\mathbf{z})$ and a finite constant $c_n > 0$ such that $p_{\mathbf{z}}$ can be written as

$$\forall \mathbf{z} \in \mathbb{R}^n \quad , \quad p_{\mathbf{z}}(\mathbf{z}) = c_n\, e^{-\Phi(\mathbf{z})} . \tag{4.13}$$

It is assumed that $\mathbf{z} \mapsto \Phi(\mathbf{z})$: (i) is continuous on $\mathbb{R}^n$, (ii) is such that $\mathbf{z} \mapsto \|\nabla_{\mathbf{z}}\Phi(\mathbf{z})\|$ is a locally bounded function on $\mathbb{R}^n$, and (iii) is such that

$$\inf_{\|\mathbf{z}\|>R} \Phi(\mathbf{z}) \to +\infty \quad \text{if} \quad R \to +\infty , \tag{4.14}$$

$$\inf_{\mathbf{z}\in\mathbb{R}^n} \Phi(\mathbf{z}) = \Phi_{\min} \quad \text{with} \quad \Phi_{\min} \in \mathbb{R} , \tag{4.15}$$

$$\int_{\mathbb{R}^n} \|\nabla_{\mathbf{z}}\Phi(\mathbf{z})\|\, e^{-\Phi(\mathbf{z})}\, d\mathbf{z} < +\infty . \tag{4.16}$$

☐ *Proposition* (the elements of the proof are given after). Let $\{(\mathbf{U}(t), \mathbf{V}(t)),\ t \in \mathbb{R}^+\}$ be the $\mathbb{R}^n \times \mathbb{R}^n$-valued stochastic process defined on the probability space $(\Theta, \mathcal{T}, \mathcal{P})$, satisfying, for all $t > 0$,

$$d\mathbf{U}(t) = \mathbf{V}(t)\, dt ,$$

$$d\mathbf{V}(t) = -\nabla_{\mathbf{u}}\Phi(\mathbf{U}(t))\, dt - \frac{1}{2} f_0 \mathbf{V}(t)\, dt + \sqrt{f_0}\, d\mathbf{W}(t) , \tag{4.17}$$

with the initial condition $\mathbf{U}(0) = \mathbf{u}_0 \in \mathbb{R}^n$ and $\mathbf{V}(0) = \mathbf{v}_0 \in \mathbb{R}^n$, where $\mathbf{W}$ is the $\mathbb{R}^n$-valued normalized Wiener process defined on $(\Theta, \mathcal{T}, \mathcal{P})$, where $f_0 > 0$ (see after), and where $\mathbf{u} \mapsto \Phi(\mathbf{u})$ verifies the previous hypotheses. Then, there is a unique asymptotic stationary solution, which is a second-order diffusion Markov process admitting an invariant measure $P_S(d\mathbf{u}, d\mathbf{v}) = \rho_s(\mathbf{u}, \mathbf{v})\, d\mathbf{u}\, d\mathbf{v}$, such that

$$\rho_s(\mathbf{u}, \mathbf{v}) = c_{2n}\, \exp\{-\frac{1}{2}\|\mathbf{v}\|^2 - \Phi(\mathbf{u})\} , \tag{4.18}$$

in which $c_{2n} > 0$ is the constant of normalization, and which is such that

$$\lim_{t\to+\infty} P_{\mathbf{U}(t),\mathbf{V}(t)}(t, d\mathbf{u}, d\mathbf{v}) = P_S(d\mathbf{u}, d\mathbf{v}) . \tag{4.19}$$

Consequently, it can be deduced that

$$\lim_{t\to+\infty} P_{\mathbf{U}(t)}(t, d\mathbf{u}) = p_{\mathbf{z}}(\mathbf{u})\, d\mathbf{u} . \tag{4.20}$$

□ *Ergodic average for computing $E\{\mathbf{h}(\mathbf{Z})\}$*. Using Section 3.6.6, for any $t_0 \geq 0$,

$$E\{\mathbf{h}(\mathbf{Z})\} = \lim_{\tau \to +\infty} \frac{1}{\tau} \int_{t_0}^{t_0+\tau} \mathbf{h}(\mathbf{U}(t,\theta))\, dt\,, \qquad (4.21)$$

in which, for any θ in Θ, $\{\mathbf{U}(t,\theta), t \in \mathbb{R}^+\}$ is one realization of the stochastic process $\{\mathbf{U}(t), t \in \mathbb{R}^+\}$, which is computed by solving the ISDE,

$$d\mathbf{U}(t,\theta) = \mathbf{V}(t,\theta)\, dt\,, \qquad (4.22)$$

$$d\mathbf{V}(t,\theta) = -\boldsymbol{\nabla}_{\mathbf{u}}\Phi(\mathbf{U}(t,\theta))\, dt - \frac{1}{2}f_0\mathbf{V}(t,\theta)\, dt + \sqrt{f_0}\, d\mathbf{W}(t,\theta)\,, \quad (4.23)$$

with the initial condition

$$\mathbf{U}(0,\theta) = \mathbf{u}_0 \in \mathbb{R}^n \quad, \quad \mathbf{V}(0,\theta) = \mathbf{v}_0 \in \mathbb{R}^n\,, \qquad (4.24)$$

in which $\{\mathbf{W}(t,\theta), t \in \mathbb{R}^+\}$ is a realization of the $\mathbb{R}^n$-valued normalized Wiener process $\{\mathbf{W}(t), t \in \mathbb{R}^+\}$.

□ *Role plays by the parameter f_0 and choice of the initial conditions*. The free parameter $f_0 > 0$ allows for introducing a dissipation term in the second-order nonlinear dynamical system corresponding to the dissipative Hamiltonian dynamical system,

$$\ddot{\mathbf{U}}(t) + \frac{1}{2}f_0\dot{\mathbf{U}}(t) + \boldsymbol{\nabla}_{\mathbf{u}}\Phi(\mathbf{U}(t)) = \sqrt{f_0}\,\mathbf{N}_\infty(t)\,, \qquad (4.25)$$

in order to remove the transient part of the response induced by the initial conditions, and consequently, allows for getting more rapidly the stationary solution corresponding to the invariant measure. Concerning the initial conditions, in general, there is no information for $\mathbf{u}_0$ that is chosen as $\mathbf{u}_0 = 0$, but $\mathbf{v}_0$ can be chosen as a realization of an $\mathbb{R}^n$-valued normalized Gaussian random vector.

□ *Random generator for $\mathbf{Z}$ and Monte Carlo method for computation of $E\{\mathbf{h}(\mathbf{Z})\}$*. Let $\theta_1, \ldots, \theta_\nu$ be in Θ such that, for $\ell = 1, \ldots, \nu$, $\{\mathbf{W}(t,\theta_\ell), t \in \mathbb{R}^+\}$ is a set of ν independent realizations of $\mathbf{W}$, and $\mathbf{v}_0^\ell$ is a set of independent realizations of an $\mathbb{R}^n$-valued normalized Gaussian random vector (also independent of $\mathbf{W}$). For $\ell = 1, \ldots, \nu$, let $\{\mathbf{U}(t,\theta_\ell), t \in \mathbb{R}^+\}$ be the ν independent realizations of stochastic process $\{\mathbf{U}(t), t \in \mathbb{R}^+\}$, which are computed by solving

$$d\mathbf{U}(t,\theta_\ell) = \mathbf{V}(t,\theta_\ell)\, dt\,, \qquad (4.26)$$

$$d\mathbf{V}(t,\theta_\ell) = -\boldsymbol{\nabla}_{\mathbf{u}}\Phi(\mathbf{U}(t,\theta_\ell))\, dt - \frac{1}{2}f_0\mathbf{V}(t,\theta_\ell)\, dt + \sqrt{f_0}\, d\mathbf{W}(t,\theta_\ell)\,, \quad (4.27)$$

with the initial condition

$$\mathbf{U}(0,\theta) = 0 \quad, \quad \mathbf{V}(0,\theta_\ell) = \mathbf{v}_0^\ell\,. \qquad (4.28)$$

The generator of independent realizations $\mathbf{Z}(\theta_1), \ldots, \mathbf{Z}(\theta_\nu)$ of $\mathbf{Z}$ is such that

$$\mathbf{Z}(\theta_\ell) = \lim_{t \to +\infty} \mathbf{U}(t, \theta_\ell) \quad \text{in probability distribution} . \tag{4.29}$$

The Monte Carlo method is used for computing $E\{\mathbf{h}(\mathbf{Z})\}$,

$$E\{\mathbf{h}(\mathbf{Z})\} = \lim_{\nu \to +\infty} \frac{1}{\nu} \sum_{\ell=1}^{\nu} \mathbf{h}(\mathbf{Z}(\theta_\ell)) . \tag{4.30}$$

☐ *Elements for the proof of the proposition.* Under the hypotheses of the proposition, using theorems 4 to 7 in pages 211 to 216 of [183], in which the Hamiltonian is taken as $\mathbb{H}(\mathbf{u}, \mathbf{v}) = \|\mathbf{v}\|^2/2 + \Phi(\mathbf{u})$, it is deduced that there is a unique solution, which is an asymptotic stationary second-order diffusion Markov process whose density $\rho_s(\mathbf{u}, \mathbf{v})$ of the invariant measure $P_S(d\mathbf{u}, d\mathbf{v})$ is the unique solution of the steady-state Fokker-Planck equation

$$\sum_{j=1}^{n} \frac{\partial}{\partial u_j} \{v_j \rho_s\} + \sum_{j=1}^{n} \frac{\partial}{\partial v_j} \{(-\frac{\partial \Phi(\mathbf{u})}{\partial u_j} - \frac{f_0}{2} v_j) \rho_s\} - \frac{f_0}{2} \sum_{j=1}^{n} \frac{\partial^2 \rho_s}{\partial v_j^2} = 0,$$

with the normalization condition $\int_{\mathbb{R}^n} \int_{\mathbb{R}^n} \rho_s(\mathbf{u}, \mathbf{v}) \, d\mathbf{u} \, d\mathbf{v} = 1$. Using the results from [183] (pp. 268–272), it is proved that the steady-state FKP has the unique solution $\rho_s(\mathbf{u}, \mathbf{v})$ given in the proposition.

☐ *Discretization scheme of the ISDE.* The Störmer-Verlet scheme [26, 103], which is an efficient scheme that preserves energy for nondissipative Hamiltonian dynamical systems, is used. Let Δt be the sampling step and let $t_k = (k-1) \Delta t$ for $k = 1, \ldots, K$. Let $\mathbf{U}^k = \mathbf{U}(t_k)$, $\mathbf{V}^k = \mathbf{V}(t_k)$, and $\mathbf{W}^k = \mathbf{W}(t_k)$), with $\mathbf{U}^1 = 0$, $\mathbf{V}^1 = \mathbf{v}_0$, and $\mathbf{W}^1 = 0$. For $k = 1, \ldots, K - 1$, the Störmer-Verlet scheme yields

$$\mathbf{U}^{k+\frac{1}{2}} = \mathbf{U}^k + \frac{\Delta t}{2} \mathbf{V}^k , \tag{4.31}$$

$$\mathbf{V}^{k+1} = \frac{1-b}{1+b} \mathbf{V}^k + \frac{\Delta t}{1+b} \mathbf{L}^{k+\frac{1}{2}} + \frac{\sqrt{f_0}}{1+b} \Delta \mathbf{W}^{k+1} , \tag{4.32}$$

$$\mathbf{U}^{k+1} = \mathbf{U}^{k+\frac{1}{2}} + \frac{\Delta t}{2} \mathbf{V}^{k+1} , \tag{4.33}$$

in which $\Delta \mathbf{W}^{k+1} = \mathbf{W}^{k+1} - \mathbf{W}^k$, where $b = f_0 \Delta t / 4$, and where the random vector $\mathbf{L}^{k+\frac{1}{2}}$ is written as $\mathbf{L}^{k+\frac{1}{2}} = -\{\nabla_{\mathbf{u}} \Phi(\mathbf{u})\}_{\mathbf{u} = \mathbf{U}^{k+\frac{1}{2}}} .$

☐ *Example and analysis of the ISDE-based algorithm for the first case for which the support is the entire space $\mathbb{R}^2$.* The example presented in Section 4.2.2 for the Metropolis-Hastings algorithm is reused here. The probability density function $p_{\mathbf{z}}$ on $\mathbb{R}^2$ is then defined by

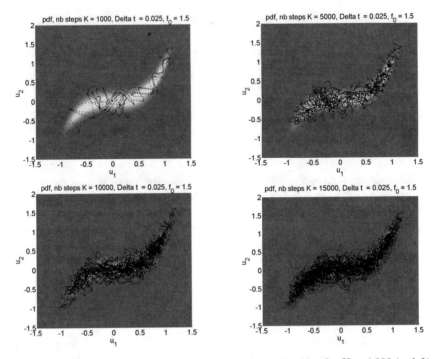

Fig. 4.6 Numerical simulation obtained with the ISDE-based algorithm for $K = 1\,000$ (up left), $5\,000$ (up right), $10\,000$ (down left), $15\,000$ (down right).

$$p_{\mathbf{z}}(\mathbf{z}) = c_0 \, \exp\{-15(z_1^3 - z_2)^2 - (z_2 - 0.3)^4\} \quad , \quad \mathbf{z} = (z_1, z_2) \in \mathbb{R}^2 \,, \quad (4.34)$$

$$\Phi(\mathbf{z}) = 15(z_1^3 - z_2)^2 + (z_2 - 0.3)^4 \,. \tag{4.35}$$

The graph of the pdf $p_{\mathbf{z}}$ on $\mathbb{R}^2$ is displayed in Figure 4.1. For $\Delta t = 0.025$ and for $f_0 = 1.5$, Figure 4.6 shows the results obtained with the ISDE-based algorithm as a function of the number K of the time steps, which is $1\,000$ (up left figure), $5\,000$ (up right figure), $10\,000$ (down left figure), and $15\,000$ (down right figure). Figure 4.7 displays the sample path $\{u_1^k, k = 1, \ldots, K\}$ of the random sequence $\{U_1^k, k = 1, \ldots, K\}$ and the sample path $\{u_2^k, k = 1, \ldots, K\}$ of the random sequence $\{U_2^k, k = 1, \ldots, K\}$ for which the number of time steps is $K = 15\,000$. The ergodic average is used for computing the estimation $m_2(K)$ of the second-order moment $E\{\|\mathbf{Z}\|^2\}$ of the random vector $\mathbf{Z}$, as a function of the number K of time steps,

$$m_2(K) = \frac{1}{K} \sum_{k=1}^{K} \|\mathbf{x}^k\|^2 \,. \tag{4.36}$$

Figure 4.8 displays the graph $K \mapsto m_2(K)$, which allows a convergence analysis to be carried out. The convergence is obtained for $9\,000$ time steps, which has to be compared to $900\,000$ realizations that are required by the Metropolis-Hastings algorithm for obtaining a good value of m_2.

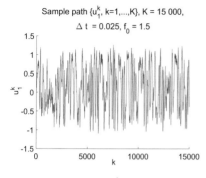

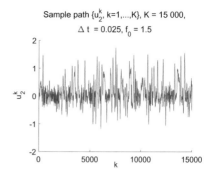

Fig. 4.7 Sample path $\{u_1^k, k = 1, \ldots, K\}$ (left figure) and $\{u_2^k, k = 1, \ldots, K\}$ (right figure).

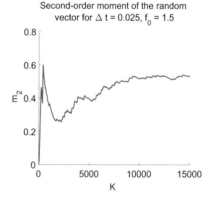

Fig. 4.8 Graph $K \mapsto m_2(K)$ for analyzing the convergence of the second-order moment.

4.3.3 Second Case: The Support Is a Known Bounded Subset of $\mathbb{R}^n$ (Rejection Method)

Let $\mathbf{Z}_b$ be an $\mathbb{R}^n$-valued random variable with a pdf $\mathbf{z} \mapsto p_{\mathbf{Z}_b}(\mathbf{z})$ on $\mathbb{R}^n$, for which its support is a known bounded subset $\mathcal{S}_n \subset \mathbb{R}^n$. For $\mathbf{z} \in \mathcal{S}_n$, there is a potential function $\mathbf{z} \mapsto \Phi_{\mathcal{S}_n}(\mathbf{z})$ defined on $\mathcal{S}_n$ and a finite constant $c_b > 0$ such that

$$\forall \mathbf{z} \in \mathcal{S}_n \subset \mathbb{R}^n, \quad , \quad p_{\mathbf{Z}_b}(\mathbf{z}) = c_b \, e^{-\Phi_{\mathcal{S}_n}(\mathbf{z})} . \tag{4.37}$$

It is assumed that function $\mathbf{z} \mapsto \Phi_{\mathcal{S}_n}(\mathbf{z})$ on $\mathcal{S}_n$ can be extended to a function $\mathbf{z} \mapsto \Phi(\mathbf{z})$ on the entire space $\mathbb{R}^n$, which verifies the hypotheses introduced in Section 4.3.2. The pdf $p_{\mathbf{Z}_b}$ can then be rewritten as

$$\forall \mathbf{z} \in \mathbb{R}^n \quad , \quad p_{\mathbf{Z}_b}(\mathbf{z}) = c_b \, \mathbb{1}_{\mathcal{S}_n}(\mathbf{z}) \, e^{-\Phi(\mathbf{z})} . \tag{4.38}$$

Let $\mathbf{Z}$ be the $\mathbb{R}^n$-valued random vector for which the pdf is written as

$$\forall \mathbf{z} \in \mathbb{R}^n \quad , \quad p_{\mathbf{Z}}(\mathbf{z}) = c_n \, e^{-\Phi(\mathbf{z})} , \tag{4.39}$$

for which its support is the entire space $\mathbb{R}^n$. We have

$$
\begin{aligned}
E\{\mathbf{h}(\mathbf{Z}_b)\} &= \int_{\mathbb{R}^n} \mathbf{h}(\mathbf{z})\, \mathbb{1}_{\mathcal{S}_n}(\mathbf{z})\, c_b\, e^{-\Phi(\mathbf{z})}\, d\mathbf{z} \\
&= \frac{c_b}{c_n} \int_{\mathbb{R}^n} \mathbf{h}(\mathbf{z})\, \mathbb{1}_{\mathcal{S}_n}(\mathbf{z})\, c_n\, e^{-\Phi(\mathbf{z})}\, d\mathbf{z} \\
&= \frac{c_b}{c_n} E\{\mathbf{h}(\mathbf{Z})\, \mathbb{1}_{\mathcal{S}_n}(\mathbf{Z})\}
\end{aligned}
\tag{4.40}
$$

Taking $\mathbf{h}(\mathbf{z}) = 1$ yields

$$
\frac{c_b}{c_n} = \frac{1}{E\{\mathbb{1}_{\mathcal{S}_n}(\mathbf{Z})\}},
\tag{4.41}
$$

and consequently, we obtain the formula that corresponds to a rejection method,

$$
E\{\mathbf{h}(\mathbf{Z}_b)\} = \frac{E\{\mathbf{h}(\mathbf{Z})\, \mathbb{1}_{\mathcal{S}_n}(\mathbf{Z})\}}{E\{\mathbb{1}_{\mathcal{S}_n}(\mathbf{Z})\}},
\tag{4.42}
$$

for which $E\{\mathbb{1}_{\mathcal{S}_n}(\mathbf{Z})\}$ and $E\{\mathbf{h}(\mathbf{Z})\, \mathbb{1}_{\mathcal{S}_n}(\mathbf{Z})\}$ are computed by using the method presented in Section 4.3.2.

4.3.4 Third Case: The Support Is a Known Bounded Subset of $\mathbb{R}^n$ (Regularization Method)

This method of regularization is detailed in [100]. Let $\mathbf{Z}_b$ be an $\mathbb{R}^n$-valued random variable with a pdf $\mathbf{z} \mapsto p_{\mathbf{Z}_b}(\mathbf{z})$ on $\mathbb{R}^n$, for which its support is a known bounded subset $\mathcal{S}_n \subset \mathbb{R}^n$. For $\mathbf{z} \in \mathcal{S}_n$, there is a potential function $\mathbf{z} \mapsto \Phi_{\mathcal{S}_n}(\mathbf{z})$ defined on $\mathcal{S}_n$ and a finite constant $c_b > 0$ such that

$$
\forall \mathbf{z} \in \mathcal{S}_n \subset \mathbb{R}^n, \quad , \quad p_{\mathbf{Z}_b}(\mathbf{z}) = c_b\, \mathbb{1}_{\mathcal{S}_n}(\mathbf{z})\, e^{-\Phi_{\mathcal{S}_n}(\mathbf{z})}.
\tag{4.43}
$$

It is assumed that function $\mathbf{z} \mapsto \Phi_{\mathcal{S}_n}(\mathbf{z})$ on $\mathcal{S}_n$ can be extended to a function $\mathbf{z} \mapsto \Phi(\mathbf{z})$ on the entire space $\mathbb{R}^n$, and the pdf $p_{\mathbf{Z}_b}$ can then be rewritten as

$$
\forall \mathbf{z} \in \mathbb{R}^n \quad , \quad p_{\mathbf{Z}_b}(\mathbf{z}) = c_b\, \mathbb{1}_{\mathcal{S}_n}(\mathbf{z})\, e^{-\Phi(\mathbf{z})} = c_b\, e^{\log(\mathbb{1}_{\mathcal{S}_n}(\mathbf{z})) - \Phi(\mathbf{z})}.
\tag{4.44}
$$

Let $\mathbf{Z}_\varepsilon$ be the $\mathbb{R}^n$-valued random vector with the pdf,

$$
\forall \mathbf{z} \in \mathbb{R}^n \quad , \quad p_{\mathbf{Z}_\varepsilon}(\mathbf{z}) = c_\varepsilon\, e^{-\Phi_\varepsilon(\mathbf{z})},
\tag{4.45}
$$

in which $\mathbf{z} \mapsto \Phi_\varepsilon(\mathbf{z})$ is the function defined on the entire space $\mathbb{R}^n$, which verifies the hypotheses introduced in Section 4.3.2, and such that

$$
\Phi_\varepsilon(\mathbf{z}) = -\log(\mathbb{1}_{\mathcal{S}_n}^\varepsilon(\mathbf{z})) + \Phi(\mathbf{z}),
\tag{4.46}
$$

where $\mathbb{1}^{\varepsilon}_{\mathcal{S}_n}$ is a positive regularization of $\mathbb{1}_{\mathcal{S}_n}$ using a convolution filter on $\mathbb{R}^n$ with an $\mathbb{R}^n$-valued normalized Gaussian kernel with a covariance matrix $\varepsilon^2[I_n]$. The method presented in Section 4.3.2 can then be used in which $\mathbf{Z}$ and $\Phi(\mathbf{z})$ are replaced by $\mathbf{Z}_{\varepsilon}$ and $\Phi_{\varepsilon}(\mathbf{z})$.

4.3.5 Fourth Case: The Support Is Unknown and a Set of Realizations Is Known. Powerful Algorithm for the High Dimensions

This method is detailed in [206].

□ *Description of data.* Let $\mathbf{Z}$ be a second-order $\mathbb{R}^n$-valued random variable with a pdf $\mathbf{z} \mapsto p_{\mathbf{Z}}(\mathbf{z})$ on $\mathbb{R}^n$, which is unknown, and consequently, for which its support is unknown. The data are made up of ν_r independent realizations, $\mathbf{z}^{r,1}, \ldots, \mathbf{z}^{r,\nu}$ of $\mathbf{Z}$, which have previously been statistically reduced by using the PCA (see Section 5.7.1). Consequently, the empirical statistical estimates, $\widehat{\mathbf{m}}_{\mathbf{Z}}$ of the mean vector $\mathbf{m}_{\mathbf{Z}}$ and $[\widehat{R}_{\mathbf{Z}}]$ of the correlation matrix $[R_{\mathbf{Z}}]$, are written as

$$\widehat{\mathbf{m}}_{\mathbf{Z}} = \frac{1}{\nu_r} \sum_{\ell=1}^{\nu_r} \mathbf{z}^{r,\ell} = \mathbf{0} \quad , \quad [\widehat{R}_{\mathbf{Z}}] = \frac{1}{\nu_r - 1} \sum_{\ell=1}^{\nu_r} \mathbf{z}^{r,\ell} (\mathbf{z}^{r,\ell})^T = [I_n]. \quad (4.47)$$

□ *Problem to be solved and methodology.* The problem to be solved is the construction of a generator of independent realizations of $\mathbf{Z}$. The methodology consists

- in constructing a representation of pdf $p_{\mathbf{Z}}$ using the multivariate kernel-density estimation of the nonparametric statistics (see Section 7.2).
- for the generator, in using the algorithm based on an ISDE for the high dimension presented in the first case for the entire space $\mathbb{R}^n$ (see Section 4.3.2).

□ *Multivariate kernel-density estimation.* The pdf $p_{\mathbf{Z}}$ is therefore estimated by using the multivariate Gaussian kernel density estimation method [25] on the basis of the knowledge of the ν_r independent realizations, $\mathbf{z}^{r,1}, \ldots, \mathbf{z}^{r,\nu_r}$ (see Section 7.2). A modification of this classical multivariate Gaussian kernel density estimation method is introduced in order that Equation (4.47) be preserved [206] (modification of the Silverman bandwidth defined by Equation (7.10)),

$$p_{\mathbf{Z}}(\mathbf{z}) = c_n \, q(\mathbf{z}), \quad (4.48)$$

$$c_n = \frac{1}{(\sqrt{2\pi}\,\widehat{s}_n)^n} \quad , \quad q(\mathbf{z}) = \frac{1}{\nu_r} \sum_{\ell=1}^{\nu_r} \exp\{-\frac{1}{2\widehat{s}_n^2}\|\frac{\widehat{s}_n}{s_n}\mathbf{z}^{r,\ell} - \mathbf{z}\|^2\}, \quad (4.49)$$

$$s_n = \left\{\frac{4}{\nu(2+n)}\right\}^{1/(n+4)} \quad , \quad \widehat{s}_n = \frac{s_n}{\sqrt{s_n^2 + \frac{\nu-1}{\nu}}}. \quad (4.50)$$

It can be verified that

$$E\{\mathbf{Z}\} = \int_{\mathbb{R}^n} \mathbf{z}\, p_{\mathbf{z}}(\mathbf{z})\, d\mathbf{z} = \frac{\widehat{s}_n}{s_n}\, \widehat{\mathbf{m}}_{\mathbf{Z}} = \mathbf{0}\,, \tag{4.51}$$

$$E\{\mathbf{Z}\mathbf{Z}^T\} = \int_{\mathbb{R}^n} \mathbf{z}\mathbf{z}^T\, p_{\mathbf{z}}(\mathbf{z})\, d\mathbf{z} = \widehat{s}_n^2\,[\,I_n\,] + (\frac{\widehat{s}_n}{s_n})^2\,\frac{(\nu-1)}{\nu}[\widehat{R}_{\mathbf{Z}}] = [\,I_n\,]\,. \tag{4.52}$$

☐ *Generator using the ISDE.* We use the algorithm based on an ISDE for the high dimensions, which has been presented in Section 4.3.2, for which the support of $p_{\mathbf{z}}$ is the entire space $\mathbb{R}^n$. The pdf $p_{\mathbf{z}}$ is rewritten as

$$\forall\, \mathbf{z} \in \mathbb{R}^n, \quad , \quad p_{\mathbf{z}}(\mathbf{z}) = c_n\, e^{-\Phi(\mathbf{z})}\,, \tag{4.53}$$

in which the potential function $\mathbf{z} \mapsto \Phi(\mathbf{z})$ is

$$\Phi(\boldsymbol{\eta}) = -\log\{q(\boldsymbol{\eta})\}\,. \tag{4.54}$$

☐ *Extension of the method for data-driven probability concentration and sampling on manifold.* The above method has been extended [210] to the more general case for which the probability distribution that is estimated from data, is concentrated on a manifold, and is constructed by using the diffusion maps [49].

Chapter 5
Fundamental Probabilistic Tools for Stochastic Modeling of Uncertainties

Abstract *This chapter is central and is devoted to the main tools that are necessary for constructing the stochastic modeling of uncertainties and for studying their propagations in computational models.*

5.1 Why a Probability Distribution Cannot Arbitrarily be Chosen for a Stochastic Modeling

5.1.1 Illustration of the Problem with the Parametric Stochastic Approach of Uncertainties

In order to explain why the probability distribution of a random vector $\mathbf{X}$, which models an uncertain parameter, cannot be chosen arbitrarily, we recall the problem presented in Section 1.6, which has to be solved and which is summarized in Figure 5.1. The probability distribution $P_{\mathbf{X}}$ of the random parameter $\mathbf{X}$ is unknown, the random observation of the system is the random vector $\mathbf{U}$, and the deterministic

© Springer International Publishing AG 2017

C. Soize, *Uncertainty Quantification*, Interdisciplinary Applied Mathematics 47, DOI 10.1007/978-3-319-54339-0_5

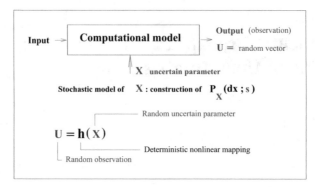

Fig. 5.1 Scheme describing the parametric stochastic approach of uncertainties.

mapping $\mathbf{h}$, such that $\mathbf{U} = \mathbf{h}(\mathbf{X})$, is assumed to given (in general, this mapping is not explicitly known and requires to construct solutions of the computational model). The problem consists in constructing the stochastic modeling of $\mathbf{X}$ (by using the direct approach or the indirect approach, see Section 5.2), and then, when $P_{\mathbf{X}}$ has been constructed, by studying the propagation of the uncertainties in the computational model, that is to say, by constructing the probability distribution $P_{\mathbf{U}}$ of $\mathbf{U}$.

5.1.2 Impact of an Arbitrary Stochastic Modeling of Uncertain Parameter X

The first two main steps in uncertainty quantification are:

– *Step 1*: the construction of a stochastic modeling of the uncertain parameter $\mathbf{X}$, that is to say, the construction of the probability distribution $P_{\mathbf{X}}(d\mathbf{x}\,;\mathbf{s})$ of random vector $\mathbf{X}$ (using the direct approach or the indirect one).

– *Step 2*: the analysis of the propagation of uncertainties by using a stochastic solver for constructing the probability distribution $P_{\mathbf{U}}(d\mathbf{u}\,;\mathbf{s})$ of the random observation vector $\mathbf{U} = \mathbf{h}(\mathbf{X})$ in which $\mathbf{h}$ is a given deterministic mapping (that is not, in general, explicitly known, but that is constructed by solving the computational model).

▷ *What is the impact of an arbitrary stochastic modeling. Even if the deterministic solver is "perfect" (computation of $\mathbf{h}$), if $P_{\mathbf{X}}$ is arbitrarily chosen (thus probably wrong), its nonlinear transformation by $\mathbf{h}$ (perfectly computed) yields a probability distribution $P_{\mathbf{U}}$ of random observation $\mathbf{U}$, which will be arbitrary (and thus probably wrong).*

5.1.3 What Is Important in UQ

□ *Case 1: no available experimental data.* Uncertainty Quantification is important for the robust analyses of predictions, for the robust optimization, and for the robust design. Consequently, Step 1, which has been defined in Section 5.1.2, is a fundamental step.

□ *Case 2: experimental data are available.*

▷ If "big data" are available, the nonparametric statistics can be used.
▷ If "big data" are not available, but only partial and limited experimental data are available (it is the framework of this course), then:
  - the parametric statistics is used, and consequently, Step 1 is a fundamental step for constructing a prior probability distribution $P_{\mathbf{X}}^{\text{prior}}(d\mathbf{x}\,;\mathbf{s})$ of the uncertain parameter $\mathbf{X}$.
  - even if the Bayesian inference (see Section 7.3.4) can be used for constructing a posterior probability distribution $P_{\mathbf{X}}^{\text{post}}(d\mathbf{x})$ of the uncertain parameter $\mathbf{X}$, in high stochastic dimension, a high quality of the prior model is required (see, for instance, Chapter 10) and thus, Step 1 is a fundamental step.

5.1.4 What Is the Objective of This Chapter

This chapter is devoted to the introduction of the main probabilistic/statistical tools that are necessary for constructing stochastic models of uncertainties.

5.2 Types of Representation for Stochastic Modeling

The following overview on the strategy is presented for a random vector but any random quantities could be used, such as a random matrix, a random field, etc.

Let $\theta \mapsto \mathbf{X}(\theta) = (X_1(\theta), \ldots, X_n(\theta))$ be a random variable, defined on a probability space $(\Theta, \mathcal{T}, \mathcal{P})$, with values in $\mathbb{R}^n$, for which the probability distribution on $\mathbb{R}^n$ is $P_{\mathbf{X}}(d\mathbf{x})$.

We recall (see Section 2.1.4) that $L^2(\Theta, \mathbb{R}^n)$ is the Hilbert space of all the second-order $\mathbb{R}^n$-valued random variables $\mathbf{X}$ such that

$$E\{\|\mathbf{X}\|^2\} = \int_{\mathbb{R}^n} \|\mathbf{x}\|^2 \, P_{\mathbf{X}}(d\mathbf{x}) < +\infty, \tag{5.1}$$

in which E is the mathematical expectation, where $\mathbf{x} = (x_1, \ldots, x_n)$, $d\mathbf{x} = dx_1 \ldots dx_n$, and $\|\mathbf{x}\|^2 = x_1^2 + \ldots + x_n^2$ is the square of the Euclidean norm of vector $\mathbf{x}$.

Two main approaches exist for constructing a (prior) probability distribution of $\mathbf{X}$: the *direct approach* and the *indirect approach*.

5.2.1 Direct Approach

It consists in directly constructing/estimating the probability density function $p_{\mathbf{X}}(\mathbf{x})$ (on $\mathbb{R}^n$ with respect to $d\mathbf{x}$) of the probability distribution $P_{\mathbf{X}}(d\mathbf{x})$ by using

- either the nonparametric statistical approach (if possible),
- or the parametric statistical approach (always possible).

□ *Nonparametric statistics if "big data" are available.* If "big data" are available (experimental data or data resulting from a numerical simulation), then an estimation of $p_{\mathbf{X}}$ can be carried out by using the multivariate kernel density estimation method (for instance, with a Gaussian kernel) (see Section 7.2).

□ *Parametric statistics and informative prior model.* It consists in constructing a parametric representation $\mathbf{x} \mapsto p_{\mathbf{X}}(\mathbf{x}, \mathbf{s})$ of the pdf (in which $\mathbf{s}$ is a vector-valued hyperparameter) by using the *Maximum Entropy Principle (MaxEnt)*, which is a very powerful and efficient tool for constructing a parametric representation of the probability distribution of random variables, random vectors, random matrices, stochastic processes, and random fields with values in any set).

- The hyperparameter $\mathbf{s}$ is related to the available information introduced as constraints for the use of the MaxEnt principle.

- If partial and limited data are available, then $\mathbf{s}$ can be estimated by solving a statistical inverse method (see Section 7.3).

- If data are not available, then $\mathbf{s}$ can be used as a parameter for carrying out a *sensitivity analysis* with respect to the statistical properties of uncertainties, such as the level of uncertainties.

5.2.2 Indirect Approach

It consists in introducing a representation $\mathbf{X} = \mathbf{f}(\mathbf{\Xi} ; \mathbf{s})$ of random vector $\mathbf{X}$, in which $\boldsymbol{\xi} \mapsto \mathbf{f}(\boldsymbol{\xi} ; \mathbf{s})$ is a deterministic nonlinear (measurable) mapping, where $\mathbf{\Xi}$ is a given $\mathbb{R}^{N_g}$-valued random variable whose probability density function $p_{\mathbf{\Xi}}(\boldsymbol{\xi})$ is given (and then is known), and where $\mathbf{s}$ is a vector-valued hyperparameter. Then, $p_{\mathbf{X}}(\mathbf{x} ; \mathbf{s})$ is the transformation of $p_{\mathbf{\Xi}}$ by the mapping $\mathbf{f}(. ; \mathbf{s})$.

Two main approaches can be used for construction, the mapping $\mathbf{f}$.

☐ *Polynomial Chaos Expansion of X.* It consists in constructing the PCE of $\mathbf{X}$ that is written (see Section 5.5) as

$$\mathbf{X} = \sum_{k=0}^{K} \mathbf{y}^k \, \Psi_{\alpha^{(k)}}(\Xi) \, .$$

In such a case, the hyperparameter is $\mathbf{s} = \{\mathbf{y}^0, \ldots, \mathbf{y}^K\}$.

☐ *Prior Algebraic Representation (PAR) of X.* It consists in writing as the prior probability model $\mathbf{X}^{\text{prior}}$ of $\mathbf{X}$,

$$\mathbf{X}^{\text{prior}} = \mathbf{f}(\Xi \, ; \mathbf{s}) \, ,$$

in which the vector-valued hyperparameter $\mathbf{s}$ has a small dimension and where $\boldsymbol{\xi} \mapsto \mathbf{f}(\boldsymbol{\xi} \, ; \mathbf{s})$ is a given nonlinear mapping (see, for instance, Chapter 10).

5.3 Maximum Entropy Principle as a Direct Approach for Constructing a Prior Stochastic Model of a Random Vector

This constructive tool is presented for a random vector but any random quantities could be used (for instance, for random matrices, as detailed in Section 5.4, etc.)

In this section, the following items are presented:

– problem definition.
– entropy as a measure of uncertainties.
– properties of the Shannon entropy.
– maximum entropy principle.
– reformulation of the optimization problem by using Lagrange multipliers and construction of a representation for the solution.
– existence and uniqueness of a solution to the MaxEnt.
– analytical examples of classical probability distributions deduced from the MaxEnt principle.
– MaxEnt as a numerical tool for constructing probability distribution in any dimension.
– computational example in high dimension for arbitrary probability distributions.

5.3.1 Problem Definition

☐ *Objective.* The objective is to construct the pdf $\mathbf{x} \mapsto p_{\mathbf{X}}(\mathbf{x}, \mathbf{s})$ of a $\mathbb{R}^n$-valued random variable $\mathbf{X}$ defined on the probability space $(\Theta, \mathcal{T}, \mathcal{P})$, by using only the *available information*, in which **s** is a vector-valued hyperparameter that must be clearly defined.

☐ *Tool/method*. The tool used is the Maximum Entropy (MaxEnt) principle, introduced by Jaynes in 1957 [112], in the framework of *Information Theory* pioneered by Shannon in 1948 [179], theory that is today an important part of the probability theory.

☐ *Simple example for illustrating what is the available information*. This aspect is very important for stochastic modeling of uncertainties in UQ. Let us consider the simple mechanical system introduced in Section 1.6.1. The problem was related to the random equation $K\,U = f$, in which the uncertain parameter is the nominal stiffness that was modeled by the random stiffness $K = \underline{k}\,X$, for which the pdf p_X on $\mathbb{R}$ of the real-valued random variable X has to be constructed. The force f is such that $0 < |f| < +\infty$. The nominal stiffness $\underline{k}$ is such that $0 < \underline{k} < +\infty)$. The random stiffness K must be positive almost surely (this is the stiffness of a stable system). It can then be deduced that X is a random variable with values in $\mathbb{R}^+$. Two sources of information must be used: the available information that is directly related to random variable X and the available information that is related to the property of the random solution U.

– *Available information directly related to random variable X*.

(C1) Since X is a random variable with values in $\mathbb{R}^+$, the support of the pdf p_X must be $\mathbb{R}^+$. The set of the values of X gives an information about the support of p_X.

(C2) We want that the mean value, $E\{K\}$, of K be equal to the nominal value $\underline{k}$, which yields $E\{X\} = \underline{x} = 1$. This is a statistical property of X.

– *Available information directly related to the property of random solution U*.

(C3) Since the mechanical system is a stable system and since the external force is finite, the statistical fluctuations of the random solution U must be finite, which means that U must be a second-order random variable, $E\{U^2\} < +\infty$. This inequality gives $E\{X^{-2}\} = c$ with $0 < c < +\infty$. This is a statistical property for X coming from a property of the random solution U.

Finally, the available information allows for defining three constraints, (C1), (C2), and (C3), for the construction of the pdf of X.

☐ *Recall of the catastrophic consequences of an arbitrary choice of pdf p_X* (illustrated with the simple mechanical example presented in Section 1.6.1 and reused in the present section).

If the pdf of X is arbitrarily chosen as a Gaussian pdf (mid-thick line in Figure 5.2), then we obtain $E\{U^2\} = +\infty$, and consequently, the constraints (C1) and (C3) are not satisfied.

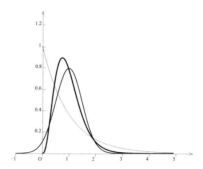

Fig. 5.2 Graph of the pdf for the Gaussian model (mid-thick line), for the exponential model (thin line), and for the probability model constructed by using the MaxEnt principle (thick line).

If the pdf of X is arbitrarily chosen as an exponential pdf (thin line in Figure 5.2), then we obtain again $E\{U^2\} = +\infty$, and consequently, the constraint (C3) is not satisfied.

If the pdf p_X of X is constructed by using the MaxEnt principle under the constraints (C1), (C2), and (C3), then we will obtain (see Equation (5.23)) a Gamma pdf, p_X^{MaxEnt} (thick line in Figure 5.2).

☐ *Reparameterization and what is the relationship between the available informa-tion and the hyperparameter* (explained with the simple mechanical example). The pdf $p_X^{\text{MaxEnt}}(x\,;c)$ that is constructed by using the MaxEnt principle under the constraints (C1), (C2), and (C3) will depend on an unknown parameter c. A question therefore comes. Can we take $s = c$ as the hyperparameter? The answer is preferably "no" for the following reason. The parameter c has been introduced to formulate the constraint (C3), but this parameter has no real physical/statistical meaning. In such a case a reparameterization can be done.

Principle of the reparameterization. If $E\{X^2\} < +\infty$ (what will be the case when the pdf p_X will be chosen as the pdf $p_X^{\text{MaxEnt}}(x\,;c)$ constructed with the MaxEnt), then the coefficient of variation δ of X can be calculated as a function of c as follows:

$$\delta = \sigma_X/m_X = \underline{x}^{-1}(E\{X^2\} - \underline{x}^2)^{1/2}, \tag{5.2}$$

$$E\{X^2\} = \int_{\mathbb{R}} x^2\, p_X^{\text{MaxEnt}}(x\,;c)\, dx, \tag{5.3}$$

which yields $\delta = g(c)$. If the inverse function $\delta \mapsto g^{-1}(\delta)$ exists, then the repa-rameterization consists in taking $c = g^{-1}(\delta)$ as a function of δ, and consequently, the hyperparameter will be chosen as $s = \delta$ (instead of $s = c$).

5.3.2 Entropy as a Measure of Uncertainties for a Vector-Valued Random Variable

Let $\mathbf{X} = (X_1, \ldots, X_n)$ be a random variable with values in any given/known subset $\mathcal{S}_n$ of $\mathbb{R}^n$ (possibly with $\mathcal{S}_n = \mathbb{R}^n$) for which the probability distribution $P_{\mathbf{X}}(d\mathbf{x})$ is defined by an unknown pdf $p_{\mathbf{X}}$ on $\mathbb{R}^n$ (with respect to $d\mathbf{x} = dx_1 \ldots dx_n$) whose support is $\mathcal{S}_n$,

$$\text{Supp } p_{\mathbf{X}} = \mathcal{S}_n \subset \mathbb{R}^n \quad \Rightarrow \quad p_{\mathbf{X}}(\mathbf{x}) = 0 \quad \text{if} \quad \mathbf{x} \notin \mathcal{S}_n . \tag{5.4}$$

The measure of uncertainties of random variable $\mathbf{X}$ is defined by the *Shannon entropy* $\mathcal{E}(p_{\mathbf{X}})$ of the pdf $p_{\mathbf{X}}$,

$$\mathcal{E}(p_{\mathbf{X}}) = - \int_{\mathbb{R}^n} p_{\mathbf{X}}(\mathbf{x}) \, \log(p_{\mathbf{X}}(\mathbf{x})) \, d\mathbf{x} = -E\{\log(p_{\mathbf{X}}(\mathbf{X}))\} . \tag{5.5}$$

5.3.3 Properties of the Shannon Entropy

□ *Set of values for the Shannon entropy.* For an $\mathbb{R}^n$-valued random variable $\mathbf{X}$, the set of values of the Shannon entropy is $\mathbb{R}$ and is not $\mathbb{R}^+$ as for the case of a discrete random variable. The Shannon entropy $\mathcal{E}(p_{\mathbf{X}})$ of $p_{\mathbf{X}}$ can then be negative and we have

$$- \infty < \mathcal{E}(p_{\mathbf{X}}) < +\infty . \tag{5.6}$$

□ *Maximum of the Shannon entropy on the set of all the pdf on $\mathbb{R}^n$ having the same compact support $K \subset \mathbb{R}^n$.* This set is thus defined by

$$\mathcal{C}_K = \{p : \mathbb{R}^n \to \mathbb{R}^+ \, , \, \text{Supp } p = K \subset \mathbb{R}^n\} , \tag{5.7}$$

in which K is any closed and bounded (compact) subset of $\mathbb{R}^n$, for which its measure (its volume) is $|K| = \int_K d\mathbf{x}$. It can easily be verified that the pdf, p_K, on $\mathbb{R}^n$, which maximizes the Shannon entropy on $\mathcal{C}_K$,

$$p_K = \arg \max_{p \in \mathcal{C}_K} \mathcal{E}(p) , \tag{5.8}$$

is the uniform pdf on $\mathbb{R}^n$ with compact support K,

$$p_K(\mathbf{x}) = \frac{1}{|K|} \, \mathbb{1}_K(\mathbf{x}) , \tag{5.9}$$

and the corresponding value of the Shannon entropy is

$$\mathcal{E}_{\max} = \mathcal{E}(p_K) = \log |K| . \tag{5.10}$$

□ *Quantification of the Shannon-entropy scale using the uniform distribution case.* Let $\{K_\ell\}_{\ell \in \mathbb{N}}$ be a sequence of compact subsets in $\mathbb{R}^n$ and let p_{K_ℓ} be the uniform pdf on $\mathbb{R}^n$ with support K_ℓ, which is thus written as $p_{K_\ell}(\mathbf{x}) = \frac{1}{|K_\ell|} \mathbb{1}_{K_\ell}(\mathbf{x})$.

- If $|K_\ell| = 1$, then $\mathcal{E}(p_{K_\ell}) = 0$.
- If $|K_\ell| \to +\infty$, then $\mathcal{E}(p_{K_\ell}) \to +\infty$.
- Let $\mathbf{x}_0$ be a fixed point in $\mathbb{R}^n$ such that $\mathbf{x}_0$ belongs to K_ℓ for all ℓ. If $|K_\ell| \to 0$, then $\mathcal{E}(p_{K_\ell}) \to -\infty$, and since the sequence $\{\frac{1}{|K_\ell|} \mathbb{1}_{K_\ell}(\mathbf{x}) \, d\mathbf{x}\}_\ell$ goes to the Dirac measure $\delta_{\mathbf{x}_0}(\mathbf{x})$ on $\mathbb{R}^n$ at point $\mathbf{x}_0$, it can be deduced that the sequence $\{\mathbf{X}_\ell\}_\ell$ (with probability distribution $p_{K_\ell}(\mathbf{x}) \, d\mathbf{x}$) converges (in probability distribution) to the deterministic value $\mathbf{x}_0$.

▷ *Summary. The maximum uncertainty is obtained when the Shannon entropy $\mathcal{E}$ is maximum. The more $\mathcal{E}$ is large, and the more uncertainty will be high. The more $\mathcal{E}$ is small and the more uncertainty will be small. The limit, $\mathcal{E} = -\infty$, corresponds to no uncertainty (deterministic case).*

5.3.4 Maximum Entropy Principle

□ *Definition of the maximum entropy principle (MaxEnt).* The Maximum entropy principle (MaxEnt), which was introduced by Jaynes [112] by using the Shannon entropy from Information Theory (Shannon [179]), allows for constructing the pdf $p_{\mathbf{X}}$ of random variable $\mathbf{X}$ by using only the available information.

Principle. The pdf $p_{\mathbf{X}}$ (that has to be constructed) is the pdf that corresponds to the largest uncertainty on the set of all the pdf that satisfy the constraints defined by the available information.

□ *Mathematical modeling of the available information.* The available information are constituted of

- the support property (mandatory),

$$\text{Supp } p_{\mathbf{X}} = \mathcal{S}_n \subset \mathbb{R}^n, \tag{5.11}$$

in which $\mathcal{S}_n$ is any subset of $\mathbb{R}^n$.

- some statistical properties on $\mathbf{X}$ and/or some properties of the stochastic solution of the computational model:

$$E\{\mathbf{g}(\mathbf{X})\} = \int_{\mathbb{R}^n} \mathbf{g}(\mathbf{x}) \, p_{\mathbf{X}}(\mathbf{x}) \, d\mathbf{x} = \mathbf{b} \in \mathbb{R}^\mu, \tag{5.12}$$

in which $\mathbf{x} \mapsto \mathbf{g}(\mathbf{x}) = (g_1(\mathbf{x}), \dots, g_\mu(\mathbf{x}))$ is a given mapping from $\mathbb{R}^n$ into $\mathbb{R}^\mu$, and where $\mathbf{b} = (b_1, \dots, b_\mu)$ is a given vector in $\mathbb{R}^\mu$.

Example. We now operate the case of the simple mechanical system introduced previously, for which $n = 1$, X is an $\mathbb{R}^+$-valued random variable, $E\{X\} = \underline{x} = 1$, and $E\{X^{-2}\} = c$. The corresponding constraints related to this available information are such that $n = 1$, $\mu = 2$, Supp $p_X = \mathcal{S}_1 = \mathbb{R}^+ \subset \mathbb{R}$, $\mathbf{g}(x) = (x, x^{-2})$, and $\mathbf{b} = (\underline{x}, c)$.

☐ *Definition of the admissible set for pdf p_X and optimization problem.* Let $\mathcal{C}_{\text{free}}$ be the subspace of all the integrable positive-valued functions defined on $\mathbb{R}^n$ for which the support is the subset $\mathcal{S}_n$ of $\mathbb{R}^n$,

$$\mathcal{C}_{\text{free}} = \{p \in L^1(\mathbb{R}^n, \mathbb{R}^+), \text{ Supp } p = \mathcal{S}_n\}. \tag{5.13}$$

We then define the admissible set $\mathcal{C}_{\text{ad}}$ such that

$$\mathcal{C}_{\text{ad}} = \left\{ p \in \mathcal{C}_{\text{free}}, \int_{\mathbb{R}^n} p(\mathbf{x}) \, d\mathbf{x} = 1, \int_{\mathbb{R}^n} \mathbf{g}(\mathbf{x}) \, p(\mathbf{x}) \, d\mathbf{x} = \mathbf{b} \right\}. \tag{5.14}$$

The pdf p_X on $\mathbb{R}^n$, which satisfies the constraints defined by the available information and constructed using the MaxEnt principle is the solution (if it exists and if it is unique) of the following optimization problem:

$$p_X = \arg \max_{p \in \mathcal{C}_{\text{ad}}} \mathcal{E}(p). \tag{5.15}$$

5.3.5 Reformulation of the Optimization Problem by Using Lagrange Multipliers and Construction of a Representation for the Solution

The optimization problem defined by Equation (5.15) is reformulated on set $\mathcal{C}_{\text{free}}$ instead of $\mathcal{C}_{\text{ad}}$ by introducing Lagrange multipliers. A representation of the solution is constructed as a function of the Lagrange multipliers.

☐ *Lagrange multipliers associated with the constraints.* A Lagrange multiplier is introduced for each constraint:

▷ $\lambda_0 \in \mathbb{R}^+$ that is associated with the constraint $\int_{\mathbb{R}^n} p(\mathbf{x}) \, d\mathbf{x} - 1 = 0$.
▷ $\boldsymbol{\lambda} \in \mathcal{C}_{\boldsymbol{\lambda}} \subset \mathbb{R}^\mu$ that is associated with the constraint $\int_{\mathbb{R}^n} \mathbf{g}(\mathbf{x}) \, p(\mathbf{x}) \, d\mathbf{x} - \mathbf{b} = \mathbf{0}$ in which the admissible set $\mathcal{C}_{\boldsymbol{\lambda}}$ is defined by

$$\mathcal{C}_{\boldsymbol{\lambda}} = \{\boldsymbol{\lambda} \in \mathbb{R}^\mu, \int_{\mathbb{R}^n} \exp(- <\boldsymbol{\lambda}, \mathbf{g}(\mathbf{x}) >) \, d\mathbf{x} < +\infty\}. \tag{5.16}$$

☐ *Expression of the Lagrangian.* For $\lambda_0 \in \mathbb{R}^+$ and $\boldsymbol{\lambda} \in \mathcal{C}_{\boldsymbol{\lambda}}$, the Lagrangian $\mathcal{L}$ is defined, for all p in $\mathcal{C}_{\text{free}}$, as

$$\mathcal{L}(p;\lambda_0,\boldsymbol{\lambda}) = \mathcal{E}(p) - (\lambda_0-1)\Big(\int_{\mathbb{R}^n} p(\mathbf{y})\,d\mathbf{y}-1\Big) - <\boldsymbol{\lambda}, \int_{\mathbb{R}^n} \mathbf{g}(\mathbf{x})p(\mathbf{x})\,d\mathbf{x}-\mathbf{b}> .$$

(5.17)

□ *Reformulation of the optimization problem and construction of a representation of the solution.* If there exists a unique solution $p_{\mathbf{X}}$ to the optimization problem $\max_{p\in\mathcal{C}_{\mathrm{ad}}} \mathcal{E}(p)$, then there exists $(\lambda_0^{\mathrm{sol}}, \boldsymbol{\lambda}^{\mathrm{sol}}) \in \mathbb{R}^+ \times \mathcal{C}_{\boldsymbol{\lambda}}$, such that the functional $(p,\lambda_0,\boldsymbol{\lambda}) \mapsto \mathcal{L}(p;\lambda_0,\boldsymbol{\lambda})$ is stationary at $p_{\mathbf{X}}$ for $\lambda_0 = \lambda_0^{\mathrm{sol}}$ and $\boldsymbol{\lambda} = \boldsymbol{\lambda}^{\mathrm{sol}}$ and is written as

$$p_{\mathbf{X}}(\mathbf{x}) = \mathbb{1}_{\mathcal{S}_n}(\mathbf{x})\, c_0^{\mathrm{sol}}\, \exp(- <\boldsymbol{\lambda}^{\mathrm{sol}}, \mathbf{g}(\mathbf{x})>) \quad, \quad \forall \mathbf{x} \in \mathbb{R}^n, \tag{5.18}$$

in which the normalization constant c_0^{sol} is written as $c_0^{\mathrm{sol}} = \exp(-\lambda_0^{\mathrm{sol}})$.

5.3.6 Existence and Uniqueness of a Solution to the MaxEnt

If the constraints are algebraically independent, that is to say, if there exists a bounded subset B of $\mathcal{S}_n$, with $\int_B d\mathbf{x} > 0$, such that for any nonzero vector $\mathbf{v}$ in $\mathbb{R}^{1+\mu}$,

$$\int_B <\mathbf{v}, \widetilde{\mathbf{g}}(\mathbf{x})>^2 d\mathbf{x} > 0 \quad \text{with} \quad \widetilde{\mathbf{g}}(\mathbf{x}) = (1, \mathbf{g}(x)) \in \mathbb{R}^{1+\mu}, \tag{5.19}$$

and under additional weak hypotheses (not crucial), it is proved in [99] (it should be noted that this complete mathematical proof is relatively difficult) that there exists a unique solution, which is written as

$$p_{\mathbf{X}}(\mathbf{x}) = \mathbb{1}_{\mathcal{S}_n}(\mathbf{x})\, c_0^{\mathrm{sol}}\, \exp(- <\boldsymbol{\lambda}^{\mathrm{sol}}, \mathbf{g}(\mathbf{x})>) \quad, \quad \forall \mathbf{x} \in \mathbb{R}^n, \tag{5.20}$$

in which $(\lambda_0^{\mathrm{sol}}, \boldsymbol{\lambda}^{\mathrm{sol}})$ is the unique solution in $\mathbb{R}^+ \times \mathcal{C}_{\boldsymbol{\lambda}}$ of the following systems of nonlinear algebraic equations,

$$\int_{\mathcal{S}_n} \exp(-\lambda_0 - <\boldsymbol{\lambda}, \mathbf{g}(\mathbf{x})>)\,d\mathbf{x} = 1, \tag{5.21}$$

$$\int_{\mathcal{S}_n} \mathbf{g}(\mathbf{x})\, \exp(-\lambda_0 - <\boldsymbol{\lambda}, \mathbf{g}(\mathbf{x})>)\,d\mathbf{x} = \mathbf{b}. \tag{5.22}$$

5.3.7 Analytical Examples of Classical Probability Distributions Deduced from the MaxEnt Principle

Four examples relative to a bounded support are presented. The first three one concerns a real-valued random variable with values in a bounded interval $[a, b]$, with values in $[0, +\infty[$, and with values in $]-\infty, +\infty[$. The fourth one in the case of a $\mathbb{R}^n$-valued random variable.

□ *Example 1: probability distribution of a real-valued random variable with values in a bounded interval.* We consider a real-valued random variable X with values in $[a,b]$ for which the pdf $p_X(x)$ is unknown. There are no additional information.

1. Available information: $\text{Supp}\, p_X = \mathcal{S}_1 = [a,b]$.
2. Maximum entropy principle yields: $p_X(x) = \mathbb{1}_{[a,b]}(x)\, \frac{1}{b-a}$.
3. Conclusion: p_X is a uniform pdf on $[a,b]$.

More generally, if a random variable is with values in a compact part $\mathcal{S}_n$ of $\mathbb{R}^n$, and if the solely available information is $\mathcal{S}_n \subset \mathbb{R}^n$, then the MaxEnt principle gives a uniform distribution on $\mathcal{S}_n$ for $\mathbf{X}$ (see Section 5.3.3 with $K = \mathcal{S}_n$).

□ *Example 2: probability distribution for a positive-valued random variable with a reparameterization.* The random variable X is with values in $]0,+\infty[$, its mean value m_X is given in $]0,+\infty[$, and it is assumed that $|E\{\log(X)\}| < +\infty$. The pdf p_X of X is unknown and there are no additional information.

1. Available information: $\text{Supp}\, p_X = \mathcal{S}_1 =]0,+\infty[$, $E\{X\} = m_X > 0$, $E\{\log(X)\} = c$ with $|c| < +\infty$.
2. Maximum entropy principle yields:

$$p_X(x) = \mathbb{1}_{]0,+\infty[}(x)\, \frac{(\delta^{-2})^{\frac{1}{\delta^2}}}{\Gamma(\delta^{-2})\, m_X}\left(\frac{x}{m_X}\right)^{\frac{1}{\delta^2}-1} \exp\left\{-\frac{x}{\delta^2 m_X}\right\}, \quad (5.23)$$

in which $\delta = \sigma_X/m_X$ is such that $0 \leq \delta < 1/\sqrt{2}$, where $\Gamma(\alpha)$ is the Gamma function, and where c, which is such that $\int_0^{+\infty} \log(x)\, p_X(x)\, dx = c$, is eliminated using δ in order to perform a reparameterization in δ instead of keeping c.
3. Conclusion: p_X corresponds to a Gamma pdf on $[0,+\infty[$.

□ *Example 3: probability distribution of a real-valued random variable for which the first two moments are given.* We consider a second-order real-valued random variable X for which the mean value m_X and the standard deviation σ_X are given. The pdf p_X of X is unknown and there are no additional information.

1. Available information: $\text{Supp}\, p_X = \mathcal{S}_1 =]-\infty,+\infty[$, mean value $E\{X\} = m_X$, and standard deviation $\{E\{X^2\} - m_X^2\}^{1/2} = \sigma_X$, which yields $E\{X^2\} = \sigma_X^2 + m_X^2$.
2. Maximum entropy principle yields:

$$p_X(x) = \frac{1}{\sqrt{2\pi}\, \sigma_X} \exp\left\{-\frac{1}{2\sigma_X^2}(x - m_X)^2\right\}. \quad (5.24)$$

3. Conclusion: p_X is a Gaussian pdf.

This result is important. It shows that, in the framework of the information theory, a Gaussian real-valued random variable corresponds to a random variable that takes its values in the real line and for which its mean value and its standard deviation are assumed to be known.

□ *Example 4: probability distribution of a vector-valued random variable for which the mean vector is given.* Let $\mathbf{X} = (X_1, \ldots, X_n)$ be a random variable with values in $]0, +\infty[\times \ldots \times]0, +\infty[\subset \mathbb{R}^n$, for which its mean value $\mathbf{m_X} = E\{\mathbf{X}\} = (m_{X_1}, \ldots, m_{X_n})$ is given in $]0, +\infty[\times \ldots \times]0, +\infty[$. The pdf $p_\mathbf{X}$ of $\mathbf{X}$ is unknown and there are no additional information.

1. Available information: $\text{Supp}\, p_\mathbf{X} = \mathcal{S}_n =]0, +\infty[\times \ldots \times]0, +\infty[\subset \mathbb{R}^n$ and $\mathbf{m_X} \in \mathcal{S}_n \subset \mathbb{R}^n$.
2. Maximum entropy principle yields:

$$p_\mathbf{X}(\mathbf{x}) = p_{X_1}(x_1) \times \ldots \times p_{X_n}(x_n), \tag{5.25}$$

in which the pdf $p_{X_j}(x_j)$ of the positive-valued random variable X_j is written as

$$p_{X_j}(x_j) = \mathbb{1}_{]0,+\infty[}(x_j) \frac{1}{m_{X_j}} \exp\left\{-\frac{x_j}{m_{X_j}}\right\}. \tag{5.26}$$

3. Conclusion: the real-valued random variables $X_1, \ldots, X_n$ are mutually independent and, for each j, the positive-valued random variable X_j has an exponential pdf. It should be noted that, although the random variable X_j^{-1} exists almost surely, this random variable is not a second-order random variable because $E\{X_j^{-2}\} = +\infty$. It should also be noted that the MaxEnt yields independent random variables $X_1, \ldots, X_n$ because there are no available information concerning some statistical moments coupling the components.

□ *Example 5: probability distribution of a vector-valued random variable for which the first two moments are given.* Let $\mathbf{X}$ be a second-order $\mathbb{R}^n$-valued random variable for which its mean value $\mathbf{m_X}$ is given in $\mathbb{R}^n$ and for which its covariance matrix $[C_X] = E\{(\mathbf{X} - \mathbf{m_X})(\mathbf{X} - \mathbf{m_X})^T\}$ is given in $\mathbb{M}_n^+(\mathbb{R})$ (therefore, matrix $[C_X]$ is invertible). Its correlation matrix (see Equations (2.31) and (2.34)) is then given by $E\{\mathbf{X}\mathbf{X}^T\} = [C_X] + \mathbf{m_X}\mathbf{m_X}^T$. The pdf $p_\mathbf{X}$ of $\mathbf{X}$ is unknown and there are no additional information.

1. Available information: $\text{Supp}\, p_\mathbf{X} = \mathcal{S}_n = \mathbb{R}^n$, $E\{\mathbf{X}\} = \mathbf{m_X} \in \mathbb{R}^n$, and $E\{\mathbf{X}\mathbf{X}^T\} = [C_X] + \mathbf{m_X}\mathbf{m_X}^T \in \mathbb{M}_n^+(\mathbb{R})$.
2. Maximum entropy principle yields:

$$p_\mathbf{X}(\mathbf{x}) = \frac{1}{\sqrt{(2\pi)^n \det[C_X]}} \exp\left\{-\frac{1}{2} <[C_X]^{-1}(\mathbf{x}-\mathbf{m_X}),(\mathbf{x}-\mathbf{m_X})>\right\}. \tag{5.27}$$

3. Conclusion: $\mathbf{X}$ is a Gaussian random vector. If the covariance matrix is not the identity matrix, then the real-valued random variables $X_1, \ldots, X_n$ are mutually dependent (see Section 2.4.2).

5.3.8 MaxEnt as a Numerical Tool for Constructing Probability Distribution in Any Dimension

☐ *What are the difficulties that are to be solved.* The use of the MaxEnt principle, under constraints defined by any available information and for the high dimensions, is a challenging problem for the following reasons.

1. The constant of normalization $c_0 = \exp(-\lambda_0)$ (that goes to zero as R^{-n} with R a "radius" of $\mathcal{S}_n$) is directly involved in the numerical calculations that are not robust for the high dimensions.
2. The second one is related to the numerical calculation of integrals in high dimension,
   – for computing the Lagrange multipliers,
   – for computing $E\{\mathbf{h}(\mathbf{X})\}$ in which $\mathbf{h}$ is a given mapping from $\mathbb{R}^n$ into $\mathbb{R}^m$.

Some advanced robust algorithms are presented hereinafter for solving numerically the problem in any dimension, in particular, for the high dimensions.

☐ *Numerical calculation of the Lagrange multiplier in any dimension.* The numerical method proposed hereinafter [19] is based on the minimization of a convex objective function [1]. The solution $p_\mathbf{X} = \arg\max_{p \in \mathcal{C}_{\mathrm{ad}}} \mathcal{E}(p)$, which has previously been constructed, can be rewritten as

$$p_\mathbf{X}(\mathbf{x}) = \mathbb{1}_{\mathcal{S}_n}(\mathbf{x})\, c_0(\boldsymbol{\lambda}^{\mathrm{sol}})\, \exp(- <\boldsymbol{\lambda}^{\mathrm{sol}}, \mathbf{g}(\mathbf{x}) >) \quad , \quad \forall \mathbf{x} \in \mathbb{R}^n, \qquad (5.28)$$

in which the constant of normalization is written, as a function of $\boldsymbol{\lambda}$, as

$$c_0(\boldsymbol{\lambda}) = \left\{ \int_{\mathcal{S}_n} \exp(- <\boldsymbol{\lambda}, \mathbf{g}(\mathbf{x}) >)\, d\mathbf{x} \right\}^{-1}. \qquad (5.29)$$

Since $\exp(-\lambda_0) = c_0(\lambda_0)$, the constraint equation, $E\{\mathbf{g}(\mathbf{X})\} = \mathbf{b}$, can be rewritten as

$$\int_{\mathcal{S}_n} \mathbf{g}(\mathbf{x})\, c_0(\boldsymbol{\lambda})\, \exp(- <\boldsymbol{\lambda}, \mathbf{g}(\mathbf{x}) >)\, d\mathbf{x} = \mathbf{b}. \qquad (5.30)$$

The optimization problem that allows $\boldsymbol{\lambda}^{\mathrm{sol}}$ to be computed in $\mathcal{C}_\lambda$ can then be rewritten as

$$\boldsymbol{\lambda}^{\mathrm{sol}} = \arg\min_{\boldsymbol{\lambda} \in \mathcal{C}_\lambda \subset \mathbb{R}^\mu} \Gamma(\boldsymbol{\lambda}), \qquad (5.31)$$

in which the objective function Γ is defined by

$$\Gamma(\boldsymbol{\lambda}) = <\boldsymbol{\lambda}, \mathbf{b}> - \log(c_0(\boldsymbol{\lambda}))\,. \tag{5.32}$$

It can be seen that the calculation does not involve constant $c_0(\boldsymbol{\lambda})$ but its Neperian logarithm $\log(c_0(\boldsymbol{\lambda}))$. Once $\boldsymbol{\lambda}^{\text{sol}}$ is calculated, c_0^{sol} is then given by $c_0^{\text{sol}} = c_0(\boldsymbol{\lambda}^{\text{sol}})$.

▷ *The numerical calculation of the constant c_0^{sol} is not mandatory, because this constant is not used for computing $E\{\mathbf{h}(\mathbf{X})\}$ with the MCMC algorithms presented in Chapter 4.*

We can now prove that the objective function Γ is strictly convex on $\mathcal{C}_\lambda \subset \mathbb{R}^\mu$. Let $\{\mathbf{X}_\lambda\}_{\lambda \in \mathcal{C}_\lambda}$ be the family of $\mathbb{R}^n$-random variables for which the pdf of the random variable $\mathbf{X}_\lambda$ is written as

$$p_{\mathbf{X}_\lambda}(\mathbf{x}) = c_0(\boldsymbol{\lambda}) \exp(- <\boldsymbol{\lambda}, \mathbf{g}(\mathbf{x})>) \quad , \quad \forall \mathbf{x} \in \mathbb{R}^n\,. \tag{5.33}$$

The gradient $\nabla\Gamma(\boldsymbol{\lambda})$ and the Hessian matrix $[\Gamma''(\boldsymbol{\lambda})]$ at point $\boldsymbol{\lambda}$ are such that

$$\nabla\Gamma(\boldsymbol{\lambda}) = \mathbf{b} - E\{\mathbf{g}(\mathbf{X}_\lambda)\}\,, \tag{5.34}$$

$$[\Gamma''(\boldsymbol{\lambda})] = E\{\mathbf{g}(\mathbf{X}_\lambda)\,\mathbf{g}(\mathbf{X}_\lambda)^T\} - E\{\mathbf{g}(\mathbf{X}_\lambda)\}\,E\{\mathbf{g}(\mathbf{X}_\lambda)\}^T\,. \tag{5.35}$$

Since the constraints are assumed to be algebraically independent, the Hessian matrix $[\Gamma''(\boldsymbol{\lambda})]$, which appears as the covariance matrix of the random vector $\mathbf{g}(\mathbf{X}_\lambda)$, is positive definite.

▷ *Consequently, function $\boldsymbol{\lambda} \mapsto \Gamma(\boldsymbol{\lambda})$ is strictly convex and reaches its minimum for $\boldsymbol{\lambda}^{\text{sol}}$ that is such that $\nabla\Gamma(\boldsymbol{\lambda}^{\text{sol}}) = \mathbf{0}$.*

We can now present an algorithm for computing $\boldsymbol{\lambda}^{\text{sol}}$ in $\mathcal{C}_\lambda \subset \mathbb{R}^\mu$. The optimization problem $\boldsymbol{\lambda}^{\text{sol}} = \arg\min_{\lambda \in \mathcal{C}_\lambda \subset \mathbb{R}^\mu} \Gamma(\boldsymbol{\lambda})$ can be solved using any minimization algorithm. Since function Γ is strictly convex, the Newton iterative method can be applied to the function $\boldsymbol{\lambda} \mapsto \nabla\Gamma(\boldsymbol{\lambda})$ for searching $\boldsymbol{\lambda}^{\text{sol}}$ such that $\nabla\Gamma(\boldsymbol{\lambda}^{\text{sol}}) = \mathbf{0}$. In order to ensure the convergence, an under-relaxation is introduced in the Newton iterative algorithm, which is written as

$$\boldsymbol{\lambda}^{i+1} = \boldsymbol{\lambda}^i - \alpha\,[\Gamma''(\boldsymbol{\lambda}^i)]^{-1}\,\nabla\Gamma(\boldsymbol{\lambda}^i)\,, \tag{5.36}$$

in which $\alpha \in]0, 1[$. In order to control the convergence, at each iteration i, the error is calculated by

$$\text{err}(i) = \frac{\|\mathbf{b} - E\{\mathbf{g}(\mathbf{X}_{\lambda^i})\}\|}{\|\mathbf{b}\|} = \frac{\|\nabla\Gamma(\boldsymbol{\lambda}^i)\|}{\|\mathbf{b}\|}\,. \tag{5.37}$$

During the iteration, for each $\boldsymbol{\lambda}^i$ in $\mathcal{C}_\lambda \subset \mathbb{R}^\mu$, the following quantities must be computed:

$$E\{\mathbf{g}(\mathbf{X}_{\lambda^i})\} \quad , \quad E\{\mathbf{g}(\mathbf{X}_{\lambda^i})\,\mathbf{g}(\mathbf{X}_{\lambda^i})^T\}\,, \tag{5.38}$$

in which the mathematical expectation is taken with respect to the $\mathbb{R}^n$-valued random variable $\mathbf{X}_{\boldsymbol{\lambda}^i}$ whose pdf is written as

$$p_{\mathbf{X}_{\boldsymbol{\lambda}^i}}(\mathbf{x}) = \mathbb{1}_{\mathcal{S}_n}(\mathbf{x})\, c_0(\boldsymbol{\lambda}^i)\, \exp(- <\boldsymbol{\lambda}^i, \mathbf{g}(\mathbf{x}) >)\,. \tag{5.39}$$

▷ *The computation of the mathematical expectations defined by Equation* (5.38) *is performed by using an MCMC method as explained in Chapter 4.*

For n large (high dimension), the algorithm based on an Itô stochastic differential equation presented in Section 4.3 can be used with a great efficiency.

☐ *Generator for the random vector and estimation of the mathematical expectation.* Once $\boldsymbol{\lambda}^{\text{sol}} \in \mathcal{C}_{\boldsymbol{\lambda}} \subset \mathbb{R}^\mu$ has been computed, the pdf of $\mathbf{X}$ is given by

$$p_{\mathbf{X}}(\mathbf{x}) = \mathbb{1}_{\mathcal{S}_n}(\mathbf{x})\, c_0^{\text{sol}}\, \exp(- <\boldsymbol{\lambda}^{\text{sol}}, \mathbf{g}(\mathbf{x}) >)\,, \tag{5.40}$$

in which $c_0^{\text{sol}} = c_0(\boldsymbol{\lambda}^{\text{sol}})$, but, as explained, c_0^{sol} is not computed since not used in the MCMC methods.

– A generator of independent realizations $\mathbf{X}(\theta_1), \ldots, \mathbf{X}(\theta_\nu)$ can then be constructed using the MCMC method as explained in Chapter 4.

– The computation of $E\{\mathbf{h}(\mathbf{X})\}$ can also be done using the MCMC methods and using either the ergodic average or the Monte Carlo method (see Chapter 4).

☐ *Example in high dimension.* We present an example in high dimension, which is taken from [19]. The objective is the construction of a generator of realizations for a non-Gaussian nonstationary centered stochastic process $\{A(t), t \in T\}$ (accelerograms [198]), for which the time sampling yields a time series $\{X_1, \ldots, X_n\}$ with $X_k = A(k\,\Delta t)$. We then have to construct the pdf $p_{\mathbf{X}}(\mathbf{x})$ of the $\mathbb{R}^n$-valued random variable $\mathbf{X} = (X_1, \ldots, X_n)$ (for which its support is $\mathcal{S}_n = \mathbb{R}^n$) and to construct the generator of realizations of the non-Gaussian random vector $\mathbf{X}$.

The Lagrange multipliers are computed by using the previous numerical method with $\mathcal{S}_n = \mathbb{R}^n$. The generator of independent realizations and the integrals in high dimension are computed by the Monte Carlo method derived from the MCMC method presented in Section 4.3.2 for the case of a support of the pdf, which is the entire space $\mathbb{R}^n$.

The following available information define $\mu = n + \kappa + 3$ constraints equations for constructing the pdf of the random vector $\mathbf{X} = (X_1, \ldots, X_n)$ by using the MaxEnt principle:

1. Supp $p_{\mathbf{X}} = \mathcal{S}_n = \mathbb{R}^n$.
2. $E\{X_j^2\} = \sigma_j^2 < +\infty$ for $j \in \{1, \ldots, n\}$, in which σ_j is the target for the standard deviation.

3. $E\{(\sum_{k=1}^{n} X_k)^2\} = 0$, which allows for imposing zero in mean-square sense for the end-velocity $V_n = \sum_{k=1}^{n} X_k$.

4. $E\{(\sum_{k=1}^{n} k\, X_k)^2\} = 0$, which allows for imposing zero in mean-square sense for the end-displacement $D_n = \sum_{k=1}^{n} (n - k + 1) X_j$.

5. $E\{(\sum_{k=1}^{n} k^2\, X_k)^2\} = 0$, which allows for imposing zero in mean-square sense for the time averaging of the displacement time series, and which therefore implies that $\mathbf{X}$ is centered.

6. $E\{s_k(\mathbf{X})\} = \underline{s}_k$ for $k \in \{1, \ldots, \kappa\}$, which allows for imposing the target for the statistical mean value of the velocity response spectrum (VRS), in which the nonlinear mapping $\mathbf{x} \mapsto s_k(\mathbf{x})$ from $\mathbb{R}^n$ into $\mathbb{R}$ is such that $s_k(\mathbf{x}) = \omega_k \max\{|y_1^k(\mathbf{x})|, \ldots, |y_n^k(\mathbf{x})|\}$ with $y_j^k(\mathbf{x}) = \{[B^j]\mathbf{x}\}_j$.

The data are the following:

$n = 1\,600$ (the pdf is then on $\mathbb{R}^{1\,600}$, problem in high dimension).

$\kappa = 20$ (number of nonlinear functions s_j).

$\mu = 1\,623$ (number of constraints equations).

$\nu = 600$ (number of integration steps for the ISDE in the MCMC method).

$\alpha = 0.3$ (Newton iterative method with under-relaxation parameter).

at each iteration of the iterative method, 900 Monte Carlo simulations are carried out.

the graph $j \mapsto \sigma_j$ that is the target of the standard deviation and the graph $k \mapsto \underline{s}_k$ that is the target of the statistical mean value of the VRS.

For θ in Θ, Figure 5.3 displays a trajectory that is constructed with the generator of the non-Gaussian time series for the accelerogram $\mathbf{X}(\theta)$, for the velocity $\mathbf{V}(\theta)$ obtained by a numerical integration of $\mathbf{X}(\theta)$, and for the displacement $\mathbf{D}(\theta)$ obtained by a numerical integration of $\mathbf{V}(\theta)$. It can be seen that the end-velocity and the end-displacement are effectively equal to zero.

The validation of the quality of the generator is illustrated with the results presented in Figure 5.4. In this figure, it can be seen a comparison of the target with the prediction for the standard deviation and for the mean VRS. In addition, for the mean VRS, the prediction computed by using an arbitrary Gaussian model is presented. It can be seen that such a Gaussian prediction is not good at all. In Figure 5.4, it can also be seen that the prediction of the confidence region of the random VRS is very good, and that the influence of the non-Gaussian modeling with respect to an arbitrary Gaussian modeling is very significant.

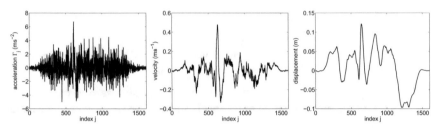

Fig. 5.3 Sample path of the time series for the accelerogram (left figure), for the velocity (central figure), and for the displacement (right figure). [Figure from [19]].

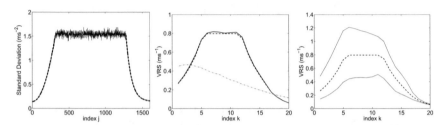

Fig. 5.4 Comparison of the target (dashed line) with the prediction (solid line), for the standard deviation (left figure) and for the mean VRS (central figure). Prediction of the mean VRS by using an arbitrary Gaussian model (mixed line in the central figure). Prediction of the confidence region of the random VRS, delimited by the upper and the lower solid lines, for a probability level $P_c = 0.95$ and mean value of the random VRS (dashed line). [Figure from [19]].

5.4 Random Matrix Theory for Uncertainty Quantification in Computational Mechanics

The random matrix theory is an important tool for constructing stochastic modeling of the system-parameter uncertainties and of the model uncertainties induced by the modeling errors in the framework of Uncertainty Quantification.

In this section, the following items are presented.

- A few words on fundamentals of the random matrix theory.
- Algebraic notations.
- Volume element and probability density function for random matrices.
- The Shannon entropy as a measure of uncertainties for a symmetric real random matrix.
- The MaxEnt Principle for symmetric real random matrices.
- A fundamental ensemble for the symmetric real random matrices with an identity matrix as a mean value.
- Fundamental ensembles for positive-definite symmetric real random matrices.
- Ensembles of random matrices for the nonparametric method in uncertainty quantification.

- Illustration of the use of ensemble SE_ε^+: Transient wave propagation in a fluid–solid multilayer.
- MaxEnt as a numerical tool for constructing ensembles of random matrices.

5.4.1 A Few Words on Fundamentals of the Random Matrix Theory

The random matrix theory was intensively studied by physicists and mathematicians in the context of nuclear physics. The theory began with Wigner in the 1950s [219], important effort in the 1960s by Wigner (1962) [220], Dyson (1963) [73], Mehta and others. In 1967 Mehta published an excellent book [133] concerning a synthesis of the random matrix theory (second edition in 1991 [134]) and third edition in 2014 [135]).

For physical applications, the most important ensemble is the *Gaussian Orthogonal Ensemble* (GOE) such that

- any random matrix in the GOE is a real symmetric random matrix,
- the entries of any random matrix in the GOE are mutually independent,
- any matrix in the GOE is invariant under orthogonal linear transformations.

Ensemble GOE corresponds to a generalization for the real symmetric matrices of the Gaussian real-valued random variables.

5.4.2 Notations in Linear Algebra

The following notations can be found in the "glossary," nevertheless, we present them for improving the readability.

□ *Vectors.*

$\mathbf{x} = (x_1, \ldots, x_n)$ is a vector in $\mathbb{R}^n$.
$\|\mathbf{x}\|$ is the Euclidean norm of $\mathbf{x}$ in $\mathbb{R}^n$.
$< \mathbf{x}, \mathbf{y} > = \sum_{j=1}^n x_j y_j$ is the Euclidean inner product in $\mathbb{R}^n$.

□ *Matrix sets.*

$\mathbb{M}_n(\mathbb{R})$ is the set of all the $(n \times n)$ real matrices.
$\mathbb{M}_n^S(\mathbb{R})$ is the set of all the $(n \times n)$ real symmetric matrices.
$\mathbb{M}_n^+(\mathbb{R})$ is the set of all the $(n \times n)$ real symmetric positive-definite matrices.
$\mathbb{M}_n^{+0}(\mathbb{R})$ is the set of all the $(n \times n)$ real symmetric positive-semidefinite matrices.

$$\mathbb{M}_n^+(\mathbb{R}) \subset \mathbb{M}_n^{+0}(\mathbb{R}) \subset \mathbb{M}_n^S(\mathbb{R}) \subset \mathbb{M}_n(\mathbb{R}).$$

□ *Operator norm of a matrix.* The operator norm of a matrix $[A]$ in $\mathbb{M}_n(\mathbb{R})$ is defined by

$$\|A\| = \sup_{\|\mathbf{x}\| \leq 1} \|[A]\mathbf{x}\| \quad , \quad \mathbf{x} \in \mathbb{R}^n .$$

□ *Frobenius norm of a matrix.* The Frobenius norm (or the Hilbert-Schmidt norm) of a matrix $[A]$ in $\mathbb{M}_n(\mathbb{R})$ is defined by

$$\|A\|_F^2 = \mathrm{tr}\{[A]^T[A]\} = \sum_{j=1}^n \sum_{k=1}^n [A]_{jk}^2 ,$$

which is such that $\|A\| \leq \|A\|_F \leq \sqrt{n}\,\|A\|$.

5.4.3 Volume Element and Probability Density Function for Random Matrices

The Euclidean space $\mathbb{M}_n(\mathbb{R})$ is equipped with the inner product and with the associated norm,

$$\ll[G],[H]\gg = \mathrm{tr}\{[G]^T[H]\} \quad , \quad \|G\|_F =\ll[G],[G]\gg^{1/2} .$$

The entries of any matrix $[G]$ in $\mathbb{M}_n(\mathbb{R})$ are denoted by $G_{jk} = [G]_{jk}$.

□ *Volume element.* The volume element dG on Euclidean space $\mathbb{M}_n(\mathbb{R})$ and the volume element $d^S G$ on Euclidean space $\mathbb{M}_n^S(\mathbb{R})$ are defined [185, 207] by

$$dG = \prod_{j,k=1}^n dG_{jk} \quad , \quad d^S G = 2^{n(n-1)/4} \prod_{1 \leq j \leq k \leq n} dG_{jk} . \tag{5.41}$$

□ *Probability density function of a symmetric real random matrix.* Let $[\mathbf{G}]$ be a random matrix, defined on $(\Theta, \mathcal{T}, \mathcal{P})$, with values in $\mathbb{M}_n^S(\mathbb{R})$ whose probability distribution $P_{[\mathbf{G}]} = p_{[\mathbf{G}]}([G])\, d^S G$ is defined by a pdf $[G] \mapsto p_{[\mathbf{G}]}([G])$ from $\mathbb{M}_n^S(\mathbb{R})$ into $\mathbb{R}^+ = [0, +\infty[$ with respect to $d^S G$ on $\mathbb{M}_n^S(\mathbb{R})$. This pdf verifies the normalization condition,

$$\int_{\mathbb{M}_n^S(\mathbb{R})} p_{[\mathbf{G}]}([G])\, d^S G = 1 . \tag{5.42}$$

□ *Support of the probability density function.* The support of pdf $p_{[\mathbf{G}]}$, denoted as $\mathrm{Supp}\, p_{[\mathbf{G}]}$, is any subset $\mathbb{S}_n$ of $\mathbb{M}_n^S(\mathbb{R})$, possibly with $\mathbb{S}_n = \mathbb{M}_n^S(\mathbb{R})$ (example: $\mathbb{S}_n = \mathbb{M}_n^+(\mathbb{R}) \subset \mathbb{M}_n^S(\mathbb{R})$). Thus, $p_{[\mathbf{G}]}([G]) = 0$ for $[G]$ not in $\mathbb{S}_n$, and we have

$$\int_{\mathbb{S}_n} p_{[\mathbf{G}]}([G])\, d^S G = 1 . \tag{5.43}$$

5.4.4 The Shannon Entropy as a Measure of Uncertainties for a Symmetric Real Random Matrix

Similarly to a random vector (see Equation (5.5)), the Shannon measure of uncertainties of a symmetric real random matrix $[\mathbf{G}]$ is defined by the entropy of information $\mathcal{E}(p_{[\mathbf{G}]})$ of the pdf $p_{[\mathbf{G}]}$ whose support is $\mathbb{S}_n \subset \mathbb{M}_n^S(\mathbb{R})$,

$$\mathcal{E}(p_{[\mathbf{G}]}) = -\int_{\mathbb{S}_n} p_{[\mathbf{G}]}([G]) \, \log\big(p_{[\mathbf{G}]}([G])\big) \, d^S G = -E\{\log\big(p_{[\mathbf{G}]}([G])\big)\} . \quad (5.44)$$

5.4.5 The MaxEnt Principle for Symmetric Real Random Matrices

Let $[\mathbf{G}]$ be an $\mathbb{M}_n^S(\mathbb{R})$- valued random matrix for which the pdf $p_{[\mathbf{G}]}$ (that is unknown with a given support $\mathbb{S}_n$) has to be constructed.

□ *Mathematical modeling of the available information.* The available information are made up of

— the support property (mandatory),

$$\text{Supp } p_{[\mathbf{G}]} = \mathbb{S}_n \subset \mathbb{M}_n^S(\mathbb{R}), \quad (5.45)$$

in which $\mathbb{S}_n$ in any subset of $\mathbb{M}_n^S(\mathbb{R})$, for instance $\mathbb{S}_n$ can be $\mathbb{M}_n^+(\mathbb{R})$.

— some statistical properties on $[\mathbf{G}]$ and/or properties of the stochastic solution of the computational model,

$$\mathbf{h}(p_{[\mathbf{G}]}) = \mathbf{0} \quad \text{in} \quad \mathbb{R}^\mu , \quad (5.46)$$

in which $p_{[\mathbf{G}]} \mapsto \mathbf{h}(p_{[\mathbf{G}]})$ is a given functional of $p_{[\mathbf{G}]}$, with values in $\mathbb{R}^\mu$.

Example. If the mean value $E\{[\mathbf{G}]\} = [\underline{G}]$ is a given matrix in $\mathbb{S}_n$, then

$$h_\alpha(p_{[\mathbf{G}]}) = \int_{\mathbb{S}_n} G_{jk} \, p_{[\mathbf{G}]}([G]) \, d^S G - \underline{G}_{jk} \quad , \quad \alpha = 1, \dots, \mu , \quad (5.47)$$

in which α is associated with (j,k) such as $1 \leq j \leq k \leq n$, and where the integer μ is such that $\mu = n(n+1)/2$.

□ *Definition of the admissible set for pdf $p_{[\mathbf{G}]}$ and optimization problem.* Let $\mathcal{C}_{\text{free}}$ be the subspace of all the integrable positive-valued functions defined on $\mathbb{M}_n^S(\mathbb{R})$ for which the support is the subset $\mathbb{S}_n$ of $\mathbb{M}_n^S(\mathbb{R})$,

$$\mathcal{C}_{\text{free}} = \{p \in L^1(\mathbb{M}_n^S(\mathbb{R}), \mathbb{R}^+) \, , \, \text{Supp}\, p = \mathbb{S}_n \, , \, \int_{\mathbb{S}_n} p([G]) \, d^S G = 1\} . \quad (5.48)$$

We then define the admissible set $\mathcal{C}_{\text{ad}}$ such that

$$\mathcal{C}_{\text{ad}} = \{p \in \mathcal{C}_{\text{free}}, \mathbf{h}(p) = \mathbf{0}\}. \tag{5.49}$$

The MaxEnt principle allows for constructing the pdf $p_{[\mathbf{G}]}$ of random matrix $[\mathbf{G}]$ by solving the optimization problem,

$$p_{[\mathbf{G}]} = \arg \max_{p \in \mathcal{C}_{\text{ad}}} \mathcal{E}(p). \tag{5.50}$$

Similarly to the case of a random vector $\mathbf{X}$ analyzed in Section 5.3, this optimization problem on set $\mathcal{C}_{\text{ad}}$ is transformed in another problem on set $\mathcal{C}_{\text{free}}$ in introducing the Lagrange multipliers associated with the constraints.

5.4.6 A Fundamental Ensemble for the Symmetric Real Random Matrices with an Identity Matrix as a Mean Value

□ *Definition of Gaussian Orthogonal Ensemble (GOE$_\delta$).* The Gaussian Orthogonal Ensemble (GOE) is the set of the random matrices $[\mathbf{G}]$, defined on $(\Theta, \mathcal{T}, \mathcal{P})$, with values in $\mathbb{M}_n^S(\mathbb{R})$, for which the probability density function $p_{[\mathbf{G}]}$ with respect to $d^S G$ is defined on $\mathbb{M}_n^S(\mathbb{R})$ and is written as

$$p_{[\mathbf{G}]}([G]) = c_G \exp\left\{-\frac{n+1}{4\delta^2} \operatorname{tr}\{[G]^2\}\right\}, \; G_{kj} = G_{jk}, \; 1 \le j \le k \le n, \tag{5.51}$$

in which c_G is the constant of normalization and where δ is a hyperparameter defined after. This ensemble of symmetric real random matrices, for which the pdf is defined by Equation (5.51), is denoted by GOE$_\delta$.

□ *Properties of symmetric real random matrices belonging to ensemble* GOE$_\delta$. The fundamental properties of the random matrices that belong to ensemble GOE$_\delta$ are the following.

– The probability distribution is invariant under any real orthogonal transformation.
– The real random variables $\{\mathbf{G}_{jk}, 1 \le j \le k \le n\}$ (the entries) are mutually independent.
– Each real-valued random variable $\mathbf{G}_{jk}$ is a Gaussian, second-order, centered random variable, with variance

$$\sigma_{jk}^2 = (1 + \delta_{jk}) \delta^2 / (n + 1). \tag{5.52}$$

The pdf with respect to dg on $\mathbb{R}$ is thus written as

$$p_{\mathbf{G}_{jk}}(g) = (\sqrt{2\pi}\sigma_{jk})^{-1} \exp\{-g^2 / (2\sigma_{jk}^2)\}. \tag{5.53}$$

☐ *Decentering and interpretation of hyperparameter δ.* The following construction is detailed in [186]. Let $[\mathbf{G}^{\mathrm{GOE}}]$ be the random matrix with values in $\mathbb{M}_n^S(\mathbb{R})$ such that

$$[\mathbf{G}^{\mathrm{GOE}}] = [I_n] + [\mathbf{G}] \quad , \quad [\mathbf{G}] \in \mathrm{GOE}_\delta . \tag{5.54}$$

Therefore $E\{[\mathbf{G}^{\mathrm{GOE}}]\} = [I_n]$ and the coefficient of variation of $[\mathbf{G}^{\mathrm{GOE}}]$ is defined by

$$\delta_{\mathrm{GOE}} = \left\{ \frac{E\{\| \mathbf{G}^{\mathrm{GOE}} - E\{\mathbf{G}^{\mathrm{GOE}}\} \|_F^2\}}{\| E\{\mathbf{G}^{\mathrm{GOE}}\} \|_F^2} \right\}^{1/2} = \left\{ \frac{1}{n} E\{\| \mathbf{G}^{\mathrm{GOE}} - I_n \|_F^2\} \right\}^{1/2} . \tag{5.55}$$

It can be verified that $\delta_{\mathrm{GOE}} = \delta$. The parameter $2\delta/\sqrt{n+1}$ can be used to specify a scale.

☐ *Generator of realizations.* For $\theta \in \Theta$, $[\mathbf{G}^{\mathrm{GOE}}(\theta)] = [I_n] + [\mathbf{G}(\theta)]$ with $\mathbf{G}_{kj}(\theta) = \mathbf{G}_{jk}(\theta)$ and $\mathbf{G}_{jk}(\theta) = \sigma_{jk} U_{jk}(\theta)$, in which $\{U_{jk}(\theta)\}_{1 \leq j \leq k \leq n}$ is the realization of $n(n+1)/2$ independent copies of a normalized (centered and unit variance) Gaussian real random variable.

☐ *Why the GOE ensemble cannot be used in uncertainty quantification if positiveness property is required.* The GOE can then be viewed as a generalization of the Gaussian real random variables to the Gaussian symmetric real random matrices. The matrices $[\mathbf{G}^{\mathrm{GOE}}]$ belonging to ensemble GOE_δ are not positive almost surely or negative almost surely (no signature), and in addition,

$$E\{\|[\mathbf{G}^{\mathrm{GOE}}]^{-1}\|_F^2\} = +\infty . \tag{5.56}$$

Consequently, this ensemble cannot be used for stochastic modeling of a symmetric real matrix for which a positiveness property and an integrability of its inverse are required. Such a situation is similar to the one that we have presented in Section 5.3.1.

☐ *Need of new ensembles of random matrices for uncertainty quantification in computational mechanics.* New ensembles of random matrices are necessary for the development of the nonparametric probabilistic approach of the model uncertainties induced by the modeling errors in computational solid mechanics and in computational fluid mechanics, which differ from the GOE and from the other known ensembles from the random matrix theory. Such ensembles of random matrices have been introduced in [140, 184, 185, 190, 203, 207] and are presented in Sections 5.4.7 and 5.4.8.

☐ *Need of advanced numerical tools for constructing any ensemble of random matrices for the high dimensions.* In addition to the fundamental ensembles of positive-definite symmetric real random matrices for which the probability distribution and the associated generator can explicitly be constructed (see Sections 5.4.7 and 5.4.8), we are in need of advanced numerical tools for constructing any ensemble of random matrices for the high dimensions by using the Max-Ent principle, for which the explicit calculation of the Lagrange multipliers cannot be performed. Such advanced numerical tools for the high dimensions have been introduced in [98, 192, 207] and are presented in Section 5.4.10.

5.4.7 Fundamental Ensembles for Positive-Definite Symmetric Real Random Matrices

The fundamental ensembles of $\mathbb{M}_n^+(\mathbb{R})$-valued random matrices that will be detailed in the present section are the following.

- Ensemble SG_0^+: the mean value is the identity matrix and the lower bound is the zero matrix (central ensemble used as random germs for the other ensembles of random matrices).
- Ensemble $\mathrm{SG}_\varepsilon^+$: the mean value is the identity matrix and there is an arbitrary positive-definite lower bound.
- Ensemble SG_b^+: the mean value is either non-given or is equal to the identity matrix and there are positive-definite lower and upper bounds.
- Ensemble SG_λ^+: the mean value is the identity matrix, the lower bound is the zero matrix, and the second-order moments of diagonal entries are imposed, what allows for imposing the variances of certain random eigenvalues.

5.4.7.1 Ensemble SG_0^+ of Positive-Definite Random Matrices with an Identity Matrix as a Mean Value [184, 185]

□ *Definition and construction of the ensemble* SG_0^+ *by using the MaxEnt principle.* A random matrix $[\mathbf{G}_0]$ in SG_0^+ is a random matrix, defined on the probability space $(\Theta, \mathcal{T}, \mathcal{P})$, with values in $\mathbb{M}_n^+(\mathbb{R}) \subset \mathbb{M}_n^S(\mathbb{R})$, which is constructed by using the MaxEnt principle with the following available information:

$$E\{[\mathbf{G}_0]\} = [I_n] \quad , \quad E\{\log(\det[\mathbf{G}_0])\} = \nu_{G_0} \quad , \quad |\nu_{G_0}| < +\infty. \quad (5.57)$$

The pdf $p_{[\mathbf{G}_0]}$ (with respect to $d^S G$) has $\mathbb{S}_n = \mathbb{M}_n^+(\mathbb{R})$ as support and writes

$$p_{[\mathbf{G}_0]}([G]) = \mathbb{1}_{\mathbb{S}_n}([G]) \, c_{G_0} \left(\det[G]\right)^{(n+1)\frac{(1-\delta^2)}{2\delta^2}} \exp\left\{-\frac{n+1}{2\delta^2} \mathrm{tr}[G]\right\}. \quad (5.58)$$

The hyperparameter δ (such that $0 < \delta < (n+1)^{1/2}(n+5)^{-1/2}$) allows for controlling the level of the statistical fluctuations of $[\mathbf{G}_0]$,

$$\delta = \left\{\frac{E\{\|\mathbf{G}_0 - E\{\mathbf{G}_0\}\|_F^2\}}{\|E\{\mathbf{G}_0\}\|_F^2}\right\}^{1/2} = \left\{\frac{1}{n}E\{\|[\mathbf{G}_0] - [I_n]\|_F^2\}\right\}^{1/2}. \quad (5.59)$$

The normalization positive constant c_{G_0} is such that

$$c_{G_0} = (2\pi)^{-n(n-1)/4}\left(\frac{n+1}{2\delta^2}\right)^{\frac{n(n+1)}{2\delta^2}}\left\{\prod_{j=1}^{n}\Gamma\left(\frac{n+1}{2\delta^2}+\frac{1-j}{2}\right)\right\}^{-1}, \quad (5.60)$$

where, for all $z > 0$, $\Gamma(z) = \int_0^{+\infty} t^{z-1}\,e^{-t}\,dt$.

Comments-1. A random matrix $[G_0]$ belonging to ensemble SG_0^+ is such that,

1. $\{[G_0]_{jk}, 1 \le j \le k \le n\}$ are dependent non-Gaussian random variables (contrary to the Gaussian ensemble GOE_δ).
2. if $(n+1)/\delta^2$ is an integer, then this pdf coincides with the Wishart probability distribution [6, 101].
3. if $(n+1)/\delta^2$ is not an integer, then this probability density function can be viewed as a particular case of the Wishart distribution, in infinite dimension, for stochastic processes [182].

☐ *Second-order moments of a random matrix belonging to ensemble* SG_0^+. A random matrix $[G_0]$, which belongs to SG_0^+, is a second-order random variable, that is to say, $E\{\|G_0\|^2\} \le E\{\|G_0\|_F^2\} < +\infty$.

1. The mean value of random matrix $[G_0]$ is the identity matrix, $E\{[G_0]\} = [I_n]$.
2. The components of the covariance tensor of random matrix $[G_0]$, which are defined by $C_{jk,j'k'} = E\{([G_0]_{jk} - [I_n]_{jk})([G_0]_{j'k'} - [I_n]_{j'k'})\}$, are such that

$$C_{jk,j'k'} = \delta^2(n+1)^{-1}\{\delta_{j'k}\delta_{jk'} + \delta_{jj'}\delta_{kk'}\}. \quad (5.61)$$

3. The variance of the random variable $[G_0]_{jk}$ is written as

$$\sigma_{jk}^2 = C_{jk,jk} = \delta^2(n+1)^{-1}(1+\delta_{jk}). \quad (5.62)$$

☐ *Invertibility and convergence property when dimension goes to infinity.* The following properties can be proved:

$$E\{\|[G_0]^{-1}\|^2\} \le E\{\|[G_0]^{-1}\|_F^2\} < +\infty. \quad (5.63)$$

$$\forall n \ge 2, \quad E\{\|[G_0]^{-1}\|^2\} \le C_\delta < +\infty, \quad (5.64)$$

in which C_δ is a positive finite constant that is independent of n but that depends on hyperparameter δ.

Comments-2.

1. The invertibility property are due to the constraint $E\{\log(\det[G_0])\} = \nu_{G_0}$ with $|\nu_{G_0}| < +\infty$.

2. This is the reason why the truncated Gaussian distribution (ensemble GOE) restricted to $\mathbb{M}_n^+(\mathbb{R})$ does not satisfy this invertibility condition that is required for stochastic modeling of uncertainties in many cases.

3. The property given by Equation (5.63) is important for proving that a stochastic equation of the type $[\mathbf{G}_0]\,\mathbf{U} = \mathbf{f}$ admits a second-order solution $\mathbf{U}$ for all $\mathbf{f}$ given in $\mathbb{R}^n$.

4. The property given by Equation (5.64) is important for interpreting ensemble SG_0^+ when dimension n goes to infinity (for instance, for proving the existence of a unique solution of stochastic elliptic partial differential equation for which a finite approximation introduces a random matrix belonging to SG_0^+ [185]) (see also Comment-5 in Section 5.4.7.2).

☐ *Probability density function of the random eigenvalues of a random matrix belonging to ensemble* SG_0^+. The pdf of the random vector $\boldsymbol{\Lambda} = (\Lambda_1, \dots, \Lambda_n)$ that is made up of the positive-valued random eigenvalues of $[\mathbf{G}_0]$ in SG_0^+, such that $[\mathbf{G}_0]\,\boldsymbol{\Phi}^j = \Lambda_j\,\boldsymbol{\Phi}^j$, is written (see [184]) as

$$p_{\boldsymbol{\Lambda}}(\boldsymbol{\lambda}) = \mathbb{1}_{[0,+\infty[^n}(\boldsymbol{\lambda})\,c_{\Lambda}\,\{\prod_{j=1}^n \lambda_j^{(n+1)\frac{(1-\delta^2)}{2\delta^2}}\}$$

$$\times\{\prod_{\alpha<\beta}|\lambda_\beta - \lambda_\alpha|\}\exp\{-\frac{(n+1)}{2\delta^2}\sum_{k=1}^n \lambda_k\}, \quad (5.65)$$

in which c_{Λ} is a constant of normalization defined by the following equation,

$$\int_0^{+\infty}\dots\int_0^{+\infty} p_{\boldsymbol{\Lambda}}(\boldsymbol{\lambda})\,d\boldsymbol{\lambda} = 1. \quad (5.66)$$

Comments-3.

1. All the random eigenvalues $\Lambda_1, \dots, \Lambda_n$ of random matrix $[\mathbf{G}_0]$ in SG_0^+ are positive almost surely, while this assertion is not true for the random eigenvalues $\Lambda_1^{\mathrm{GOE}}, \dots, \Lambda_n^{\mathrm{GOE}}$ of the random matrix $[\mathbf{G}^{\mathrm{GOE}}] = [I_n] + [\mathbf{G}]$ in which $[\mathbf{G}]$ is a random matrix belonging to the GOE_δ ensemble.

2. A comparison of the random eigenvalues for random matrices in SG_0^+ and in GOE_δ can be found in [186].

☐ *Algebraic representation and generator of realizations for a random matrix belonging to ensemble* SG_0^+. The following algebraic representation [184, 185, 190] gives an explicit generator of realizations for any random matrix $[\mathbf{G}_0]$ that belongs to ensemble SG_0^+ for which the pdf is defined by Equation (5.58),

$$[\mathbf{G}_0] = [\mathbf{L}]^T\,[\mathbf{L}], \quad (5.67)$$

in which $[\mathbf{L}]$ is an upper triangular random matrix with values in $\mathbb{M}_n(\mathbb{R})$ such that

1. the random variables $\{[\mathbf{L}]_{jj'}, j \leq j'\}$ are mutually independent.
2. for $j < j'$, we have $[\mathbf{L}]_{jj'} = \sigma U_{jj'}$ in which $\sigma = \delta(n+1)^{-1/2}$ and where $U_{jj'}$ is a real-valued Gaussian random variable with zero mean and with a variance that is equal to 1.
3. for $j = j'$, we have $[\mathbf{L}]_{jj} = \sigma\sqrt{2V_j}$ where V_j is a positive-valued Gamma random variable whose probability density function with respect to dv is written as

$$p_{V_j}(v) = \mathbb{1}_{\mathbb{R}^+}(v)\frac{1}{\Gamma\left(\frac{n+1}{2\delta^2} + \frac{1-j}{2}\right)} v^{\frac{n+1}{2\delta^2} - \frac{1+j}{2}} e^{-v}. \tag{5.68}$$

Comments-4.

1. The algebraic representation defined by Equation (5.67) shows that although the entries $\{[\mathbf{L}]_{jj'}, j \leq j'\}$ of $[\mathbf{L}]$ are mutually independent, the entries $\{[\mathbf{G}_0]_{jj'}, j \leq j'\}$ of $[\mathbf{G}_0]$ are mutually dependent.
2. The diagonal entries $[\mathbf{L}]_{jj}, j = 1, \ldots, n$ of random matrix $[\mathbf{L}]$ depend on j.

5.4.7.2 Ensemble $\mathrm{SG}_\varepsilon^+$ of Positive-Definite Random Matrices with an Identity Matrix as a Mean Value and with an Arbitrary Positive-Definite Lower Bound [203, 207]

☐ *Definition and construction of the ensemble* $\mathrm{SG}_\varepsilon^+$. The ensemble $\mathrm{SG}_\varepsilon^+$ is the subset of SG_0^+, for which the mean value is the identity matrix and for which there is an arbitrary lower bound that is a positive-definite matrix controlled by an arbitrary positive number ε, defined by

$$[\mathbf{G}] = \frac{1}{1+\varepsilon}\{[\mathbf{G}_0] + \varepsilon[I_n]\} \quad , \quad [\mathbf{G}_0] \in \mathrm{SG}_0^+ \quad , \quad \varepsilon > 0, \tag{5.69}$$

$$0 < [\mathbf{G}_\ell] < [\mathbf{G}] \quad \text{a.s.}, \tag{5.70}$$

in which the lower bound is the positive-definite matrix,

$$[\mathbf{G}_\ell] = c_\varepsilon[I_n] \quad \text{with} \quad c_\varepsilon = \varepsilon/(1+\varepsilon). \tag{5.71}$$

For $\varepsilon = 0$, $\mathrm{SG}_\varepsilon^+ = \mathrm{SG}_0^+$ and then, $[\mathbf{G}] = [\mathbf{G}_0]$.

☐ *Properties of random matrices belonging to* $\mathrm{SG}_\varepsilon^+$. For all $\varepsilon > 0$, we have

$$E\{[\mathbf{G}]\} = [I_n] \quad , \quad E\{\log(\det([\mathbf{G}] - [\mathbf{G}_\ell]))\} = \nu_{G_\varepsilon}, \tag{5.72}$$

with $\nu_{G_\varepsilon} = \nu_{G_0} - n\log(1+\varepsilon)$.

☐ *Coefficient of variation* δ_G *of random matrix* $[\mathbf{G}]$. This coefficient of variation, which is defined by

$$\delta_G = \left\{ \frac{E\{\|\mathbf{G} - E\{\mathbf{G}\}\|_F^2\}}{\|E\{\mathbf{G}\}\|_F^2} \right\}^{1/2} = \left\{ \frac{1}{n} E\{\|[\mathbf{G}] - [I_n]\|_F^2\} \right\}^{1/2}, \quad (5.73)$$

is written as

$$\delta_G = \frac{\delta}{1 + \varepsilon}, \quad (5.74)$$

where δ is the hyperparameter defined for ensemble SG_0^+ (see Equation (5.59)).

□ *Lower bound and invertibility of random matrix* $[\mathbf{G}]$. Let $L^2(\Theta, \mathbb{R}^n)$ be the Hilbert space of all the $\mathbb{R}^n$-vector valued random variables, equipped with the inner product $<\mathbf{X}, \mathbf{Y}>_\Theta = E\{<\mathbf{X}, \mathbf{Y}>\}$ and the associated norm $\|\mathbf{X}\|_\Theta$. For all $\varepsilon > 0$, the bilinear form $b(\mathbf{X}, \mathbf{Y}) = <[\mathbf{G}]\mathbf{X}, \mathbf{Y}>_\Theta$ on $L^2(\Theta, \mathbb{R}^n) \times L^2(\Theta, \mathbb{R}^n)$ is such that

$$b(\mathbf{X}, \mathbf{X}) \geq c_\varepsilon \|\mathbf{X}\|_\Theta^2. \quad (5.75)$$

Random matrix $[\mathbf{G}]$ is invertible almost surely and its inverse $[\mathbf{G}]^{-1}$ is a second-order random variable,

$$E\{\|[\mathbf{G}]^{-1}\|_F^2\} < +\infty. \quad (5.76)$$

Comment-5. This ensemble will be useful for stochastic modeling of elliptic operators [152, 203, 208] (see also point 3 of Comment-2 in Section 5.4.7.1).

5.4.7.3 Ensemble SG_b^+ of Positive-Definite Random Matrices with Given Positive-Definite Lower and Upper Bounds, and with a Given Mean Value [96, 207]

□ *Available information for constructing* SG_b^+ *by the MaxEnt principle*. The ensemble $\mathrm{SG}_b^+ \subset \mathrm{SG}_0^+$ is constituted of random matrices $[\mathbf{G}_b]$, such that

$$[0] < [G_\ell] < [\mathbf{G}_b] < [G_u] \text{ with } 0 < [G_\ell] < [G_u] \text{ given in } \mathbb{M}_n^+(\mathbb{R}). \quad (5.77)$$

The available information are the following.

1. Supp $p_{[\mathbf{G}_b]} = \mathbb{S}_n = \{ [G] \in \mathbb{M}_n^+(\mathbb{R}) \mid [G_\ell] < [G] < [G_u] \} \subset \mathbb{M}_n^+(\mathbb{R})$.
2. $[\underline{G}_b] = E\{[\mathbf{G}_b]\}$ with $[\underline{G}_b]$ given such that $[G_\ell] < [\underline{G}_b] < [G_u]$.
3. $E\{\log(\det([\mathbf{G}_b] - [G_\ell]))\} = \nu_\ell$ and $E\{\log(\det([G_u] - [\mathbf{G}_b]))\} = \nu_u$ with $|\nu_\ell| < +\infty$ and $|\nu_u| < +\infty$.

□ *Construction of* SG_b^+ *with the MaxEnt*. Under the constraints defined by the available information, the MaxEnt principle yields the following result. The random matrix $[\mathbf{A}_0]$ with valued in $\mathbb{M}_n^+(\mathbb{R})$, which is defined by

$$[\mathbf{A}_0] = ([\mathbf{G}_b] - [G_\ell])^{-1} - [G_{\ell u}]^{-1}, \quad (5.78)$$

in which $[G_{\ell u}] = [G_u] - [G_\ell]$ belongs to $\mathbb{M}_n^+(\mathbb{R})$, can be written as

$$[\mathbf{A}_0] = [\underline{L}_0]^T [\mathbf{G}_0] [\underline{L}_0], \tag{5.79}$$

in which $[\mathbf{G}_0]$ belongs to ensemble SG_0^+ (which depends on the hyperparameter δ) and where $[\underline{L}_0] \in \mathbb{M}_n(\mathbb{R})$ is the upper triangular matrix such that $[\underline{A}_0] = E\{[\mathbf{A}_0]\} = [\underline{L}_0]^T [\underline{L}_0]$. The inversion yields

$$[\mathbf{G}_b] = [G_\ell] + \left([\underline{L}_0]^T [\mathbf{G}_0] [\underline{L}_0] + [G_{\ell u}]^{-1}\right)^{-1}, \tag{5.80}$$

and for any arbitrary small $\varepsilon_0 > 0$ (for instance, $\varepsilon_0 = 10^{-6}$),

$$\| E\{([\mathbf{A}_0] + [G_{\ell u}]^{-1})^{-1}\} + [G_\ell] - [\underline{G}_b] \|_F \; \leq \; \varepsilon_0 \|\underline{G}_b\|_F. \tag{5.81}$$

An efficient algorithm for the construction of $[\underline{L}_0]$ is required in order that $E\{[\mathbf{G}_b]\} = [\underline{G}_b]$, and we propose an efficient algorithm hereinafter.

□ *Algorithm for the construction of* $[\underline{L}_0]$. Let $\mathbb{M}_n^U(\mathbb{R})$ be the set of all the upper triangular real $(n \times n)$ matrices with positive diagonal entries. Therefore, $[\underline{L}_0]$ belongs to $\mathbb{M}_n^U(\mathbb{R})$. For a fixed value of δ, and for a given target value of $[\underline{G}_b]$, the value $[\underline{L}_0^{\text{opt}}]$ of $[\underline{L}_0]$ is calculated by solving the optimization problem

$$[\underline{L}_0^{\text{opt}}] = \arg \min_{[\underline{L}_0] \in \mathbb{M}_n^U(\mathbb{R})} \mathcal{F}([\underline{L}_0]). \tag{5.82}$$

$$\mathcal{F}([\underline{L}_0]) = \| E\{([\underline{L}_0]^T [\mathbf{G}_0] [\underline{L}_0] + [G_{\ell u}]^{-1})^{-1}\} + [G_\ell] - [\underline{G}_b] \|_F / \|\underline{G}_b\|_F. \tag{5.83}$$

□ *Coefficient of variation* δ_b *for controlling the level of statistical fluctuations of* $[\mathbf{G}_b]$. The coefficient δ_b is defined by

$$\delta_b = \left\{ \frac{E\{\| \mathbf{G}_b - \underline{G}_b \|_F^2\}}{\|\underline{G}_b\|_F^2} \right\}^{1/2}. \tag{5.84}$$

5.4.7.4 Ensemble SG_λ^+ of Positive-Definite Random Matrices with an Identity Matrix as a Mean Value, with a Zero Lower Bound, and with Imposed Second-Order Moments [140]

□ *Available information*. The ensemble SG_λ^+ of random matrices is the subset of SG_0^+ defined by the MaxEnt principle under the constraints defined by the following available information.

1. Supp $p_{[\mathbf{G}_\lambda]} = \mathbb{S}_n = \mathbb{M}_n^+(\mathbb{R})$.
2. $E\{[\mathbf{G}_\lambda]\} = [I_n]$.
3. $E\{\log(\det[\mathbf{G}_\lambda])\} = \nu_{G_\lambda}$ with $|\nu_{G_\lambda}| < +\infty$.
4. $E\{[\mathbf{G}_\lambda]_{jj}^2\} = s_j^2$, $j = 1, \ldots m$ with $m < n$ and $s_1^2, \ldots, s_m^2$ given positive constants.

□ *Construction of* SG_λ^+ *using the MaxEnt principle.* The pdf $p_{[\mathbf{G}_\lambda]}$ (with respect to $d^S G$) whose support is $\mathbb{S}_n = \mathbb{M}_n^+(\mathbb{R})$ is written as

$$p_{[\mathbf{G}_\lambda]}([G]) = \mathbb{1}_{\mathbb{S}_n}([G]) \times C_{G_\lambda} \times \left(\det[G]\right)^{\alpha-1}$$

$$\times \exp\{-\mathrm{tr}\{[\mu]^T[G]\} - \sum_{j=1}^m \tau_j G_{jj}^2\}, \quad (5.85)$$

in which C_{G_λ} is the normalization constant, α is a parameter such that $n + 2\alpha - 1 > 0$, $[\mu]$ is a diagonal real $(n \times n)$ matrix such that $\mu_{jj} = (n + 2\alpha - 1)/2$ for $j > m$, and where $\mu_{11}, \ldots, \mu_{mm}$ and $\tau_1, \ldots, \tau_m$ are $2m$ positive parameters, which are expressed as a function of α and $s_1^2, \ldots, s_m^2$.

□ *Expression of the coefficient of variation δ that controls the global level of statistical fluctuations of* $[\mathbf{G}_\lambda]$. The coefficient of variation δ is defined by

$$\delta = \left\{ \frac{E\{\|\mathbf{G}_\lambda - E\{\mathbf{G}_\lambda\}\|_F^2\}}{\|E\{\mathbf{G}_\lambda\}\|_F^2} \right\}^{1/2} = \left\{ \frac{1}{n} E\{\|[\mathbf{G}_\lambda] - [I_n]\|_F^2\} \right\}^{1/2}, \quad (5.86)$$

and is written as

$$\delta^2 = \frac{1}{n} \sum_{j=1}^m s_j^2 + \frac{n + 1 - (m/n)(n + 2\alpha - 1)}{n + 2\alpha - 1}. \quad (5.87)$$

5.4.8 Ensembles of Random Matrices for the Nonparametric Method in Uncertainty Quantification

The objective of this section is to present the construction of ensembles SE_0^+, SE_ε^+, SE^{+0}, SE^{rect}, and SE^{HT} of random matrices, which are useful for performing the stochastic modeling of matrices encountered in uncertainty quantification of computational models for many fields of applications, such as

- structural dynamics, acoustics, vibroacoustics, fluid–structure interaction, unsteady aeroelasticity, soil–structure interaction, etc.,
- elasticity in solid mechanics (elasticity tensors of random elastic continuous media, matrix-valued random fields for heterogeneous microstructures, etc),
- thermic (thermal conductivity tensor), electromagnetism (dielectric tensor), etc.

These ensembles of random matrices result from some transformations of the fundamental ensembles introduced in Section 5.4.7. In a few words, these ensembles of random matrices have the following characteristics.

- Ensemble SE_0^+: this ensemble is similar to SG_0^+ but the given mean value differs from the identity matrix.

- Ensemble SE_ε^+: this ensemble is similar to SG_ε^+ but the given mean value differs from the identity matrix and with a positive-definite lower bound.
- Ensemble SE^{+0}: this ensemble is similar to ensemble SG_0^+ but is constituted of positive-semidefinite $(m \times m)$ real random matrices for which the mean value is any given positive-semidefinite matrix.
- Ensemble SE^{rect}: this ensemble is made up of rectangular random matrices for which the mean value is a given rectangular matrix and which is constructed using ensemble SE_ε^+.
- Ensemble SE^{HT}: this is the set of random functions with values in the set of the complex matrices such that the real part and the imaginary part are positive-definite random matrices that are constrained by an underlying Hilbert transform induced by a causality property.

5.4.8.1 Ensemble SE_0^+ of Positive-Definite Random Matrices with a Given Mean Value [184, 185]

□ *Definition of ensemble* SE_0^+. Let $[A]$ be a deterministic matrix given in $\mathbb{M}_n^+(\mathbb{R})$ and representing the given mean value that is different from the identity matrix. Any random matrix $[\mathbf{A}_0]$ in SE_0^+ is with values in $\mathbb{M}_n^+(\mathbb{R})$, and is such that

$$E\{\mathbf{A}_0\} = [A] \in \mathbb{M}_n^+(\mathbb{R}) \ , \ \ E\{\log(\det[\mathbf{A}_0])\} = \nu_{A_0} \ , \ \ |\nu_{A_0}| < +\infty. \quad (5.88)$$

Random matrix $[\mathbf{A}_0]$ can then be written as

$$[\mathbf{A}_0] = [L_A]^T [\mathbf{G}_0] [L_A] \ \ , \ \ \ [\mathbf{G}_0] \in SG_0^+ \ , \quad (5.89)$$

in which $[A] = [L_A]^T [L_A]$ (Cholesky factorization of $[A] \in \mathbb{M}_n^+(\mathbb{R})$). It should be noted that matrix $[A]$ can also be written as $[A] = [A]^{1/2} [A]^{1/2}$, and consequently, matrix $[\mathbf{A}_0]$ can also be rewritten as

$$[\mathbf{A}_0] = [A]^{1/2} [\mathbf{G}_0] [A]^{1/2} . \quad (5.90)$$

□ *Properties of random matrix* $[\mathbf{A}_0]$ *and its inverse.* The second-order moment of $[\mathbf{A}_0]$ and of $[\mathbf{A}_0]^{-1}$ are finite,

$$E\{\|\mathbf{A}_0\|^2\} \le E\{\|\mathbf{A}_0\|_F^2\} < +\infty , \quad (5.91)$$

$$E\{\|[\mathbf{A}_0]^{-1}\|^2\} \le E\{\|[\mathbf{A}_0]^{-1}\|_F^2\} < +\infty . \quad (5.92)$$

The components of the covariance tensor of random matrix $[\mathbf{A}_0]$, which are defined by $C_{jk,j'k'} = E\{([\mathbf{A}_0]_{jk} - A_{jk})([\mathbf{A}_0]_{j'k'} - A_{j'k'})\}$, are such that

$$C_{jk,j'k'} = \frac{\delta^2}{n+1} \{A_{j'k} A_{jk'} + A_{jj'} A_{kk'}\} . \quad (5.93)$$

The variance $\sigma_{jk}^2 = C_{jk,jk}$ of random variable $[\mathbf{A}_0]_{jk}$ is such that

$$\sigma_{jk}^2 = \frac{\delta^2}{n+1}\left\{A_{jk}^2 + A_{jj}A_{kk}\right\}.\tag{5.94}$$

□ *Coefficient of variation.* The coefficient of variation δ_{A_0} of random matrix $[\mathbf{A}_0]$ is defined by

$$\delta_{A_0} = \left\{\frac{E\{\|\mathbf{A}_0 - A\|_F^2\}}{\|A\|_F^2}\right\}^{1/2},\tag{5.95}$$

and, since $E\{\|\mathbf{A}_0 - A\|_F^2\} = \sum_{j=1}^{n}\sum_{k=1}^{n}\sigma_{jk}^2$, it can be written as

$$\delta_{A_0} = \frac{\delta}{\sqrt{n+1}}\left\{1 + \frac{(\operatorname{tr}[A])^2}{\|A\|_F^2}\right\}^{1/2}.\tag{5.96}$$

5.4.8.2 Ensemble $\mathrm{SE}_\varepsilon^+$ of Positive-Definite Random Matrices with a Given Mean Value and with a Positive-Definite Lower Bound [203, 207]

□ *Definition of ensemble* $\mathrm{SE}_\varepsilon^+$. Let $[A]$ be a deterministic matrix given in $\mathbb{M}_n^+(\mathbb{R})$ and representing the given mean value that can be different or equal to the identity matrix. For fixed $\varepsilon > 0$, any random matrix $[\mathbf{A}]$ in $\mathrm{SE}_\varepsilon^+$ is a random matrix with values in $\mathbb{M}_n^+(\mathbb{R})$, which is written as

$$[\mathbf{A}] = [L_A]^T[\mathbf{G}][L_A] \quad , \quad [\mathbf{G}] \in \mathrm{SG}_\varepsilon^+ ,\tag{5.97}$$

in which $[A] = [L_A]^T[L_A]$ (Cholesky factorization of $[A] \in \mathbb{M}_n^+(\mathbb{R})$). Consequently, any random matrix $[\mathbf{A}]$ belonging to $\mathrm{SE}_\varepsilon^+$ can be written as

$$[\mathbf{A}] - [A_\ell] = \frac{1}{1+\varepsilon}[\mathbf{A}_0] > 0 \quad \text{a.s.} \quad \text{with} \quad [\mathbf{A}_0] \in \mathrm{SE}_0^+ ,\tag{5.98}$$

where $[A_\ell] \in \mathbb{M}_n^+(\mathbb{R})$ is the positive-definite lower bound defined by

$$[A_\ell] = c_\varepsilon[A] \quad \text{with} \quad c_\varepsilon = \varepsilon/(1+\varepsilon).\tag{5.99}$$

If $\varepsilon = 0$, then ensemble $\mathrm{SE}_\varepsilon^+$ coincides with SE_0^+. For $\varepsilon > 0$, this ensemble allows for introducing a positive-definite lower bound $[A_\ell]$ that is arbitrarily constructed for the case where it is known that a lower bound exists but for which there are no available information for identifying it.

□ *Properties of random matrix* $[\mathbf{A}]$ *and its inverse.* Any random matrix $[\mathbf{A}]$ that belongs to ensemble $\mathrm{SE}_\varepsilon^+$ has the following properties.

1. The inequality $0 < [A_\ell] < [\mathbf{A}]$ holds almost surely.
2. The second-order moment exists, $E\{\|\mathbf{A}\|^2\} \le E\{\|\mathbf{A}\|_F^2\} < +\infty$.
3. The mean value is the matrix $[A]$ given in $\mathbb{M}_n^+(\mathbb{R})$: $E\{[\mathbf{A}]\} = [A]$.
4. The pdf of random matrix $[\mathbf{A}]$ goes to zero when $[A] - [A_\ell] \to [0_+]$. We have $E\{\log(\det([\mathbf{A}]-[A_\ell]))\} = \nu_A$ with $|\nu_A| < +\infty$ and $\nu_A = \nu_{A_0} - n\log(1+\varepsilon)$.

5. For all random vector $\mathbf{X}$ in $L^2(\Theta, \mathbb{R}^n)$, we have $b_A(\mathbf{X}, \mathbf{X}) \geq c_\varepsilon <$ $[\mathbf{A}]\mathbf{X}, \mathbf{X}>_\Theta = c_\varepsilon \|[L_A]\mathbf{X}\|^2_\Theta$. Consequently, random matrix $[\mathbf{A}]$ is invertible almost surely and $[\mathbf{A}]^{-1}$ is a second-order random matrix,

$$E\{\|[\mathbf{A}]^{-1}\|^2\} \leq E\{\|[\mathbf{A}]^{-1}\|^2_F\} < +\infty. \tag{5.100}$$

□ *Coefficient of variation.* The coefficient of variation δ_A of random matrix $[\mathbf{A}]$, which is defined by

$$\delta_A = \left\{ \frac{E\{\|\mathbf{A} - A\|^2_F\}}{\|A\|^2_F} \right\}^{1/2}, \tag{5.101}$$

is written as

$$\delta_A = \frac{1}{1+\varepsilon} \delta_{A_0}, \tag{5.102}$$

in which δ_{A_0} is the coefficient of variation of the random matrix $[A_0]$ that belongs to SE_0^+.

5.4.8.3 Ensemble SE^{+0} of Positive-Semidefinite Random Matrices with a Given Positive-Semidefinite Mean Value [190, 207]

□ *Algebraic structure of random matrices belonging to* SE^{+0}. The ensemble SE^{+0} is constituted of random matrices with values in $\mathbb{M}_m^{+0}(\mathbb{R})$ (positive-semidefinite real $(m \times m)$ matrices). The mean value $[A] = E\{[\mathbf{A}]\}$ is a deterministic matrix $[A]$ that is given in $\mathbb{M}_m^{+0}(\mathbb{R})$. The dimension of the null space, $\mathrm{null}([A])$, of matrix $[A]$ is μ_{null} such that $1 \leq \mu_{\mathrm{null}} < m$. There is a rectangular matrix $[R_A]$ in $\mathbb{M}_{n,m}(\mathbb{R})$, with $n = m - \mu_{\mathrm{null}}$, such that

$$[A] = [R_A]^T [R_A]. \tag{5.103}$$

Such a factorization is performed using a classical algorithms. The ensemble SE^{+0} is then defined as follows. The null space of any random matrix $[\mathbf{A}]$ that belongs to SE^{+0} coincides with the null space of $[A]$, which means that $\mathrm{null}([\mathbf{A}])$ is a deterministic subspace of $\mathbb{R}^m$ with a fixed dimension $\mu_{\mathrm{null}} < m$ such that $\mathrm{null}([\mathbf{A}]) = \mathrm{null}([A])$.

□ *Definition and construction of ensemble* SE^{+0}. The ensemble SE^{+0} is defined as the subset of all the second-order random matrices $[\mathbf{A}]$ with values in $\mathbb{M}_m^{+0}(\mathbb{R})$ such that

$$[\mathbf{A}] = [R_A]^T [\mathbf{G}] [R_A] \quad, \quad [\mathbf{G}] \in \mathrm{SG}_\varepsilon^+, \tag{5.104}$$

in which $[\mathbf{G}]$ (that is a random matrix in $\mathrm{SG}_\varepsilon^+$) is with values in $\mathbb{M}_n^+(\mathbb{R})$ with $n = m - \mu_{\mathrm{null}}$. The coefficient of variation δ_G of random matrix $[\mathbf{G}]$ is defined by Equation (5.74), which has been expressed as a function of hyperparameter δ

that is defined by Equation (5.59), and which allows for controlling the level of statistical fluctuations of random matrix $[\mathbf{A}]$.

5.4.8.4 Ensemble SE^{rect} of Rectangular Random Matrices with a Given Mean Value [190, 207]

□ *Decomposition of the mean value of a random matrix belonging to* SE^{rect}. Any random rectangular matrix $[\mathbf{A}]$ in SE^{rect} is a second-order random matrix with values in $\mathbb{M}_{m,n}(\mathbb{R})$, whose mean value $[A] = E\{[\mathbf{A}]\}$ is given in $\mathbb{M}_{m,n}(\mathbb{R})$ and for which its null space is reduced to $\{0\}$ ($[A]\mathbf{x} = \mathbf{0}$ yields $\mathbf{x} = \mathbf{0}$). Such a rectangular matrix $[A]$ can be factorized as follows [90]:

$$[A] = [U][T], \tag{5.105}$$

in which the square matrix $[T]$ and the rectangular matrix $[U]$ are such that

$$[T] \in \mathbb{M}_n^+(\mathbb{R}) \quad \text{and} \quad [U] \in \mathbb{M}_{m,n}(\mathbb{R}) \quad \text{such that} \quad [U]^T[U] = [I_n]. \tag{5.106}$$

Such a factorization is directly deduced from the SVD of $[A]$.

□ *Definition of ensemble* SE^{rect}. For $\varepsilon > 0$ fixed, any $\mathbb{M}_{m,n}(\mathbb{R})$-valued random matrix $[\mathbf{A}]$ in SE^{rect} is written as

$$[\mathbf{A}] = [U][\mathbf{T}], \tag{5.107}$$

in which the random $(n \times n)$ matrix $[\mathbf{T}]$ belongs to ensemble SE$_\varepsilon^+$ and is then written as

$$[\mathbf{T}] = [L_T]^T[\mathbf{G}][L_T] \quad , \quad [\mathbf{G}] \in \text{SG}_\varepsilon^+, \tag{5.108}$$

in which $[L_T]$ is the upper triangular $(n \times n)$ real matrix given by the Cholesky factorization $[T] = [L_T]^T[L_T]$ of matrix $[T] \in \mathbb{M}_n^+(\mathbb{R})$. The coefficient of variation δ_G of random matrix $[\mathbf{G}]$ is defined by Equation (5.74), which has been expressed as a function of hyperparameter δ that is defined by Equation (5.59), and which allows for controlling the level of statistical fluctuations of random matrix $[\mathbf{T}]$ and consequently, of random matrix $[\mathbf{A}]$.

5.4.8.5 Ensemble SE^{HT} of a Pair of Positive-Definite Matrix-Valued Random Functions Related by a Hilbert Transform [204, 207, 37]

In many problems such as in linear structural dynamics and in linear vibroacoustics, formulated in the frequency domain, and for which a reduced-order computational model is constructed, it appears a frequency-dependent complex matrix that is written as $[Z(\omega)] = [K(\omega)] + i\omega[D(\omega)]$ in which $[D(\omega)]$ is a positive-definite real matrix and where $[K(\omega)]$ is a symmetric real matrix. For instance,

- in linear viscoelastic structural dynamics [154, 155, 156], $[D(\omega)]$ is the reduced damping matrix and $[K(\omega)]$ is the reduced stiffness matrix that is positive-definite (if the possible rigid body displacements are removed).
- in vibroacoustics, for the exterior Neumann problem related to the Helmholtz equation in an unbounded domain [154, 156], $(i\omega)^{-1}[Z(\omega)]$ is the boundary element matrix related to the acoustic impedance boundary operator for the external fluid-structure interface, $[D(\omega)]$ is the reduced damping matrix induced by the acoustic radiation at infinity (the Sommerfeld radiation), and $[K(\omega)]$ is symmetric but, in general, is not a positive matrix.

The ensemble SE^{HT} is a set of random functions (stochastic processes) indexed by $\mathbb{R}$, with values in a subset of $\mathbb{M}_n(\mathbb{C})$ (set of all the $(n \times n)$ complex random matrices). A frequency-dependent random matrix $[\mathbf{Z}(\omega)]$ that belongs to SE^{HT}, is written as

$$\omega \mapsto [\mathbf{Z}(\omega)] = [\mathbf{K}(\omega)] + i\omega \, [\mathbf{D}(\omega)] \,, \tag{5.109}$$

in which $[\mathbf{K}(\omega)]$ belongs to $\mathbb{M}_n^S(\mathbb{R})$, where $[\mathbf{D}(\omega)]$ belongs to $\mathbb{M}_n^+(\mathbb{R})$, and which are assumed to be constrained by an underlying Hilbert transform induced by the existence of a causality property. Consequently, the stochastic processes $[\mathbf{K}]$ and $[\mathbf{D}]$ cannot independently be modeled and are statistically dependent stochastic processes. In order to simplify the presentation, it is assumed that the frequency-dependent matrix $[\mathbf{K}(\omega)]$ is positive-definite matrix (case of the linear viscoelastic structural dynamics, for example). The extension for which $[\mathbf{K}(\omega)]$ is not positive but is only symmetric is straightforward.

□ *Defining the deterministic matrix problem.* We thus consider the case for which, for all $\omega \in \mathbb{R}$, the complex matrix $[Z(\omega)] = i\omega \, [D(\omega)] + [K(\omega)] \in \mathbb{M}_n(\mathbb{C})$ is such that

1. the frequency-dependent matrices $[D(\omega)]$ and $[K(\omega)]$ belong to $\mathbb{M}_n^+(\mathbb{R})$.
2. for all ω in $\mathbb{R}$, $[D(-\omega)] = [D(\omega)]$ and $[K(-\omega)] = [K(\omega)]$.
3. for all $\omega \geq 0$, $\omega \, [D(\omega)]$ is written as

$$\omega \, [D(\omega)] = [\widehat{N}^I(\omega)] \quad , \quad [K(\omega)] = [K_0] + [\widehat{N}^R(\omega)] \,, \tag{5.110}$$

in which $[K_0]$ is a matrix given in $\mathbb{M}_n^+(\mathbb{R})$ and where the real matrices $[\widehat{N}^R(\omega)]$ and $[\widehat{N}^I(\omega)]$ are defined by

$$[\widehat{N}^R(\omega)] = \Re\mathfrak{e}\{[\widehat{N}(\omega)]\} \quad , \quad [\widehat{N}^I(\omega)] = \Im\mathfrak{m}\{[\widehat{N}(\omega)]\} \,. \tag{5.111}$$

The function $\omega \mapsto [\widehat{N}(\omega)]$ from $\mathbb{R}$ into $\mathbb{M}_n(\mathbb{C})$ is defined as the Fourier transform,

$$[\widehat{N}(\omega)] = \int_{\mathbb{R}} e^{-i\omega t}[N(t)] \, dt \quad , \quad \forall \omega \in \mathbb{R} \,, \tag{5.112}$$

of an integrable causal function $t \mapsto [N(t)]$ from $\mathbb{R}$ into $\mathbb{M}_n(\mathbb{R})$, that is to say, is such that

$$\{t \mapsto [N(t)]\} \in L^1(\mathbb{R}, \mathbb{M}_n(\mathbb{R})) \quad , \quad [N(t)] = [0] \, , \, \forall t < 0 \, . \tag{5.113}$$

Therefore, $\omega \mapsto [\widehat{N}^R(\omega)]$ and $\omega \mapsto [\widehat{N}^I(\omega)]$ are continuous functions on $\mathbb{R}$, which goes to $[0]$ as $|\omega| \to +\infty$, and which are related by the Hilbert transform,

$$[\widehat{N}^R(\omega)] = \frac{1}{\pi} \text{p.v} \int_{-\infty}^{+\infty} \frac{1}{\omega - \omega'} [\widehat{N}^I(\omega')] \, d\omega' \, , \tag{5.114}$$

in which p.v denotes the Cauchy principal value. The real matrix $[K_0]$ that belongs to $\mathbb{M}_n^+(\mathbb{R})$ can then be written as

$$[K_0] = [K(0)] + \frac{2}{\pi} \int_0^{+\infty} [D(\omega)] \, d\omega \, . \tag{5.115}$$

Consequently, for all $\omega \geq 0$, we have the following equation that relates $[K(\omega)]$ and $[D(\omega)]$,

$$[K(\omega)] = [K(0)] + \frac{\omega}{\pi} \text{p.v} \int_{-\infty}^{+\infty} \frac{1}{\omega - \omega'} [D(\omega')] \, d\omega' \, , \tag{5.116}$$

or equivalently,

$$[K(\omega)] = [K(0)] + \frac{2\,\omega^2}{\pi} \text{p.v} \int_0^{+\infty} \frac{1}{\omega^2 - \omega'^2} [D(\omega')] \, d\omega' \, . \tag{5.117}$$

□ *Construction of ensemble* SE^{HT} *of stochastic processes by using a nonparametric stochastic approach.* The ensemble SE^{HT} of stochastic processes $\{[\mathbf{Z}(\omega), \omega \in \mathbb{R}\}$ is constructed by using the stochastic modeling principle corresponding to the nonparametric stochastic modeling of model uncertainties presented in Chapter 8. It consists, for all real ω, in modeling $[D(\omega)]$ and $[K(\omega)]$ by random matrices $[\mathbf{D}(\omega)]$ and $[\mathbf{K}(\omega)]$ such that, for all ω in $\mathbb{R}$,

$$[\mathbf{D}(-\omega)] = [\mathbf{D}(\omega)] \quad , \quad [\mathbf{K}(-\omega)] = [\mathbf{K}(\omega)] \quad \text{a.s.} \, , \tag{5.118}$$

for which the mean values are the deterministic matrices introduced before,

$$E\{[\mathbf{D}(\omega)]\} = [D(\omega)] \quad , \quad E\{[\mathbf{K}(\omega)]\} = [K(\omega)] \, . \tag{5.119}$$

These random matrices are constructed, for all $\omega \geq 0$, as follows.

▷ *Stochastic modeling of random matrix* $[\mathbf{D}(\omega)]$. For all fixed $\omega \geq 0$, the Cholesky factorization of the mean value $[D(\omega)]$ in $\mathbb{M}_n^+(\mathbb{R})$ yields

$$[D(\omega)] = [L_D(\omega)]^T [L_D(\omega)] \, , \tag{5.120}$$

and then the random matrix $[\mathbf{D}(\omega)]$ is constructed in $\mathrm{SE}_\varepsilon^+$ as

$$[\mathbf{D}(\omega)] = [L_D(\omega)]^T [\mathbf{G}_D] [L_D(\omega)] \quad , \quad [\mathbf{G}_D] \in \mathrm{SG}_\varepsilon^+ . \tag{5.121}$$

The hyperparameter δ_{G_D} of the random matrix $[\mathbf{G}_D]$ that allows the level of uncertainties to be controlled is written as $\delta_{G_D} = \delta_D/(1 + \varepsilon)$ (see Equation (5.74)).

▷ *Stochastic modeling of the random matrix* $[\mathbf{K}(0)]$. The deterministic matrix $[K(0)]$, which is given in $\mathbb{M}_n^+(\mathbb{R})$, is thus replaced by a random matrix $[\mathbf{K}(0)]$ with values in $\mathbb{M}_n^+(\mathbb{R})$ such that $E\{[\mathbf{K}(0)]\} = [K(0)]$. The Cholesky factorization of $[K(0)]$ is written as

$$[K(0)] = [L_{K(0)}]^T [L_{K(0)}], \tag{5.122}$$

and then $[\mathbf{K}(0)]$ is constructed in ensemble $\mathrm{SE}_\varepsilon^+$ of random matrices such that

$$[\mathbf{K}(0)] = [L_{K(0)}]^T [\mathbf{G}_{K(0)}] [L_{K(0)}] \quad , \quad [\mathbf{G}_{K(0)}] \in \mathrm{SG}_\varepsilon^+ . \tag{5.123}$$

The hyperparameter $\delta_{G_{K(0)}}$ of the random matrix $[\mathbf{G}_{K(0)}]$ that allows the level of uncertainties to be controlled is written as $\delta_{G_{K(0)}} = \delta_K/(1+\varepsilon)$ (see Equation (5.74)).

▷ *Stochastic modeling of random matrix* $[\mathbf{K}(\omega)]$. From Equation (5.117), for fixed $\omega \geq 0$, the random matrix $[\mathbf{K}(\omega)]$ is constructed by

$$[\mathbf{K}(\omega)] = [\mathbf{K}(0)] + \frac{2\,\omega^2}{\pi} \mathrm{p.v} \int_0^{+\infty} \frac{1}{\omega^2 - \omega'^2} [\mathbf{D}(\omega')] \, d\omega' , \tag{5.124}$$

which can also be rewritten (recommended for computation because the singularity in $u = 1$ is independent of ω),

$$[\mathbf{K}(\omega)] = [\mathbf{K}(0)] + \frac{2\,\omega}{\pi} \mathrm{p.v} \int_0^{+\infty} \frac{1}{1 - u^2} [\mathbf{D}(\omega u)] \, du ,$$

$$= [\mathbf{K}(0)] + \frac{2\,\omega}{\pi} \lim_{\eta \to 0} \{ \int_0^{1-\eta} + \int_{1+\eta}^{+\infty} \} . \tag{5.125}$$

The ingredients for a very efficient computation of the integral in the right-hand side of Equation (5.125) can be found in [37].

Sufficient condition for the positiveness of random matrix $[\mathbf{K}(\omega)]$. In general, if $[\mathbf{K}(0)]$ is positive definite almost surely and if $[\mathbf{D}(\omega)]$ is positive definite almost surely for all ω, these properties do not imply that $[\mathbf{K}(\omega)]$ is positive definite almost surely for all ω. Nevertheless, there is a sufficient condition on $[\mathbf{D}(\omega)]$ for that the property holds. If for all real vector $\mathbf{y} = (y_1, \ldots, y_n)$, and if almost surely, the random function $\omega \mapsto\, < [\mathbf{D}(\omega)] \, \mathbf{y} , \mathbf{y} >$ is decreasing in ω for $\omega \geq 0$, then, for all $\omega \geq 0$, $[\mathbf{K}(\omega)]$ is a positive-definite random matrix (see the proof in [204]).

5.4.9 Illustration of the Use of Ensemble SE_ε^+: Transient Wave Propagation in a Fluid-Solid Multilayer [65, 66]

The problem consists in using the axial transmission technique for the identification of the mechanical properties of the cortical bone. A transmitter applies an impulse in the ultrasonic range and a receiver records the signals. The first arrival signal is due to a guided wave in the cortical bone. Its velocity is estimated and the mechanical properties of the cortical bone are deduced by solving an inverse problem related to a computational model with uncertainties in order to take into account the variabilities of the measurements. The geometry and the physical properties of the multilayer "acoustic fluid" - "elastic solid" system are displayed in Figure 5.5. The upper layer represents the coupling gel, the skin, and the muscles, and is modeled by a deterministic acoustic fluid layer. The central layer represents the cortical bone and is modeled by a random homogeneous anisotropic elastic solid layer. The fourth-order elasticity tensor of the homogeneous anisotropic elastic solid random layer is modeled by a (6×6) random matrix belonging to SE_ε^+ (anisotropic statistical fluctuations) whose mean value corresponds to a homogeneous transversely isotropic elastic medium. The lower layer is made up of the bone marrow and is modeled by a deterministic acoustic fluid layer. The prediction of the pdf of the random wave velocity of the first arrival signal is performed with the computational stochastic model. Figure 5.6 displays the comparison of the pdf that is predicted and the one estimated using 600 measurements performed in vivo. The comparison is good.

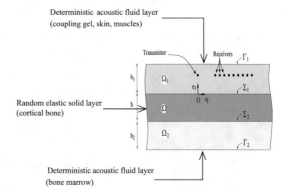

Fig. 5.5 Scheme describing the geometry and the physical properties of the multilayer "acoustic fluid"—"elastic solid" system.

Fig. 5.6 Computational pdf
estimated with the stochastic
model (blue line), compared
with the experimental pdf
estimated from 600 measurements performed in vivo (red
line). [Figure from [65]].

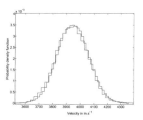

5.4.10 MaxEnt as a Numerical Tool for Constructing Ensembles of Random Matrices

For each ensemble of random matrices that we have constructed by using the Max-
Ent principle and that we have presented in Sections 5.4.7 and 5.4.8, an explicit
solution has been obtained for calculating the pdf and for constructing the associated generator (through the development of difficult mathematical calculations that
have not been shown).

For certain available information, the Lagrange multipliers cannot explicitly be calculated. This is the case for instance for the positive-definite matrices having a
symmetry class induced by material symmetries in the framework of elasticity, as
presented in Chapter 10. Such a situation is extremely challenging for the high dimensions. The method proposed hereinafter [192, 207] consists in introducing an
adapted parameterization of the random matrices under construction in order to use
the advanced methodology presented in Section 5.3.8, which is devoted to the use
of the MaxEnt principle as a numerical tool for constructing the pdf of a random
vector in any dimension.

☐ *Description of the random matrix under construction.* Let $[\mathbf{A}]$ be a random matrix
with values in $\mathbb{S}_n \subset \mathbb{M}_n^S(\mathbb{R})$, possibly with $\mathbb{S}_n = \mathbb{M}_n^S(\mathbb{R})$. For instance, $\mathbb{S}_n$ can
be the set $\mathbb{S}_n = \mathbb{M}_n^+(\mathbb{R})$. Let $p_{[\mathbf{A}]}$ be the pdf of $[\mathbf{A}]$ with respect to $d^S A$ on
$\mathbb{M}_n^S(\mathbb{R})$, such that $\mathrm{Supp}\, p_{[\mathbf{A}]} = \mathbb{S}_n$ (thus, $p_{[\mathbf{A}]}([A]) = 0$ for $[A]$ not in $\mathbb{S}_n$). The
normalization condition is then written as $\int_{\mathbb{S}_n} p_{[\mathbf{A}]}([A])\, d^S A = 1$.

☐ *Available information.* The available information is written as

$$E\{\mathcal{G}([\mathbf{A}])\} = \mathbf{f} \quad , \quad \text{on} \quad \mathbb{R}^\mu , \tag{5.126}$$

in which $\mathbf{f}$ is given in $\mathbb{R}^\mu$ with $\mu \geq 1$ and where $[A] \mapsto \mathcal{G}([A])$ is a given mapping
from $\mathbb{S}_n$ into $\mathbb{R}^\mu$.

Example. For instance, the mapping $\mathcal{G}$ can be defined by the mean value $E\{[\mathbf{A}]\} = [\underline{A}]$ in which $[\underline{A}]$ is a given matrix in $\mathbb{S}_n$ and by the condition $E\{\|[\mathbf{A}]^{-1}\|^2\} = c_A$ in which $|c_A| < +\infty$.

□ *Parameterization.* A parameterization of ensemble $\mathbb{S}_n$ is introduced such that any matrix $[A]$ belonging to $\mathbb{S}_n \subset \mathbb{M}_n^S(\mathbb{R})$ is written as

$$[A] = [\mathcal{A}(\mathbf{x})] \quad , \quad \mathbf{x} \in \mathcal{S}_N \subset \mathbb{R}^N \quad \text{with} \quad N \leq n(n+1)/2 , \qquad (5.127)$$

where $\mathbf{x} \mapsto [\mathcal{A}(\mathbf{x})]$ is a given mapping from $\mathcal{S}_N$ into $\mathbb{S}_n$. Let $\mathbf{x} \mapsto \mathbf{g}(\mathbf{x})$ be the mapping from $\mathcal{S}_N \subset \mathbb{R}^N$ into $\mathbb{R}^\mu$ such that

$$\mathbf{g}(\mathbf{x}) = \mathcal{G}([\mathcal{A}(\mathbf{x})]) \quad , \quad \mathbf{x} \in \mathcal{S}_N . \qquad (5.128)$$

Example (from [97]). For the positive-definite symmetric $(n \times n)$ real matrices, a given symmetry class is defined by a subset $\mathbb{M}_n^{\text{sym}}(\mathbb{R})$ of $\mathbb{M}_n^+(\mathbb{R})$. Let $\mathbb{S}_n = \mathbb{M}_n^{\text{sym}}(\mathbb{R}) \subset \mathbb{M}_n^+(\mathbb{R})$ such that for all $[A] \in \mathbb{S}_n$,

$$[A] = \exp_{\mathbb{M}}(\sum_{j=1}^{N} x_j \, [E_j^{\text{sym}}]) \quad , \quad [E_j^{\text{sym}}] \in \mathbb{M}_n^S(\mathbb{R}) , \qquad (5.129)$$

in which $\exp_{\mathbb{M}}$ denotes the exponential of the symmetric real matrices, where $\mathbf{x} = (x_1, \ldots, x_N) \in \mathbb{R}^N$, and where $\{[E_j^{\text{sym}}], j = 1, \ldots, N\}$ is the matrix algebraic basis of $\mathbb{M}_n^{\text{sym}}(\mathbb{R})$ (such as the Walpole tensor basis in elasticity), and where $\mathcal{S}_N = \mathbb{R}^N$.

□ *Stochastic representation using the parameterization.* Let $\mathbf{X}$ be an $\mathbb{R}^N$-valued second-order random variable for which the probability distribution on $\mathbb{R}^N$ is represented by the pdf $\mathbf{x} \mapsto p_{\mathbf{X}}(\mathbf{x})$ from $\mathbb{R}^N$ into $\mathbb{R}^+ = [0, +\infty[$ with respect to $d\mathbf{x} = dx_1 \ldots dx_N$. The support of function $p_{\mathbf{X}}$ is $\mathcal{S}_N$ (possibly $\mathcal{S}_N = \mathbb{R}^N$ as in the previous example). Function $p_{\mathbf{X}}$ satisfies the normalization condition,

$$\int_{\mathcal{S}_N} p_{\mathbf{X}}(\mathbf{x}) \, d\mathbf{x} = 1 . \qquad (5.130)$$

For random vector $\mathbf{X}$, the available information is written as

$$E\{\mathbf{g}(\mathbf{X})\} = \mathbf{f} . \qquad (5.131)$$

The advanced computational tool presented in Section 5.3.8 can directly be used.

5.5 Polynomial Chaos Representation as an Indirect Approach for Constructing a Prior Probability Distribution of a Second-Order Random Vector

The Polynomial Chaos Expansion is an important tool for Uncertainty Quantification, in particular for constructing a stochastic modeling in the framework of the indirect approach, in particular for constructing stochastic models of non-Gaussian random fields (see Sections 10.1, 10.3, 10.4, and 10.7.6), and for constructing stochastic solvers for boundary value problems using a Galerkin method (see Section 6.3).

In this section, the following items are presented.

- What is a polynomial chaos expansion of a second-order random vector.
- Polynomial chaos (PC) expansion with deterministic coefficients [85, 123, 188, 206, 222].
- Computational aspects for constructing realizations of polynomial chaos with high degrees, for an arbitrary probability distribution with a separable or a non-separable pdf [162, 199].
- Polynomial chaos expansion with random coefficients [196].

5.5.1 What Is a Polynomial Chaos Expansion of a Second-Order Random Vector

☐ *Introducing the random germ.* Let $N_g \geq 1$ be a given integer. Let $\boldsymbol{\xi} = (\xi_1, \ldots, \xi_{N_g})$ be in $\mathbb{R}^{N_g}$ and let $d\boldsymbol{\xi} = d\xi_1 \ldots d\xi_{N_g}$ be the Lebesgue measure. Let $\boldsymbol{\Xi} = (\Xi_1, \ldots, \Xi_{N_g})$ be a given $\mathbb{R}^{N_g}$-valued random variable, defined on $(\Theta, \mathcal{T}, \mathcal{P})$, with a given pdf $p_{\boldsymbol{\Xi}}(\boldsymbol{\xi})$ with respect to $d\boldsymbol{\xi}$, with support Supp $p_{\boldsymbol{\Xi}} = s_g \subset \mathbb{R}^{N_g}$ (possibly with $s_g = \mathbb{R}^{N_g}$), and such that

$$\int_{\mathbb{R}^{N_g}} \|\boldsymbol{\xi}\|^m \, p_{\boldsymbol{\Xi}}(\boldsymbol{\xi}) \, d\boldsymbol{\xi} < +\infty \quad , \quad \forall m \in \mathbb{N}. \tag{5.132}$$

The random variable $\boldsymbol{\Xi}$ will be called the *random germ*.

Example 1. The canonical Gaussian pdf on $\mathbb{R}^{N_g}$ (for which $s_g = \mathbb{R}^{N_g}$, see Equation (2.47)) is written as

$$p_{\boldsymbol{\Xi}}(\boldsymbol{\xi}) = \frac{1}{(2\pi)^{N_g/2}} \exp\{-\frac{1}{2} \|\boldsymbol{\xi}\|^2\}, \tag{5.133}$$

satisfies Equation (5.132).

Example 2. The uniform pdf on $\mathbb{R}^{N_g}$ with compact support $s_g = [-1, 1]^{N_g}$ is written as

$$p_\Xi(\xi) = \frac{1}{|s_g|} \mathbb{1}_{s_g}(\xi), \tag{5.134}$$

satisfies Equation (5.132).

□ *Introducing the random vector* **X** *as a deterministic nonlinear mapping of the random germ* Ξ. Let $\xi \mapsto \mathbf{f}(\xi)$ be a deterministic nonlinear mapping from $\mathbb{R}^{N_g}$ into $\mathbb{R}^n$ such that $\mathbf{X} = \mathbf{f}(\Xi)$ is a second-order $\mathbb{R}^n$-valued random variable, which means that **X** belongs to $L^2(\Theta, \mathbb{R}^n)$.

□ *Algebraic form of a polynomial chaos expansion.* The polynomial chaos expansion (PCE) of $\mathbf{X} = \mathbf{f}(\Xi)$ is written as

$$\mathbf{X} = \sum_{k=0}^{+\infty} \mathbf{y}^k \, \Psi_{\alpha^{(k)}}(\Xi), \tag{5.135}$$

in which, for $\{\Psi_{\alpha^{(k)}}(\Xi), k \in \mathbb{N}\}$, are the multivariate polynomial chaos and where $\{\mathbf{y}^k \in \mathbb{R}^n, k \in \mathbb{N}\}$ are the vector-valued coefficients, which completely define the mapping **f**. The convergence of the series in the right-hand side member of Equation (5.135) is in $L^2(\Theta, \mathbb{R}^n)$ (mean-square convergence).

5.5.2 Polynomial Chaos (PC) Expansion with Deterministic Coefficients

□ *Preamble.* Since $\mathbf{X} = \mathbf{f}(\Xi) = (f_1(\Xi), \dots, f_n(\Xi))$ belongs to $L^2(\Theta, \mathbb{R}^n)$, then for all j, $X_j = f_j(\Xi)$ belongs to $L^2(\Theta, \mathbb{R})$, and thus

$$E\{X_j^2\} = E\{f_j(\Xi)^2\} = \int_{\mathbb{R}^{N_g}} f_j(\xi)^2 \, p_\Xi(\xi) \, d\xi < +\infty, \tag{5.136}$$

which shows that function $\xi \mapsto f_j(\xi)$ from $\mathbb{R}^{N_g}$ into $\mathbb{R}$ is square integrable with respect to the probability distribution $p_\Xi(\xi) \, d\xi$.

□ *Hilbert space* $\mathbb{H}$. We thus introduce the Hilbert space $\mathbb{H}$ of all the $\mathbb{R}$-valued functions that are square integrable on $\mathbb{R}^{N_g}$ with respect to the probability distribution $p_\Xi(\xi) \, d\xi$, equipped with the inner product and the associated norm, such that, for all g and h in $\mathbb{H}$,

$$< g, h >_{\mathbb{H}} = \int_{\mathbb{R}^{N_g}} g(\xi) \, h(\xi) \, p_\Xi(\xi) \, d\xi = E\{g(\Xi) \, h(\Xi)\}, \tag{5.137}$$

$$\|g\|_{\mathbb{H}} = < g, g >_{\mathbb{H}}^{1/2} = \{E\{g(\Xi)^2\}\}^{1/2}. \tag{5.138}$$

□ *Multi-index notation.* Let $\alpha = (\alpha_1, \ldots, \alpha_{N_g}) \in \mathcal{A} = \mathbb{N}^{N_g}$ be the multi-indices, and let $|\alpha| = \alpha_1 + \ldots + \alpha_{N_g}$ be the length of α.

□ *Hilbert basis of* $\mathbb{H}$. Let $\{\Psi_\alpha, \alpha \in \mathcal{A}\}$ be a Hilbert basis of $\mathbb{H}$, which is such that

1. for all α in $\mathcal{A}$, function $\boldsymbol{\xi} \mapsto \Psi_\alpha(\boldsymbol{\xi})$ belongs to $\mathbb{H}$.
2. $\{\Psi_\alpha, \alpha \in \mathcal{A}\}$ is an orthonormal family in $\mathbb{H}$, which means that $< \Psi_\alpha, \Psi_\beta >_\mathbb{H} = \delta_{\alpha\beta}$, and consequently,

$$E\{\Psi_\alpha(\Xi)\,\Psi_\beta(\Xi)\} = \delta_{\alpha\beta}\,. \tag{5.139}$$

3. If $< g, \Psi_\alpha >_\mathbb{H} = 0$ for all $\alpha \in \mathcal{A}$, then $g = 0$, which means that the family is total in $\mathbb{H}$.

□ *Use of the classical theorem of Hilbert's bases.* For all $j = 1, \ldots, n$, function $\mathbf{x} \mapsto f_j(\boldsymbol{\xi})$ belongs to $\mathbb{H}$, and consequently, can be written as

$$f_j(\boldsymbol{\xi}) = \sum_{\alpha \in \mathcal{A}} y_j^{(\alpha)}\,\Psi_\alpha(\boldsymbol{\xi})\,, \tag{5.140}$$

in which the coefficients are such that

$$y_j^{(\alpha)} = < f_j, \Psi_\alpha >_\mathbb{H}$$
$$= \int_{\mathbb{R}^{N_g}} f_j(\boldsymbol{\xi})\,\Psi_\alpha(\boldsymbol{\xi})\,p_\Xi(\boldsymbol{\xi})\,d\boldsymbol{\xi} = E\{f_j(\Xi)\,\Psi_\alpha(\Xi)\}$$
$$= E\{X_j\,\Psi_\alpha(\Xi)\}\,. \tag{5.141}$$

The series in the right-hand side of Equation (5.141) is convergent in $\mathbb{H}$ (for the norm),

$$E\{X_j^2\} = \|f_j\|_\mathbb{H}^2 = E\{f_j(\Xi)^2\} = \sum_{\alpha \in \mathcal{A}} (y_j^{(\alpha)})^2 < +\infty\,. \tag{5.142}$$

□ *Expansion of random vector* $\mathbf{X} = \mathbf{f}(\Xi)$. The expansion of random vector $\mathbf{X} = \mathbf{f}(\Xi) \in L^2(\Theta, \mathbb{R}^n)$ can immediately be deduced,

$$\mathbf{X} = \mathbf{f}(\Xi) = \sum_{\alpha \in \mathcal{A}} \mathbf{y}^{(\alpha)}\,\Psi_\alpha(\Xi)\,, \tag{5.143}$$

in which the $\mathbb{R}^n$-valued coefficients $\mathbf{y}^{(\alpha)}$ are given by

$$\mathbf{y}^{(\alpha)} = \int_{\mathbb{R}^{N_g}} \mathbf{f}(\boldsymbol{\xi})\,\Psi_\alpha(\boldsymbol{\xi})\,p_\Xi(\boldsymbol{\xi})\,d\boldsymbol{\xi} = E\{\mathbf{f}(\Xi)\,\Psi_\alpha(\Xi)\}$$
$$= E\{\mathbf{X}\,\Psi_\alpha(\Xi)\}\,, \tag{5.144}$$

and the mean-square norm of $\mathbf{X}$ is written as

$$E\{\|\mathbf{X}\|^2\} = E\{\|\mathbf{f}(\Xi)\|^2\}$$
$$= \sum_{\alpha \in \mathcal{A}} \|\mathbf{y}^{(\alpha)}\|^2 < +\infty\,. \tag{5.145}$$

□ *Finite approximation of the expansion of random vector* $\mathbf{X} = \mathbf{f}(\Xi)$. For $N_d \geq 1$, let $\mathcal{A}_{N_d} \subset \mathcal{A}$ be the finite subset of multi-indices, which is defined by

$$\mathcal{A}_{N_d} = \{\boldsymbol{\alpha} = (\alpha_1, \ldots, \alpha_{N_g}) \in \mathbb{N}^{N_g} \mid 0 \leq \alpha_1 + \ldots + \alpha_{N_g} \leq N_d\}. \tag{5.146}$$

The $1 + K_{N_d}$ elements of $\mathcal{A}_{N_d}$ are denoted by $\boldsymbol{\alpha}^{(0)}, \ldots \boldsymbol{\alpha}^{(K_{N_d})}$ with $\boldsymbol{\alpha}^{(0)} = (0, \ldots, 0)$, and where the integer K_{N_d} is defined as a function of N_d by

$$K_{N_d} = \frac{(N_g + N_d)!}{N_g!\, N_d!} - 1\,. \tag{5.147}$$

It can then be deduced that

$$\mathbf{X} = \lim_{N_d \to +\infty} \mathbf{X}^{(K_{N_d})}\,, \tag{5.148}$$

with the convergence in $L^2(\Theta, \mathbb{R}^n)$ (convergence in norm), where, for each integer K_{N_d}, the $\mathbb{R}^n$-valued random variable $\mathbf{X}^{(K_{N_d})}$ is defined by

$$\mathbf{X}^{(K_{N_d})} = \sum_{k=0}^{K_{N_d}} \mathbf{y}^k\, \Psi_{\boldsymbol{\alpha}^{(k)}}(\Xi) \quad , \quad \mathbf{y}^k = E\{\mathbf{X}\, \Psi_{\boldsymbol{\alpha}^{(k)}}(\Xi)\} \in \mathbb{R}^n\,. \tag{5.149}$$

For a given K_{N_d}, the error induced by the use of $\mathbf{X}^{(K_{N_d})}$ instead of $\mathbf{X}$ can then be calculated by

$$E\{\|\mathbf{X} - \mathbf{X}^{(K_{N_d})}\|^2\} = E\{\|\mathbf{X}\|^2\} - \sum_{k=0}^{K_{N_d}} \|\mathbf{y}^k\|^2\,. \tag{5.150}$$

□ *Polynomial chaos expansion of random vector* $\mathbf{X} = \mathbf{f}(\Xi)$ *for a separable pdf.* Let us assume that the support s_g of p_Ξ is written as $s_g = s_1 \times \ldots \times s_{N_g}$ in which s_j is a subset of $\mathbb{R}$ (possibly with $s_j = \mathbb{R}$) for all $j = 1, \ldots, N_g$. Let us also assume that p_Ξ is written as $p_\Xi(\boldsymbol{\xi}) = p_1(\xi_1) \times \ldots \times p_{N_g}(\xi_{N_g})$ in which, for all $j = 1, \ldots, N_g$, p_j is a pdf on $\mathbb{R}$ for which its support is s_j. This means that the pdf is separable and, consequently, the real-valued random variables $\Xi_1, \ldots, \Xi_{N_g}$ are mutually independent. For all $j = 1, \ldots, N_g$, let $\{\psi_m^j(\xi_j), m \in \mathbb{N}\}$ be the family of orthonormal polynomials such that

$$\int_{s_j} \psi_m^j(\xi_j)\, \psi_{m'}^j(\xi_j)\, p_j(\xi_j)\, d\xi_j = \delta_{mm'}\,. \tag{5.151}$$

Then the family $\{\Psi_{\boldsymbol{\alpha}}, \boldsymbol{\alpha} \in \mathcal{A}\}$ of multivariate polynomials defined by

$$\Psi_{\boldsymbol{\alpha}}(\boldsymbol{\xi}) = \psi^1_{\alpha_1}(\xi_1) \times \ldots \times \psi^{N_g}_{\alpha_{N_g}}(\xi_{N_g}), \tag{5.152}$$

is a Hilbert basis of $\mathbb{H}$, and Equations (5.148) to (5.150) constitute the *polynomial chaos expansion (PCE)* of $\mathbf{X}$. It can be seen that the degree of multivariate polynomial $\Psi_{\boldsymbol{\alpha}}(\boldsymbol{\xi})$ is $|\boldsymbol{\alpha}|$, and that for all $\boldsymbol{\alpha}$ in $\mathcal{A}_{N_d}$, the maximum degree of the polynomials is N_d.

Fundamental example: the Gaussian polynomial chaos expansion. Let $p(\xi) = \frac{1}{\sqrt{2\pi}} \exp\{\frac{1}{2}\xi^2\}$ be the canonical Gaussian pdf on $\mathbb{R}$. Let $\{\psi_m(\xi), m \in \mathbb{N}\}$ be the family of the normalized Hermite polynomials that are generated by the following recurrence:

$$\psi_{n+1}(\xi) = \frac{1}{\sqrt{n+1}} \left(\xi \, \psi_n(\xi) - \frac{d\psi_n(\xi)}{d\xi} \right) \quad , \quad n \geq 0, \tag{5.153}$$

with $\psi_0(\xi) = 1$. The first ones are written as

$\psi_0(\xi) = 1$
$\psi_1(\xi) = \xi$
$\psi_2(\xi) = (\xi^2 - 1)/\sqrt{2}$
$\psi_3(\xi) = (\xi^3 - 3\xi)/\sqrt{3!},$
$\psi_4(\xi) = (\xi^4 - 6\xi^2 + 3)/\sqrt{4!},$
$\psi_5(\xi) = (\xi^5 - 10\xi^3 + 15\xi)/\sqrt{5!},$
etc.

Let $p_{\boldsymbol{\Xi}} = p(\xi_1) \times \ldots \times p(\xi_{N_g})$ be the pdf of $\boldsymbol{\Xi}$ and let $\{\Psi_{\boldsymbol{\alpha}}, \boldsymbol{\alpha} \in \mathcal{A}\}$ be the family of the multivariate normalized Hermite polynomials defined by $\Psi_{\boldsymbol{\alpha}}(\boldsymbol{\xi}) = \psi_{\alpha_1}(\xi_1) \times \ldots \times \psi_{\alpha_{N_g}}(\xi_{N_g})$. The polynomial chaos expansion defined by Equations (5.148) to (5.150), for which the multivariate polynomials are the multivariate normalized Hermite polynomials, is called the Gaussian polynomial chaos expansion of the second-order $\mathbb{R}^n$-valued random variable $\mathbf{X} = \mathbf{f}(\boldsymbol{\Xi})$.

☐ *Polynomial chaos expansion of random vector* $\mathbf{X} = \mathbf{f}(\boldsymbol{\Xi})$ *for a nonseparable pdf.* It is now assume that the pdf $p_{\boldsymbol{\Xi}}$ of $\boldsymbol{\Xi}$ is not separable. Consequently, the components $\Xi_1, \ldots, \Xi_{N_g}$ of $\boldsymbol{\Xi}$ are mutually dependent and the multivariate orthonormal polynomials $\{\Psi_{\boldsymbol{\alpha}}, \boldsymbol{\alpha} \in \mathcal{A}\}$ cannot be written as $\Psi_{\boldsymbol{\alpha}}(\boldsymbol{\xi}) = \psi_{\alpha_1}(\xi_1) \times \ldots \times \psi_{\alpha_{N_g}}(\xi_{N_g})$. It should be noted that a pdf that would be written as $p_{\boldsymbol{\Xi}}(\boldsymbol{\xi}) = \mathbb{1}_{s_g}(\boldsymbol{\xi}) \times q_1(\xi_1) \times \ldots \times q_{N_g}(\xi_{N_g})$, for which the support $s_g \subset \mathbb{R}^{N_g}$ would not be separable, that is to say, could not be written as $s_g = s_1 \times \ldots \times s_{N_g}$, would not be separable and, therefore, the components $\Xi_1, \ldots, \Xi_{N_g}$ of $\boldsymbol{\Xi}$ would be mutually dependent.

From a theoretical point of view, it is known that the family $\{\Psi_{\boldsymbol{\alpha}}, \boldsymbol{\alpha} \in \mathcal{A}_{N_d}\}$ of orthonormal polynomials is constructed using a Gram-Schmidt orthonormalization algorithm of the multivariate monomials defined, for $\boldsymbol{\alpha} = (\alpha_1, \ldots, \alpha_n) \in \mathcal{A}_{N_d}$ and for $\boldsymbol{\xi} = (\xi_1, \ldots, \xi_{N_g}) \in s_g \subset \mathbb{R}^{N_g}$, by

$$\mathcal{M}_{\boldsymbol{\alpha}}(\boldsymbol{\xi}) = \xi_1^{\alpha_1} \times \ldots \times \xi_{N_g}^{\alpha_{N_g}} . \tag{5.154}$$

Nevertheless, the Gram-Schmidt algorithm cannot directly be used for computing realizations of the multivariate orthonormal polynomials as it is explained in Section 5.5.3.

5.5.3 Computational Aspects for Constructing Realizations of Polynomial Chaos with High Degrees, for an Arbitrary Probability Distribution with a Separable or a Nonseparable pdf

☐ *Objective of the computational aspects.* We consider the case of K multivariate polynomial chaos $\{\Psi_{\boldsymbol{\alpha}^{(k)}}(\boldsymbol{\Xi}), k = 1, \ldots, K\}$ with high degrees for which the pdf $p_{\boldsymbol{\Xi}}(\boldsymbol{\xi})$ on $\mathbb{R}^{N_g}$ of $\boldsymbol{\Xi}$ is separable or is not separable.

Let us assume that $K \leq \nu$. The objective is to compute the matrix $[\Psi_\nu]$ in $\mathbb{M}_{K,\nu}(\mathbb{R})$ of ν independent realizations $\{\Psi_{\boldsymbol{\alpha}^{(k)}}(\boldsymbol{\Xi}(\theta_\ell)); k = 1, \ldots, K; \ell = 1, \ldots, \nu\}$ with $\theta_1, \ldots, \theta_\nu$ in Θ, of the multivariate polynomial chaos $\{\Psi_{\boldsymbol{\alpha}^{(k)}}(\boldsymbol{\Xi}), k = 1, \ldots, K\}$, in preserving the orthogonality properties defined by Equation (5.139). This matrix is thus defined by

$$[\Psi_\nu] = \begin{bmatrix} \Psi_{\boldsymbol{\alpha}^{(1)}}(\boldsymbol{\Xi}(\theta_1)) & \ldots & \Psi_{\boldsymbol{\alpha}^{(1)}}(\boldsymbol{\Xi}(\theta_\nu)) \\ \cdot & \ldots & \cdot \\ \Psi_{\boldsymbol{\alpha}^{(K)}}(\boldsymbol{\Xi}(\theta_1)) & \ldots & \Psi_{\boldsymbol{\alpha}^{(K)}}(\boldsymbol{\Xi}(\theta_\nu)) \end{bmatrix} . \tag{5.155}$$

The orthonormality property of the multivariate polynomial chaos, which is defined by Equation (5.139) and written as $E\{\Psi_{\boldsymbol{\alpha}^{(k)}}(\boldsymbol{\Xi}) \, \Psi_{\boldsymbol{\alpha}^{(k')}}(\boldsymbol{\Xi})\} = \delta_{kk'}$, is estimated by using the empirical statistical estimator as

$$\lim_{\nu \to +\infty} \frac{1}{\nu - 1} [\Psi_\nu] [\Psi_\nu]^T = [I_K] . \tag{5.156}$$

☐ *Why the computational aspects must be carefully taken into account.* For a separable pdf, the problem is not trivial at all. The use of the explicit algebraic formulas (constructed with a symbolic Toolbox) or the use of computational recurrence relations with respect to the degree induces important numerical noise and the orthogonality property defined by Equation (5.156) is lost as soon as the degree N_d of the polynomials increases. This difficulty is also encountered for a nonseparable pdf. If a global orthogonalization was done to correct this problem, then the independence of the realizations would be lost.

The method proposed hereinafter allows for preserving the orthogonality properties and the independence of the realizations and consists in the following steps.

1. Construction of the ν realizations of the multivariate monomials using a generator of independent realizations of the random germ Ξ for which the given pdf $p_\Xi(\xi)$ on $\mathbb{R}^{N_g}$ is arbitrary.
2. Doing an orthogonalization of the ν realizations of the multivariate monomials with an algorithm that is different from the Gram-Schmidt orthogonalization algorithm, which is not stable in presence of high degrees.

☐ *Algorithm.* The multivariate monomials are defined by

$$\mathcal{M}_{\alpha^{(k)}}(\xi) = \xi_1^{\alpha_1^{(k)}} \times \ldots \times \xi_{N_g}^{\alpha_{N_g}^{(k)}} \quad , \quad k = 1, \ldots, K. \tag{5.157}$$

Let $[M_\nu]$ be the matrix in $\mathbb{M}_{K,\nu}(\mathbb{R})$ of the ν independent realizations, which is defined by

$$
\begin{aligned}
[M_\nu] &= [\ \mathcal{M}(\Xi(\theta_1)) \ \ldots \ \mathcal{M}(\Xi(\theta_\nu))] \\
&= \begin{bmatrix} \mathcal{M}_{\alpha^{(1)}}(\Xi(\theta_1)) & \ldots & \mathcal{M}_{\alpha^{(1)}}(\Xi(\theta_\nu)) \\ \cdot & \ldots & \cdot \\ \mathcal{M}_{\alpha^{(K)}}(\Xi(\theta_1)) & \ldots & \mathcal{M}_{\alpha^{(K)}}(\Xi(\theta_\nu)) \end{bmatrix}.
\end{aligned} \tag{5.158}
$$

Therefore, matrix $[\Psi_\nu]$ can be rewritten as $[\Psi_\nu] = [A_\nu][M_\nu]$, in which $[A_\nu]$ is a matrix in $\mathbb{M}_K(\mathbb{R})$. It is assumed that $K \leq \nu$ is such that $[A_\nu]$ in invertible. Let $[R]$ be the matrix in $\mathbb{M}_K(\mathbb{R})$ defined by

$$
\begin{aligned}
[R] &= E\{\mathcal{M}(\Xi)\,\mathcal{M}(\Xi)^T\} \\
&= \lim_{\nu \to +\infty} \frac{1}{\nu-1}\,[M_\nu]\,[M_\nu]^T \\
&= \lim_{\nu \to +\infty} [A_\nu]^{-1}[A_\nu]^{-T}.
\end{aligned} \tag{5.159}
$$

The algorithm can then be summarized as follows:

1. computing matrix $[M_\nu]$ and then $[R] \simeq \frac{1}{\nu-1}[M_\nu][M_\nu]^T$ for ν sufficiently high.
2. computing $[A_\nu]^{-T}$ that corresponds to the Cholesky decomposition of $[R]$.
3. computing the lower triangular matrix $[A_\nu]$.
4. computing $[\Psi_\nu] = [A_\nu][M_\nu]$.

☐ *Numerical validation* [199]. We consider the case of the Gaussian multivariate normalized Hermite polynomials with $N_g = 1$. For $K = N_d = 1, \ldots, 30$ and for $\nu = 10^6$ realizations, Figure 5.7 displays the graphs of the relative error that measures the loss of the orthonormality property with respect to the known theoretical result when the explicit algebraic formula is used, when the recurrence equation is used, when the proposed computational method is used, and when the theoretical result is used (in this case, the error is exactly zero). It can be seen that the proposed method coincides with the theoretical result (the error is the numerical zero). The explicit algebraic formula and the recurrence equation give a big loss of orthogonality when the degree N_d of the polynomials increases.

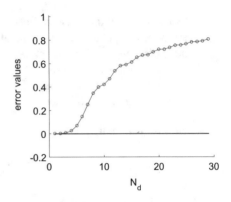

Fig. 5.7 For $N_g = 1$, graphs of the relative error as a function of $K = N_d = 1, \ldots, 30$ with $\nu = 10^6$: explicit algebraic formula and recurrence equations (thin line with circles); proposed computational method and theory (thick line). [Figure from [199]].

5.5.4 Polynomial Chaos Expansion with Random Coefficients

The polynomial chaos expansion, with random coefficients, of a second-order $\mathbb{R}^n$-valued random variable X is an important tool in the framework of statistical inverse problems. Let us assume that X corresponds to a stochastic modeling of an uncertain vector-valued parameter in a computational model, for which the probability distribution P_X is unknown and must be identified using experimental measurements. For the high dimensions (big value of n), the indirect approach is used consisting in introducing the PCE of X for which the coefficients must be identified but for which the dimension $N_g \leq n$ must also be identified. For such a problem it is interesting to introduce a PCE with random coefficients (see Step 7 of Section 10.4).

The formulation of the polynomial chaos expansion of a second-order $\mathbb{R}^n$-valued random vector $\mathbf{X} = \mathbf{f}(\Xi)$ with random coefficients can be formulated as follows.

Let us consider the PCE of $\mathbf{X}$ that has been introduced in Section 5.5.2, which is written as $\mathbf{X} = \sum_{\alpha} \mathbf{y}^{(\alpha)} \Psi_{\alpha}(\Xi)$ in which the deterministic coefficients $\mathbf{y}^{(\alpha)}$ that belong to $\mathbb{R}^n$ are such that $\mathbf{y}^{(\alpha)} = E\{\mathbf{X}\,\Psi_{\alpha}(\Xi)\}$, and where $\Xi = (\Xi_1, \ldots, \Xi_{N_g})$ is a given $\mathbb{R}^{N_g}$-random variable whose pdf $\boldsymbol{\xi} \mapsto p_{\Xi}(\boldsymbol{\xi})$ is known.

For $1 \leq N_g' < N_g$, let $\Xi' = (\Xi_1, \ldots, \Xi_{N_g'})$ be the random variable with values in $\mathbb{R}^{N_g'}$ and let $\Xi'' = (\Xi_{N_g'+1}, \ldots, \Xi_{N_g})$ be the random variable with values in $\mathbb{R}^{N_g - N_g'}$, such that $\Xi = (\Xi', \Xi'')$. It is assumed that the pdf of Ξ is written as $p_{\Xi}(\boldsymbol{\xi}) = p_{\Xi'}(\boldsymbol{\xi}') \times p_{\Xi''}(\boldsymbol{\xi}'')$ in which $\boldsymbol{\xi} = (\boldsymbol{\xi}', \boldsymbol{\xi}'')$, which means that the random variables Ξ' and Ξ'' are independent.

The PCE, with random coefficients, of $\mathbf{X}$, with respect to the $\mathbb{R}^{N'_g}$-valued random variable $\boldsymbol{\Xi}'$, can be written as

$$\mathbf{X} = \sum_{\alpha'} \mathbf{Y}^{\alpha'} \Psi_{\alpha'}(\boldsymbol{\Xi}') \quad , \quad \mathbf{Y}^{\alpha'} = E_{\boldsymbol{\Xi}'}\{\mathbf{X}\Psi_{\alpha'}(\boldsymbol{\Xi}')\}, \tag{5.160}$$

where $\{\mathbf{Y}^{\alpha'}\}_{\alpha'}$ is a family of mutually dependent $\mathbb{R}^n$-valued random vectors, which is independent of $\boldsymbol{\Xi}'$, and which is such that

$$E\{\mathbf{Y}^0\} = E\{\mathbf{X}\} \quad , \quad E\{\sum_{\alpha'} \|\mathbf{Y}^{\alpha'}\|^2\} = E\{\|\mathbf{X}\|^2\}. \tag{5.161}$$

5.6 Prior Algebraic Representation with a Minimum Number of Hyperparameters

As we have explained in Sections 5.1 to 5.4, a prior stochastic model of uncertain parameters (for random vectors and for random matrices) must be constructed either for carrying out a sensitivity analysis with respect to the level of uncertainties controlled by the hyperparameters (in order to perform a robust analysis with the computational model) or for performing the identification of such a stochastic model by solving a statistical inverse problem with data coming from numerical simulations or from experiments. These aspects will be developed in Chapter 7 and will be illustrated in Chapters 8 to 10 for stochastic models in high dimension.

For constructing a prior algebraic representation (with a minimum number of hyperparameters) of a prior stochastic model $\mathbf{X}^{\text{prior}}$ of a random variable $\mathbf{X}$, an efficient method consists in constructing an informative prior model by using the MaxEnt principle under the constraints defined by the available information (as explained in Sections 5.3 and 5.4). Such a prior stochastic model is written as

$$\mathbf{X}^{\text{prior}} = \mathbf{f}(\boldsymbol{\Xi}\,;\mathbf{s})\,, \tag{5.162}$$

in which

1. $\boldsymbol{\Xi}$ is the stochastic germ, which is a given random vector whose probability density function $p_{\boldsymbol{\Xi}}$ is known.
2. $\mathbf{s}$ is a vector-valued hyperparameter that should have a relative small dimension.
3. $\boldsymbol{\xi} \mapsto \mathbf{f}(\boldsymbol{\xi}\,;\mathbf{s})$ is a given nonlinear mapping, which has to be constructed.

Several illustrations of such a construction are given in Chapter 10.

5.7 Statistical Reduction (PCA and KL Expansion)

The statistical reduction is a very important tool for reducing the stochastic dimension of the random quantities used in stochastic modeling of uncertain parameters and of the random observations in a computational model, but also for the stochastic modeling of data coming from experimental measurements or coming from numerical simulations.

Two main tools can be used,

1. the principal component analysis (PCA) for the statistical reduction of a random vector $\mathbf{X}$ in finite dimension.
2. the Karhunen-Loève expansion for the statistical reduction of a random field $\mathbf{U}$ in infinite dimension.

These statistical reductions can be done only for second-order random variables or for random fields for which the covariance matrix or the covariance operator is known. If not, these types of statistical reductions cannot be done.

5.7.1 Principal Component Analysis (PCA) of a Random Vector X

□ *Second-order description.* Let $\mathbf{X}$ be a second-order random vector, defined on a probability space $(\Theta, \mathcal{T}, \mathcal{P})$, with values in $\mathbb{R}^n$, for which the mean vector $\mathbf{m_X}$ and the covariance matrix $[C_\mathbf{X}]$ are given,

$$\mathbf{m_X} = E\{\mathbf{X}\} \quad , \quad [C_\mathbf{X}] = E\{(\mathbf{X} - \mathbf{m_X})(\mathbf{X} - \mathbf{m_X})^T\}. \tag{5.163}$$

□ *Eigenvalue problem for the covariance matrix.* We consider the following value problem for the covariance matrix $[C_\mathbf{X}]$ that belongs either to $\mathbb{M}_n^{+0}$ (the matrix is not invertible) or to $\mathbb{M}_n^+$ (the matrix is invertible),

$$[C_\mathbf{X}]\, \varphi^i = \lambda_i\, \varphi^i. \tag{5.164}$$

The eigenvalues are real and such that

$$\lambda_1 \geq \lambda_2 \geq \ldots \geq \lambda_n \geq 0. \tag{5.165}$$

If $[C_\mathbf{X}] \in \mathbb{M}_n^+$, then $\lambda_n > 0$ and consequently, all the eigenvalues are strictly positive. The associated eigenvectors $\varphi^1, \ldots, \varphi^n$ constitute an orthonormal basis of $\mathbb{R}^n$,

$$<\varphi^i, \varphi^{i'}> = \delta_{ii'}. \tag{5.166}$$

□ *Statistical reduction by using the principal component analysis.* For $1 \leq m \leq n$, the reduced statistical model $\mathbf{X}^{(m)}$ of $\mathbf{X}$ is written as

$$\mathbf{X}^{(m)} = \mathbf{m_X} + \sum_{i=1}^{m} \sqrt{\lambda_i}\, \eta_i\, \varphi^i\,, \tag{5.167}$$

in which $\boldsymbol{\eta} = (\eta_1, \ldots, \eta_m)$ is the second-order random vector with values in $\mathbb{R}^m$, such that

$$\eta_i = \frac{1}{\sqrt{\lambda_i}} < \mathbf{X} - \mathbf{m_X}, \varphi^i > \quad,\quad i = 1, \ldots, m\,. \tag{5.168}$$

The components $\eta_1, \ldots, \eta_m$ of random vector $\boldsymbol{\eta}$ are such that

$$E\{\eta_i\} = 0 \quad,\quad E\{\eta_i \eta_{i'}\} = \delta_{ii'}\,. \tag{5.169}$$

Consequently, the mean vector $\mathbf{m_{\eta}}$ in $\mathbb{R}^m$ and the covariance matrix $[C_{\eta}]$ in $\mathbb{M}_m^+(\mathbb{R})$ of the $\mathbb{R}^m$-valued random variable $\boldsymbol{\eta}$ are such that

$$\mathbf{m_{\eta}} = \mathbf{0} \quad,\quad [C_{\eta}] = [I_m]\,. \tag{5.170}$$

Comments.

1. If m is chosen equal to n, then Equation (5.167) does not correspond to a statistical reduction but is simply a change of basis in $\mathbb{R}^n$. If $m < n$, then there is a statistical reduction.
2. In Equation (5.170), the first equation means that the second-order random vector $\boldsymbol{\eta}$ is centered.
3. In Equation (5.170), the second equation means that the components $\eta_1, \ldots, \eta_m$ of the second-order random vector $\boldsymbol{\eta}$ are not correlated because the covariance matrix $[C_{\eta}]$ is the identity matrix (see Section 2.2.3).
4. The property $[C_{\eta}] = [I_m]$ does not mean that the real-valued random variables $\eta_1, \ldots, \eta_m$ are mutually independent. In general, the random variables $\eta_1, \ldots, \eta_m$ are mutually dependent. Nevertheless, if $\mathbf{X}$ is a Gaussian random vector, then $\boldsymbol{\eta}$ is a Gaussian random vector and since $[C_{\eta}] = [I_m]$, it can then be deduced that $\eta_1, \ldots, \eta_m$ are mutually independent (see Chapter 2), but this result does not hold for a non-Gaussian random vector $\mathbf{X}$.

□ *Convergence analysis and calculation of the reduced order* m. The convergence in $L^2(\Theta, \mathbb{R}^n)$ of the reduced statistical model defined by Equation (5.167) is controlled by

$$E\{\|\mathbf{X} - \mathbf{X}^{(m)}\|^2\} = \sum_{m+1}^{n} \lambda_i = E\{\|\mathbf{X} - \mathbf{m_X}\|^2\} - \sum_{i=1}^{m} \lambda_i\,. \tag{5.171}$$

Since $E\{\|\mathbf{X} - \mathbf{m_X}\|^2\} = \mathrm{tr}[C_{\mathbf{X}}]$, a relative error function $m \mapsto \mathrm{err}(m)$ defined on $\{1, \ldots, n-1\}$ with values in $\mathbb{R}^+$, based on Equation (5.171), is introduced such that

$$\mathrm{err}(m) = 1 - \frac{\sum_{i=1}^{m} \lambda_i}{\mathrm{tr}[C_{\mathbf{X}}]}\,. \tag{5.172}$$

Note that $m \mapsto \text{err}(m)$ is monotonically decreasing and that $\text{err}(n) = 0$. For a given covariance matrix $[C_{\mathbf{X}}]$ and for a given tolerance $\varepsilon > 0$, the optimal value m_ε of the reduced order m in Equation (5.167) is calculated in order that

$$m_\varepsilon = \inf\{m \in \{1, \ldots, n-1\} \mid \text{err}(m) \leq \varepsilon, \ \forall m \geq m_\varepsilon\}. \qquad (5.173)$$

□ *Numerical aspects.* Often, the covariance matrix $[C_{\mathbf{X}}]$ is not exactly known but is estimated from the knowledge of ν independent realizations $\mathbf{x}^1, \ldots, \mathbf{x}^\nu$ of $\mathbf{X}$ (coming from experimental data or from numerical simulations). The empirical estimates of the mean vector $\mathbf{m}_{\mathbf{X}} = E\{\mathbf{X}\}$ and of the covariance matrix $[C_{\mathbf{X}}] = E\{(\mathbf{X} - \mathbf{m}_{\mathbf{X}})(\mathbf{X} - \mathbf{m}_{\mathbf{X}})^T\}$ are usually written as

$$\widehat{\mathbf{m}}_{\mathbf{X}}^\nu = \frac{1}{\nu} \sum_{\ell=1}^\nu \mathbf{x}^\ell \quad , \quad [\widehat{C}_{\mathbf{X}}^\nu] = \frac{1}{\nu-1} \sum_{\ell=1}^\nu (\mathbf{x}^\ell - \widehat{\mathbf{m}}_{\mathbf{X}}^\nu)(\mathbf{x}^\ell - \widehat{\mathbf{m}}_{\mathbf{X}}^\nu)^T. \qquad (5.174)$$

If n is big (high dimension), for instance if $n = 10^6$, then the storing of the symmetric matrix $[\widehat{C}_{\mathbf{X}}^\nu]$ requires $4\,000$ GB (Gigabytes). Let us assume that $\nu < n$. For such a case, an efficient algorithm derived from [7, 83, 107] is presented hereinafter for computing the first m largest eigenvalues and eigenvectors without assembling and without storing the big matrix $[\widehat{C}_{\mathbf{X}}^\nu]$.

For $\ell = 1, \ldots, \nu$, let $\mathbf{y}^\ell = \mathbf{x}^\ell - \widehat{\mathbf{m}}_{\mathbf{X}}^\nu$ be the centered realizations. Let $[\mathsf{y}] = [\mathbf{y}^1 \ldots \mathbf{y}^\nu]$ be the matrix in $\mathbb{M}_{n,\nu}(\mathbb{R})$ whose columns are the centered realizations. The estimate $[\widehat{C}_{\mathbf{X}}^\nu]$ defined by Equation (5.174) can then be rewritten as

$$[\widehat{C}_{\mathbf{X}}^\nu] = \frac{1}{\nu-1}[\mathsf{y}][\mathsf{y}]^T. \qquad (5.175)$$

It should be noted that, since it is assumed that $\nu < n$, then the rank of matrix $[\widehat{C}_{\mathbf{X}}^\nu]$ is less or equal to ν. Consequently, the nonzero eigenvalues will be obtained for $m \leq \nu$. The eigenvalue problem defined by Equation (5.164) is rewritten as

$$[\widehat{C}_{\mathbf{X}}^\nu][\varPhi] = [\varPhi][\varLambda], \qquad (5.176)$$

in which $[\varLambda]$ is the diagonal $(\nu \times \nu)$ matrix of the positive eigenvalues $\lambda_1 \geq \ldots \geq \lambda_\nu > 0$ and where the ν columns of the matrix $[\varPhi]$ in $\mathbb{M}_{n,\nu}(\mathbb{R})$ are the associated eigenvectors. Matrix $[\varPhi]$ is such that

$$[\varPhi]^T[\varPhi] = [I_\nu]. \qquad (5.177)$$

Let us consider the "thin SVD" [90] ("economy size" decomposition with MATLAB) of matrix $[\mathsf{y}] \in \mathbb{M}_{n,\nu}(\mathbb{R})$, which is written as

$$[\mathsf{y}] = [\mathsf{U}][\varSigma][\mathsf{V}]^T. \qquad (5.178)$$

It can then be seen that

$$[\Phi] = [\mathbb{U}] \quad , \quad [\Lambda] = \frac{1}{\nu - 1} [\Sigma]^2 . \tag{5.179}$$

5.7.2 Karhunen-Loève Expansion of a Random Field U

The Karhunen-Loève (KL) expansion of a random field (or a stochastic process) **U** *is similar to the principal component analysis (PCA) for a random vector* **X**, *but in infinite dimension. We will also explain the connection between the KL expansion of random field* **U**, *and the PCA of a random vector* **X** *that corresponds to a sampling of* **U**.

☐ *Probabilistic description of the random field* **U**. A second-order random field $\{\mathbf{U}(\zeta), \zeta \in \Omega\}$, defined on the probability space $(\Theta, \mathcal{T}, \mathcal{P})$, indexed by any part $\Omega \subset \mathbb{R}^d$ $(d \geq 1)$, with values in $\mathbb{R}^\mu$, is an uncountable family of second-order $\mathbb{R}^\mu$-valued random variables. If $d = 1$, **U** is usually called a "stochastic process" and for $d \geq 2$, **U** is usually called a "random field." Nevertheless, a random field can also be viewed as a stochastic process indexed by a set whose dimension d is greater or equal to 2. The second-order quantities of random field **U** are defined as follows:

1. *Mean function.* The mean function of random field **U** is the function $\zeta \mapsto \mathbf{m}_\mathbf{U}(\zeta)$ from Ω into $\mathbb{R}^\mu$ such that

$$\mathbf{m}_\mathbf{U}(\zeta) = E\{\mathbf{U}(\zeta)\} . \tag{5.180}$$

2. *Autocorrelation function.* The autocorrelation function of random field **U** is the function $(\zeta, \zeta') \mapsto [R_\mathbf{U}(\zeta, \zeta')]$ from $\Omega \times \Omega$ into $\mathbb{M}_\mu(\mathbb{R})$ such that

$$[R_\mathbf{U}(\zeta, \zeta')] = E\{\mathbf{U}(\zeta)\, \mathbf{U}(\zeta')^T\} . \tag{5.181}$$

Since **U** has been assumed to be a second-order random field, we have

$$E\{\|\mathbf{U}(\zeta)\|^2\} = \mathrm{tr}[R_\mathbf{U}(\zeta, \zeta)] < +\infty \quad , \quad \forall \zeta \in \Omega . \tag{5.182}$$

3. *Covariance function.* The covariance function of random field **U** is the function $(\zeta, \zeta') \mapsto [C_\mathbf{U}(\zeta, \zeta')]$ from $\Omega \times \Omega$ into $\mathbb{M}_\mu(\mathbb{R})$ such that

$$[C_\mathbf{U}(\zeta, \zeta')] = E\{(\mathbf{U}(\zeta) - \mathbf{m}_\mathbf{U}(\zeta))\, (\mathbf{U}(\zeta') - \mathbf{m}_\mathbf{U}(\zeta'))^T\} . \tag{5.183}$$

It can easily be seen that

$$[C_\mathbf{U}(\zeta, \zeta')] = [R_\mathbf{U}(\zeta, \zeta')] - \mathbf{m}_\mathbf{U}(\zeta)\, \mathbf{m}_\mathbf{U}(\zeta')^T . \tag{5.184}$$

In addition, it will be assumed that the covariance function satisfies the following property

$$\int_{\Omega} \int_{\Omega} \| [C_U(\zeta, \zeta')] \|_F^2 \, d\zeta' \, d\zeta' < +\infty. \tag{5.185}$$

The hypothesis defined by Equation (5.185) is introduced for obtaining the existence of the KL expansion involving a countable family of random variables.

□ *What is a KL expansion and what is its interest.* Mathematically, and under the hypotheses defined by Equations (5.182) and (5.185), the Karhunen-Loève expansion consists in representing the uncountable family of second-order random variables $\{U(\zeta), \zeta \in \Omega\}$ by a countable family $\{\eta_i, i \in \mathbb{N}^*\}$ in the mean-square sense,

$$U(\zeta) = m_U(\zeta) + \sum_{i=1}^{\infty} \sqrt{\lambda_i} \, \eta_i \, \phi^i(\zeta), \tag{5.186}$$

in which, for all ζ in Ω, the series is mean-square convergent, that is to say, is convergent in $L^2(\Theta, \mathbb{R}^\mu)$. The KL expansion, defined by Equation (5.186), will allow for constructing a finite representation $U^{(m)}$ that approximates U with a finite number of terms and which will be called the statistical reduced representation of random field U.

□ *Construction of a Hilbert basis for the KL expansion.* Let $L^2(\Omega, \mathbb{R}^\mu)$ be the Hilbert space of the square integrable functions defined on Ω with values in $\mathbb{R}^\mu$, equipped with the inner product and the associated norm,

$$<u, v>_{L^2} = \int_{\Omega} <u(\zeta), v(\zeta)> \, d\zeta \ , \quad \|u\|_{L^2} = \{<u, u>_{L^2}\}^{1/2}. \tag{5.187}$$

The hypothesis defined by Equation (5.185) for the covariance function implies that the covariance linear operator C_U defined by

$$\{C_U \phi\}(\zeta) = \int_{\Omega} [C_U(\zeta, \zeta')] \, \phi(\zeta') \, d\zeta', \tag{5.188}$$

is a Hilbert-Schmidt operator in $L^2(\Omega, \mathbb{R}^\mu)$, which means that the eigenvalue problem

$$C_U \phi = \lambda \phi, \tag{5.189}$$

admits a decreasing sequence of positive eigenvalues $\lambda_1 \geq \lambda_2 \geq \ldots \to 0$ such that $\sum_{i=1}^{+\infty} \lambda_i^2 < +\infty$ and the family $\{\phi^i, i \in \mathbb{N}^*\}$ of eigenvectors constitutes a Hilbert basis of $L^2(\Omega, \mathbb{R}^\mu)$. Therefore, the family $\{\phi^i\}_i$ is an orthonormal family in $L^2(\Omega, \mathbb{R}^\mu)$,

$$<\phi^i, \phi^{i'}>_{L^2} = \delta_{ii'}. \tag{5.190}$$

□ *The KL expansion as a representation of* **U** *with the Hilbert basis*. Applying the theorem of Hilbert's basis to $\mathbf{U} - \mathbf{m}_\mathbf{U}$ yields the following KL expansion of random field **U**,

$$\mathbf{U}(\zeta) = \mathbf{m}_\mathbf{U}(\zeta) + \sum_{i=1}^\infty \sqrt{\lambda_i}\, \eta_i\, \phi^i(\zeta), \tag{5.191}$$

in which, for all ζ in Ω, the series is convergent in $L^2(\Theta, \mathbb{R}^\mu)$, and where, for all i, η_i is a second-order real-valued random variable that is given by

$$\eta_i = \frac{1}{\sqrt{\lambda_i}} \int_\Omega <\mathbf{U}(\zeta) - \mathbf{m}_\mathbf{U}(\zeta), \phi^i(\zeta)>\, d\zeta. \tag{5.192}$$

The second-order real-valued random variables $\{\eta_i, i \in \mathbb{N}^*\}$ are centered and not correlated,

$$E\{\eta_i\} = 0 \quad , \quad E\{\eta_i\, \eta_{i'}\} = \delta_{ii'}. \tag{5.193}$$

Similarly to the comment formulated for the PCA, the second-order real-valued random variables $\{\eta_i, i \in \mathbb{N}^*\}$ are not correlated but are mutually dependent. If the random field **U** is Gaussian, then the random variables $\{\eta_i, i \in \mathbb{N}^*\}$ are mutually independent, but this property is not true for a non-Gaussian random field.

□ *Statistical reduction using the KL expansion of* **U**. For a fixed value of the integer m such that $1 \leq m$, the reduced statistical model of the random field $\{\mathbf{U}(\zeta), \zeta \in \Omega\}$ is written as $\{\mathbf{U}^{(m)}(\zeta), \zeta \in \Omega\}$ and is defined by

$$\mathbf{U}^{(m)}(\zeta) = \mathbf{m}_\mathbf{U}(\zeta) + \sum_{i=1}^m \sqrt{\lambda_i}\, \eta_i\, \phi^i(\zeta). \tag{5.194}$$

Let $\eta = (\eta_1, \ldots, \eta_m)$ be the second-order $\mathbb{R}^m$-valued random variable. From Equation (5.193), it can be deduced that the mean value $\mathbf{m}_\eta$ in $\mathbb{R}^m$ and the co-variance matrix $[C_\eta]$ in $\mathbb{M}_m^+(\mathbb{R})$ of the random vector η are such that

$$\mathbf{m}_\eta = \mathbf{0} \quad , \quad [C_\eta] = [I_m]. \tag{5.195}$$

□ *Convergence analysis and calculation of the reduced order* m. The convergence of the reduced statistical model defined by Equation (5.194) is controlled by

$$E\{\|\mathbf{U} - \mathbf{U}^{(m)}\|_{L^2}^2\} = \sum_{i=m+1}^\infty \lambda_i = E\{\|\mathbf{U} - \mathbf{m}_\mathbf{U}\|_{L^2}^2\} - \sum_{i=1}^m \lambda_i, \tag{5.196}$$

in which, taking into account Equation (5.187),

$$\|\mathbf{U} - \mathbf{U}^{(m)}\|_{L^2}^2 = \int_\Omega \|\mathbf{U}(\zeta) - \mathbf{U}^{(m)}(\zeta)\|^2\, d\zeta.$$

A relative error function $m \mapsto \mathrm{err}(m)$ defined on $\mathbb{N}^*$ with values in $\mathbb{R}^+$, based on Equation (5.196), is introduced such that

$$\mathrm{err}(m) = 1 - \frac{\sum_{i=1}^{m} \lambda_i}{E\{\|\mathbf{U} - \mathbf{m_U}\|_{L^2}^2\}} . \tag{5.197}$$

Since $E\{\|\mathbf{U} - \mathbf{m_U}\|_{L^2}^2\} = \sum_{i=1}^{+\infty} \lambda_i < +\infty$ and since $\lambda_1 \geq \lambda_2 \geq \ldots \to 0$, $m \mapsto \mathrm{err}(m)$ is monotonically decreasing and $\mathrm{err}(+\infty) = 0$.

For a given covariance function $(\zeta, \zeta') \mapsto [C_{\mathbf{U}}(\zeta, \zeta')]$ and for a given tolerance $\varepsilon > 0$, the optimal value m_ε of the reduced order m in Equation (5.194) is calculated in order that

$$m_\varepsilon = \inf\{m \in \mathbb{N}^* \mid \mathrm{err}(m) \leq \varepsilon , \ \forall m \geq m_\varepsilon\} . \tag{5.198}$$

☐ *Numerical aspects and what is the connection with a PCA.* The construction of Hilbert basis $\{\phi^i\}_i$ requires to solve the eigenvalue problem defined by Equation (5.189) and, in general, an explicit solution cannot be obtained. Consequently, computational methods must then be applied for constructing a numerical approximation of Equation (5.189), which allows for obtaining a numerical approximation of the eigenvalues $\lambda_1 \geq \ldots \geq \lambda_m$ and of the eigenfunctions $\{\phi^i\}_{i=1,\ldots,m}$.

Instead of numerically approximating the integral equation related to the eigenvalue problem defined by Equation (5.189), another approach consists in introducing a finite family $\{\mathbf{U}(\zeta^1), \ldots, \mathbf{U}(\zeta^N)\}$ of the random field $\mathbf{U}$, which corresponds to the sampling of $\mathbf{U}$ at the points $\zeta^1, \ldots, \zeta^N$ in Ω, and which correctly approximates in $L^2(\Theta, \mathbb{R}^\mu)$ the uncountable family $\{\mathbf{U}(\zeta), \zeta \in \Omega\}$. The PCA of the random vector $\mathbf{X} = (\mathbf{U}(\zeta^1), \ldots, \mathbf{U}(\zeta^N))$ with value in $(\mathbb{R}^\mu)^N$ can then be done for constructing a statistical reduction as explained in Section 5.7.1.

Example. Let us assume that the random field corresponds to a stochastic modeling of the coefficients of a partial differential equation of a boundary value problem on a bounded set of $\mathbb{R}^d$. Let us also assume that the computational model is constructed by using the finite element method for performing the spatial discretization of the boundary value problem. The sampling points $\zeta^1, \ldots, \zeta^N$ in Ω can then be chosen as the union of the Gauss-Legendre quadrature points of all the finite elements of the mesh of domain Ω.

Chapter 6
Brief Overview of Stochastic Solvers for the Propagation of Uncertainties

Abstract *The course is focused on the stochastic modeling of uncertainties and their identification by solving statistical inverse problems. The course is not devoted to the mathematical developments related to the stochastic solvers (there are many textbooks devoted to this subject). In this context, this chapter is limited to a list of principal approaches and to a brief description of the Galerkin method (spectral approach) and to the Monte Carlo method.*

6.1 Why a Stochastic Solver Is Not a Stochastic Modeling

☐ *Illustration by using the parametric stochastic approach of uncertainties.*
Figure 6.1 displays a scheme that describes the two fundamental steps in the framework of the use of the parametric stochastic approach of uncertainties. In this figure, it can be seen,

- the random vector $\mathbf{X}$ that represents the uncertain parameters in the computational model, for which its stochastic model consists in constructing its prior probability distribution $P_{\mathbf{X}}^{\text{prior}}$.
- the random observation $\mathbf{U}$ that is the transformation of $\mathbf{X}$ by a deterministic nonlinear mapping $\mathbf{h}$ such that $\mathbf{U} = \mathbf{h}(\mathbf{X})$. Mapping $\mathbf{h}$ is not explicitly known and is numerically constructed by using the computational model.
- the probability distribution $P_{\mathbf{U}}^{\text{prior}}$ of random vector $\mathbf{U}$ that must be estimated.

This figure also shows that there are two mains steps that are

1. the *stochastic modeling* of $\mathbf{X}$ consisting in constructing $P_{\mathbf{X}}^{\text{prior}}$.

© Springer International Publishing AG 2017

C. Soize, *Uncertainty Quantification*, Interdisciplinary Applied Mathematics 47,
DOI 10.1007/978-3-319-54339-0_6

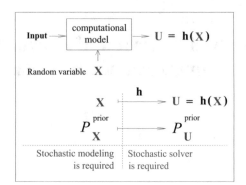

Fig. 6.1 Scheme describing
the parametric stochastic
approach of uncertainties.

2. the *stochastic solver* that allows for constructing P_U^{prior} knowing that P_X^{prior} and
   h are given.

□ *Fundamental notions/problems.* The previous analysis, illustrated by Figure 6.1,
allows us for stating the following important conclusions:

- the stochastic modeling of uncertainties must not be confused with the
  stochastic solver that is required to analyze the propagation of uncertainties.
- the stochastic modeling of uncertainties is then absolutely required and is a
  fundamental problem that has to carefully be analyzed and studied.
- this means that the probability distribution P_X^{prior} must objectively be constructed and not be arbitrarily chosen (as illustrated in Section 1.6). If the
  stochastic model is arbitrary and therefore, probably erroneous, even if the
  stochastic solver is "perfect," the result, P_U^{prior}, will be erroneous.

6.2 Brief Overview on the Types of Stochastic Solvers

There are two main classes of techniques for constructing stochastic solvers,

- the techniques that are related to Galerkin-type projections introduced in the
  weak formulations of partial differential equations for stochastic boundary value
  problems (spectral stochastic methods).
- the techniques that are related to a sampling technique that corresponds to a direct
  simulation method.

□ *First class of techniques related to Galerkin-type projections introduced in the
weak formulations of partial differential equations for boundary value problems.*

- *The spectral stochastic method based on the use of PCE.* In this framework, a
  popular and efficient technique is the spectral stochastic method that is based
  on the use of polynomial chaos expansion of the solution, which has been initiated by Roger Ghanem in 1990-1991 [84, 85]. Such an approach has been

developed and applied in many fields of computational sciences and engineering for the cases of uncertain computational models, in particular in computational solid mechanics and in computational fluid dynamics [123]. Numerical aspects related to the spectral method for stochastic boundary value problem can also be found in [223].

— *Alternative methods.* Some alternative methods to the spectral stochastic methods based on the use of the PCE are the methods based on the construction of optimal separated representations of the solution such as
  — the Karhunen-Loeve expansion of the solution of the stochastic boundary value problem[130].
  — the generalized spectral decomposition method [152].

— *Addition bibliographical comments concerning the spectral stochastic method using the PCE.* A systematic construction of PCE of second-order random fields and their use for analyzing the uncertainty propagation in boundary value problems were initiated in [84, 85]. Polynomial chaos expansions and their use for solving stochastic boundary value problems and some associated statistical inverse problems have rapidly grown [10, 11, 15, 56, 59, 63, 69, 70, 81, 82, 86, 87, 123, 145, 151, 153, 200, 211, 222, 223]). Several extensions have been proposed concerning the generalized chaos expansions, the PCE for an arbitrary probability measure, the PCE with random coefficients [12, 74, 126, 162, 188, 196, 217], and the construction of a basis adaptation in homogeneous chaos spaces [213]. Significant work has also been devoted to the acceleration of stochastic convergence of the PCE [23, 24, 88, 114, 129, 206, 213].

☐ *Second class of techniques related to sampling techniques that correspond to a direct simulation method.* The second class of techniques is related to a sampling technique for which the most efficient and the most popular is the Monte Carlo simulation method [172]. The speed of convergence is independent of the stochastic dimension (consequently, there is not a curse of dimensionality), but the speed of convergence is in $1/\sqrt{\nu}$ in which ν is the number of independent realizations (as shown in Section 2.8). Nevertheless, the speed of convergence can be improved by using one of the following techniques,

— the advanced Monte Carlo simulation procedures [177].
— the subset simulation techniques [13, 14].
— the important sampling for the high dimensions [13].
— the local domain Monte Carlo simulation [166].

6.3 Spectral Stochastic Approach by Using the PCE

In this section, we present a simple example for illustrating the methodology of the spectral stochastic approach by using the polynomial chaos expansion in order to solve a stochastic boundary value problem.

6.3.1 Description in a Simple Framework for Easy Understanding

For simplifying the presentation, it is assumed the following.

1. The boundary value problem is relative to a system of partial differential equations (described in the physical space) for which the space discretization is carried out by a usual method (finite difference method, finite volume method, finite element method, etc.). We then deduce a computational model in finite dimension, which can correspond to a reduced-order computational model if a reduction process has been applied.
2. The computational model (possibly reduced) that depends on an uncertain parameter $\mathbf{x}$ with values in an admissible set $\mathcal{S}_n \subset \mathbb{R}^n$ is written as

$$\mathcal{D}(\mathbf{u}; \mathbf{x}) = \mathcal{F}(\mathbf{x}) \quad \text{on} \quad \mathbb{R}^m \quad \text{with} \quad \mathbf{u} \in \mathbb{R}^m. \tag{6.1}$$

   For all $\mathbf{x}$ in $\mathcal{S}_n$, it is assumed that there exists a unique solution $\mathbf{u} = \mathbf{h}(\mathbf{x})$ in $\mathbb{R}^m$.
3. The parametric stochastic approach of uncertainties is used. Therefore, uncertain parameter $\mathbf{x}$ is modeled by a random variable $\mathbf{X}$, defined on $(\Theta, \mathcal{T}, \mathcal{P})$, with values in $\mathbb{R}^n$, for which the support of the probability distribution $P_{\mathbf{X}}(d\mathbf{x})$ is $\mathcal{S}_n$. In addition, it is assumed that the stochastic modeling of $\mathbf{X}$ is such that the stochastic solution $\mathbf{U}$ is a second-order $\mathbb{R}^m$-valued random variable,

$$\mathbf{U} = \mathbf{h}(\mathbf{X}) \in L^2(\Theta, \mathbb{R}^m). \tag{6.2}$$

A parameterization of $\mathbf{X}$ is introduced, which is written as $\mathbf{X} = \mathbf{f}(\Xi)$ where Ξ is a second-order $\mathbb{R}^{N_g}$-valued random variable with a given pdf, $p_\Xi(\xi)$ with respect to $d\xi$, and where $\mathbf{f}$ is a given mapping from $\mathbb{R}^{N_g}$ into $\mathcal{S}_n \subset \mathbb{R}^n$, which transforms the probability distribution $p_\Xi(\xi) \, d\xi$ of Ξ into the probability distribution $P_{\mathbf{X}}(d\mathbf{x})$ of $\mathbf{X}$.

6.3.2 Illustration of the Spectral Stochastic Approach with PCE

Two formulations are presented.

1. The intrusive approach that requires to modify the solvers in the computational softwares for implementing the method. Such an approach can then be relatively difficult to implement in the commercial softwares.

2. The nonintrusive approach that corresponds to a post-processing of the outputs
   of computational softwares and consequently, which does not require to modify
   the solvers in the computational softwares.

☐ *Intrusive approach.* The stochastic equation to be solved is $\mathcal{D}(\mathbf{U};\mathbf{X}) = \mathcal{F}(\mathbf{X})$,
  which is rewritten as

$$\mathcal{D}(\mathbf{U};\mathbf{f}(\boldsymbol{\Xi})) = \mathcal{F}(\mathbf{f}(\boldsymbol{\Xi})) . \tag{6.3}$$

The finite approximation of the PCE of the second-order random vector $\mathbf{U}$ (see
Equation (5.149) for which the notations are adapted to $\mathbf{U}$) is written as

$$\mathbf{U}^{(K_{N_d})} = \sum_{k=0}^{K_{N_d}} \mathbf{u}^k \, \Psi_{\boldsymbol{\alpha}^{(k)}}(\boldsymbol{\Xi}) \quad , \quad \mathbf{u}^k = E\{\mathbf{U}\Psi_{\boldsymbol{\alpha}^{(k)}}(\boldsymbol{\Xi})\} \in \mathbb{R}^m . \tag{6.4}$$

The unknown deterministic coefficients (vectors) $\mathbf{u}^0, \mathbf{u}^1, \ldots, \mathbf{u}^{K_{N_d}}$ are then the
solution of the following system of $1 + K_{N_d}$ deterministic vectorial equations,
such that, for $k' = 0, 1, \ldots, K_{N_d}$,

$$E\{\Psi_{\boldsymbol{\alpha}^{(k')}}(\boldsymbol{\Xi}) \, \mathcal{D}(\mathbf{U}^{(K_{N_d})};\mathbf{f}(\boldsymbol{\Xi}))\} = E\{\Psi_{\boldsymbol{\alpha}^{(k')}}(\boldsymbol{\Xi}) \, \mathcal{F}(\mathbf{f}(\boldsymbol{\Xi}))\} . \tag{6.5}$$

It should be noted that if $\mathbf{u} \mapsto \mathcal{D}(\mathbf{u};\mathbf{x})$ is a linear mapping, then the system of
deterministic vectorial equations defined by Equation (6.5) is linear, while it is
a nonlinear one if $\mathbf{u} \mapsto \mathcal{D}(\mathbf{u};\mathbf{x})$ is a nonlinear mapping. Concerning the convergence of the method and the computational aspects, the following comments can
be done.

– The method is intrusive with respect to the commercial softwares.
– The method is accurate and the convergence with respect to the number K_{N_d}
  of the polynomial chaos can be controlled during the computation. However,
  for a nonlinear problem $\boldsymbol{\Xi} \mapsto \mathbf{U}$ (that is generally the case), all the terms of
  the system of equations must be recomputed for each value of K_{N_d} (contrary
  to a linear case).
– The speed of convergence depends on the dimension N_g of random vector $\boldsymbol{\Xi}$
  (curse of dimensionality).
– For the high dimensions (N_g not very small), the mathematical expectations
  that appear in Equation (6.5) for evaluating the coefficients of the linear algebraic system of equations must be estimated by using the Monte Carlo method
  (see Chapters 2 and 4).

☐ *Nonintrusive approach.* Instead of solving Equation (6.5), the unknown deter-
  ministic vectors $\mathbf{u}^0, \mathbf{u}^1, \ldots, \mathbf{u}^K$ are computed by using the following Monte
  Carlo approach (see Section 6.4)

$$\mathbf{u}^k = E\{\mathbf{U}\Psi_{\boldsymbol{\alpha}^{(k)}}(\boldsymbol{\Xi})\} \simeq \frac{1}{\nu} \sum_{\ell=1}^{\nu} \mathbf{U}(\theta_\ell) \, \Psi_{\boldsymbol{\alpha}^{(k)}}(\boldsymbol{\Xi}(\theta_\ell)) , \tag{6.6}$$

in which, for $\ell = 1, \ldots, \nu$, $\mathbf{U}(\theta_\ell)$ is the solution of the deterministic equation,

$$\mathcal{D}(\mathbf{U}(\theta_\ell); \mathbf{X}(\theta_\ell)) = \mathcal{F}(\mathbf{X}(\theta_\ell)) \quad , \quad \mathbf{X}(\theta_\ell) = \mathbf{f}(\Xi(\theta_\ell)), \tag{6.7}$$

where $\Xi(\theta_1)), \ldots, \Xi(\theta_\nu))$ are ν independent realizations that are constructed by using the generator of Ξ.

This nonintrusive approach appears as a posttreatment of the Monte Carlo method for constructing a PCE of the stochastic solution $\mathbf{U}$ for which ν independent realizations $\mathbf{U}(\theta_\ell), \ell = 1, \ldots, \nu$ are known. The CPU time induced by the PCE is negligible and in this case, the stochastic solver is thus the Monte Carlo numerical method presented in Section 6.4.

6.4 Monte Carlo Numerical Simulation Method

The methodology of a stochastic solver based on the use of the Monte Carlo numerical simulation is summarized by the scheme displayed in Figure 6.2. The mathematical foundations are those presented in Section 2.8 (concerning the use of the central limit theorem and its application to the computation of integrals for the high dimensions) and in Chapter 4 (concerning the Markov Chain Monte Carlo methods for generating realizations and for estimating the mathematical expectation of nonlinear mappings of random vectors).

The first row of the array shown in Figure 6.2 summarizes the problem that must be solved.

1. The probability distribution $P_{\mathbf{X}}^{\text{prior}}$ of the random vector $\mathbf{X}$ is known. This means that an objective construction of $P_{\mathbf{X}}^{\text{prior}}$ by using the direct approach or the indirect approach must be done as previously explained.

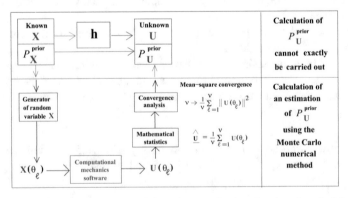

Fig. 6.2 Scheme summarizing the methodology for a stochastic solver based on the Monte Carlo method.

2. Only an approximation of the deterministic mapping $\mathbf{h}$ can be constructed with the computational model using a software, such as a finite element code.
3. The objective is to estimate the probability distribution $P_{\mathbf{U}}^{\text{prior}}$ of random observation $\mathbf{U} = \mathbf{h}(\mathbf{X})$, which cannot be exactly calculated.

The second row of the array shown in Figure 6.2 summarizes the methodology of the Monte Carlo solver.

1. The first step consists in generating ν independent realizations $\{\mathbf{X}(\theta_\ell), \ell = 1, \ldots, \nu\}$ of $\mathbf{X}$ for which its probability distribution $P_{\mathbf{X}}^{\text{prior}}$ is known. In general, such a generator is based on the Markov Chain Monte Carlo methods presented in Chapter 4.
2. The second step consists in computing the ν independent realizations $\{\mathbf{U}(\theta_\ell), \ell = 1, \ldots, \nu\}$ of $\mathbf{U}$ such that, for each ℓ, $\mathbf{U}(\theta_\ell) = \mathbf{h}(\mathbf{X}(\theta_\ell))$. This is a deterministic computation performed with a computational software. This step can easily massively be parallelized.
3. The third step consists in estimating the quantity of interest by using the mathematical statistics and by controlling the convergence of the estimates. The control of the convergence can be done in mean-square, or better, by using the central limit theorem as explained in Section 2.8.

Concerning the convergence of the Monte Carlo method and the computational aspects, the following comments can be done.

− The Monte Carlo method is not intrusive with respect to the commercial softwares.
− The method is adapted to parallel computation without any software developments.
− The convergence can be controlled with respect to the number ν of realizations and the distance to the unknown exact solution can be estimated during the computation thanks to the central limit theorem (see Section 2.8).
− The speed of convergence is independent of the dimension N_g of Ξ (not curse of dimensionality).

Chapter 7
Fundamental Tools for Statistical Inverse Problems

7.1. Fundamental Methodology
7.2. Multivariate Kernel Density Estimation Method in Nonparametric Statistics
7.3. Statistical Tools for the Identification of Stochastic Models of Uncertainties
7.4. Application of the Bayes Method to the Output-Prediction-Error Method

Abstract *This chapter deals with the fundamental statistical tools for solving statistical inverse problems that allow for identifying the stochastic models of uncertainties through the computational models. This part of the mathematical statistics is very well developed and there are a huge number of textbooks with which one can easily get lost when trying to learn. We have voluntarily limited the presentation to the basic ideas of the statistical inversion theory. Consequently, the nonstationary inverse problems such as the Bayesian filtering leading, for instance, to the linear Kalman filters and to the extended Kalman filters will not be presented.*

7.1 Fundamental Methodology

7.1.1 Description of the Problem

□ *What is an inverse problem in the context of computational models.* It consists in identifying some parameters of a computational model by using data related to observations of quantities of interest that can be calculated with the computational model. In general, such a deterministic inverse problem, which consists to inverse the direct problem, is an ill-posed problem. However, the same inverse problem reformulated in the framework of the probability theory becomes, almost always a well-posed problem.

□ *What is a statistical inverse problem.* It consists in solving an inverse problem in a probabilistic framework by using mathematical statistical tools.

© Springer International Publishing AG 2017
C. Soize, *Uncertainty Quantification*, Interdisciplinary Applied Mathematics 47,
DOI 10.1007/978-3-319-54339-0_7

☐ *What are the sources of data.* A statistical inverse problem can only be solved if some data are available. The data can come

   – either from experimental observations, which will be called "experimental data"
   – or from numerical simulations performed with computational models, which will be called "simulated data."

☐ *What are the main situations that can be encountered?*

   – If there are no available data, then the statistical inverse problem is an ill-posed problem, which cannot be solved.
   – If data are available, then the statistical inverse problem is generally a well-posed problem, but the quality of the solution is related to the quality of the convergence of the statistical estimates with respect to the size of data (case of limited and/or partial data).

7.1.2 Before Presenting the Statistical Tools, What Are the Fundamental Strategies Used in Uncertainty Quantification?

7.1.2.1 No Data Are Available

In such a case, any statistical inverse problem is an ill-posed problem. Nevertheless uncertainties can be and must be taken into account in order to perform a robust analysis, a robust optimization, or a robust design with the computational model, as a function of the level of uncertainties.

▷ *Consequently, a prior stochastic model of both the system-parameter uncertainties and the model uncertainties induced by modeling errors must carefully be constructed. The unknown vector-valued hyperparameter* **s** *of such a prior stochastic model is then used for performing a sensitivity analysis with respect to the level of uncertainties and to the sources of uncertainties.*

7.1.2.2 Data Are Available

Two methods can then be used for performing the identification by solving the statistical inverse problem.

☐ *First method: a least-square method or the maximum likelihood method.*

   1. The first step consists in carefully constructing a prior stochastic model $P^{\text{prior}}(\mathbf{s})$ of uncertainties (both the model-parameter uncertainties and the model uncertainties induced by the modeling errors), which depends on an unknown vector-valued hyperparameter **s**.

2. The second step consists in identifying an optimal value $\mathbf{s}^{opt}$ of $\mathbf{s}$ by solving a statistical inverse problem using a least-square method or the maximum likelihood method. Such an identification allows for obtaining an optimal prior model

$$P^{prior,opt} = P^{prior}(\mathbf{s}^{opt}), \qquad (7.1)$$

with respect to the level of uncertainties and to the sources of uncertainties.

☐ *Second method: the Bayesian method.* With the first method, the prior model $P^{prior}(\mathbf{s})$ is parameterized by a hyperparameter $\mathbf{s}$. When $\mathbf{s}$ runs through its admissible set $\mathcal{C}_{ad}$, then the family $\{P^{prior}(\mathbf{s}), \mathbf{s} \in \mathcal{C}_{ad}\}$ of prior models spans a subset of the entire set of all the possible probability distributions, and consequently $P^{prior,opt} = P^{prior}(\mathbf{s}^{opt})$ is optimal for representing data in this subset of probability distributions. The Bayesian method allows for estimating a *posterior stochastic model* P^{post} which is optimal for representing data in the entire set of probability distributions. The Bayesian method corresponds to an updating P^{post} (the posterior model) with the data of a prior model P^{prior}. The prior stochastic model can be chosen as:

– a noninformative prior model: P^{prior} is arbitrarily chosen (not recommended, above all for the high dimensions with limited data).
– an informative prior model: the prior model $P^{prior}(\mathbf{s}^0)$ for a fixed value $\mathbf{s}^0$ of $\mathbf{s}$.
– an informative optimal prior model: the optimal prior model $P^{prior}(\mathbf{s}^{opt})$. If the same set of data is used for calculating $\mathbf{x}^{opt}$ using the first method and for calculating P^{post} with the Bayesian method, this means that P^{post} is constructed in the region of all the probability distributions that are in the neighborhood of $P^{prior}(\mathbf{s}^{opt})$.

☐ *Summary of the identification problem.* Figure 7.1 displays a summary of the identification problem for which experimental data are available. In the right part of this figure, it can be seen the notion of variabilities of the real system (that has been introduced and commented in Section 1.2) and the notion of the measurement noise that requires additional explanations. For certain fields of applications, such that, for instance, for medical image processing or for astrophysical image processing, the measurement noise can be not negligible and must absolutely be taken into account in the statistical inverse problems. For the identification of computational models developed for predicting the behavior (and also for the design and optimization) of complex mechanical systems, the measurement noise is, in general, negligible with respect to the variabilities induced by the experimental configuration that is not precisely known or induced by the fluctuations due to the manufacturing process.

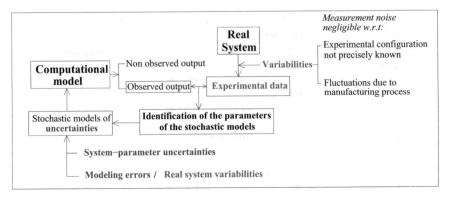

Fig. 7.1 Scheme describing the identification procedure from experimental observations.

7.2 Multivariate Kernel Density Estimation Method in Nonparametric Statistics

For many problems related to the construction of generators of realizations of a random vector (see, for instance, Section 4.3.5) or related to the methodologies for solving statistical inverse problems (problem analyzed in the present Chapter), it is necessary to construct an estimate of the probability density function of a random vector from a known set of its realizations. If no additional information are available (this means that only the set of realizations is known), the way for solving this problem is to use the nonparametric density estimate and multivariate smoothing, which constitute a powerful statistical tool [25, 89, 106]. In this framework, the most popular and efficient method is the multivariate kernel density method, for which the Gaussian kernel is often well adapted and is very efficient.

7.2.1 Problem Definition

Let $\mathbf{U}$ be an $\mathbb{R}^m$-valued random variable, defined on the probability space $(\Theta, \mathcal{T}, \mathcal{P})$, whose probability distribution $P_{\mathbf{U}}(d\mathbf{u})$ is defined by an unknown pdf $\mathbf{u} \mapsto p_{\mathbf{U}}(\mathbf{u})$ on $\mathbb{R}^m$ with respect to $d\mathbf{u}$. The support $\mathcal{S}_m \subset \mathbb{R}^m$ of $p_{\mathbf{U}}$ is also unknown. A set of $\nu > m$ independent realizations $\{\mathbf{u}^{r,1}, \ldots, \mathbf{u}^{r,\nu}\}$ of $\mathbf{U}$ is available (experimental data or simulated data). The problem to be solved consists in estimating $\widehat{p}_{\mathbf{U}}(\mathbf{u}^0)$ of the value of the pdf $\mathbf{u} \mapsto p_{\mathbf{U}}(\mathbf{u})$ at the point $\mathbf{u}^0$ given in $\mathbb{R}^m$, using only the data set $\{\mathbf{u}^{r,1}, \ldots, \mathbf{u}^{r,\nu}\}$ of independent realizations of $\mathbf{U}$.

7.2.2 Multivariate Kernel Density Estimation Method

The method consists in three steps.

1. Estimation of the mean vector and of the covariance matrix of random vector $\mathbf{U}$ by using the data set $\{\mathbf{u}^{r,1}, \ldots, \mathbf{u}^{r,\nu}\}$.
2. Rotation of the coordinates.
3. Multivariate kernel density estimation in the rotated coordinates.

□ *Estimation of the mean vector and of the covariance matrix of random vector* $\mathbf{U}$. Let $\widehat{\mathbf{m}}_{\mathbf{U}}$ in $\mathbb{R}^m$ and let $[\widehat{C}_{\mathbf{U}}]$ in $\mathbb{M}_m^+$ be the empirical estimates of the unknown mean vector $\mathbf{m}_{\mathbf{U}} = E\{\mathbf{U}\}$ and of the unknown covariance matrix $[C_{\mathbf{U}}] = E\{(\mathbf{U} - \mathbf{m}_{\mathbf{U}})(\mathbf{U} - \mathbf{m}_{\mathbf{U}})^T\}$, which are such that

$$\widehat{\mathbf{m}}_{\mathbf{U}} = \frac{1}{\nu} \sum_{\ell=1}^{\nu} \mathbf{u}^{r,\ell}, \tag{7.2}$$

$$[\widehat{C}_{\mathbf{U}}] = \frac{1}{\nu - 1} \sum_{\ell=1}^{\nu} (\mathbf{u}^{r,\ell} - \widehat{\mathbf{m}}_{\mathbf{U}})(\mathbf{u}^{r,\ell} - \widehat{\mathbf{m}}_{\mathbf{U}})^T. \tag{7.3}$$

□ *Rotation of the coordinates.* The eigenvalue problem is solved for the positive-definite matrix $[\widehat{C}_{\mathbf{U}}]$,

$$[\widehat{C}_{\mathbf{U}}][\Phi] = [\Phi][\Lambda], \tag{7.4}$$

in which $[\Lambda]$ is the diagonal matrix of the positive eigenvalues $\lambda_1 \geq \lambda_2 \geq \ldots \geq \lambda_m > 0$ and where $[\Phi] \in \mathbb{M}_m(\mathbb{R})$ is the matrix of the eigenvectors such that $[\Phi][\Phi]^T = [\Phi]^T[\Phi] = [I_m]$. The $\mathbb{R}^m$-valued random vector $\mathbf{Q}$ of the random coordinates in the principal axes defined by the eigenvectors is written as

$$\mathbf{Q} = [\Phi]^T(\mathbf{U} - \widehat{\mathbf{m}}_{\mathbf{U}}). \tag{7.5}$$

The transformation of data set $\{\mathbf{u}^{r,1}, \ldots, \mathbf{u}^{r,\nu}\}$ is therefore written as

$$\forall \ell = 1, \ldots, \nu \quad , \quad \mathbf{q}^{r,\ell} = [\Phi]^T(\mathbf{u}^{r,\ell} - \widehat{\mathbf{m}}_{\mathbf{U}}). \tag{7.6}$$

Since $p_{\mathbf{U}}(\mathbf{u})\,d\mathbf{u} = p_{\mathbf{Q}}(\mathbf{q})\,d\mathbf{q}$ and since $d\mathbf{q} = |\det([\Phi]^T)|\,d\mathbf{u} = d\mathbf{u}$, we have $p_{\mathbf{U}}(\mathbf{u}) = p_{\mathbf{Q}}(\mathbf{q})$, and therefore, the estimate at point $\mathbf{u}^0$ given in $\mathbb{R}^m$, is written as

$$\widehat{p}_{\mathbf{U}}(\mathbf{u}^0) = \widehat{p}_{\mathbf{Q}}(\mathbf{q}^0) \quad , \quad \mathbf{q}^0 = [\Phi]^T(\mathbf{u}^0 - \widehat{\mathbf{m}}_{\mathbf{U}}). \tag{7.7}$$

□ *Multivariate kernel density estimation in the rotated coordinates.* The multivariate kernel density estimate $\widehat{p}_{\mathbf{Q}}(\mathbf{q}^0)$ of the pdf $p_{\mathbf{Q}}$ at point $\mathbf{q}^0$ given in $\mathbb{R}^m$, by using the data set $\{\mathbf{q}^{r,1}, \ldots, \mathbf{q}^{r,\nu}\}$, is written as

$$\widehat{p}_{\mathbf{Q}}(\mathbf{q}^0) = \frac{1}{\nu} \sum_{\ell=1}^{\nu} \prod_{k=1}^{m} \left\{ \frac{1}{h_k} \mathcal{K} \left(\frac{q_k^{r,\ell} - q_k^0}{h_k} \right) \right\}, \tag{7.8}$$

in which $h_1, \ldots, h_m$ are the smoothing parameters, where $\mathbf{q}^0 = (q_1^0, \ldots, q_m^0)$, where $\mathbf{q}^{r,\ell} = (q_1^{r,\ell}, \ldots, q_m^{r,\ell})$, and where $\mathcal{K}$ is a kernel on $\mathbb{R}$, which satisfies the following properties:

1. $q \mapsto \mathcal{K}(q)$ is an integrable function from $\mathbb{R}$ into $\mathbb{R}^+$.
2. $\operatorname{Supp} \mathcal{K} = \mathcal{S}_1 \subset \mathbb{R}$ (possibly with $\mathcal{S}_1 = \mathbb{R}$).
3. $\int_{\mathcal{S}_1} \mathcal{K}(q) \, dq = 1$.

These properties imply that, for all q_k^0 fixed in $\mathcal{S}_1$,

$$\lim_{h_k \to 0} \frac{1}{h_k} \mathcal{K} \left(\frac{q - q_k^0}{h_k} \right) \, dq = \delta_0(q - q_k^0), \tag{7.9}$$

in which δ_0 is the Dirac measure at point 0 in $\mathbb{R}$.

▷ *Case of the Gaussian kernel.* For the multivariate Gaussian kernel density estimation, h_k can be chosen as the Silverman smoothing parameter that is defined by

$$h_k = \sqrt{[\lambda]_{kk}} \left\{ \frac{4}{\nu(2+m)} \right\}^{\frac{1}{4+m}} \quad , \quad \mathcal{K}(q) = \frac{1}{\sqrt{2\pi}} e^{-q^2/2}. \tag{7.10}$$

▷ *Comment about Equation* (7.8). It can easily be seen that the value $p_{\mathbf{Q}}(\mathbf{q}^0)$ of the pdf of random vector $\mathbf{Q}$, at point $\mathbf{q}^0$ fixed in $\mathbb{R}^m$, is such that

$$p_{\mathbf{Q}}(\mathbf{q}^0) = \int_{\mathbb{R}^m} p_{\mathbf{Q}}(\mathbf{q}) \, \delta_0(\mathbf{q} - \mathbf{q}^0) = E\{\delta_0(\mathbf{Q} - \mathbf{q}^0)\}, \tag{7.11}$$

in which $\delta_0(\mathbf{q}) = \otimes_{k=1}^{m} \delta_0$ is the Dirac measure on $\mathbb{R}^m$ at point $\mathbf{0} = (0, \ldots, 0) \in \mathbb{R}^m$. For $\sup_k h_k$ sufficiently small, using the following estimate of the mathematical expectation,

$$E\{\delta_0(\mathbf{Q} - \mathbf{q}^0)\} \simeq \frac{1}{\nu} \sum_{\ell=1}^{\nu} \prod_{k=1}^{m} \delta_0(q_k^{r,\ell} - q_k^0), \tag{7.12}$$

and using Equation (7.9) for evaluating the right-hand side of Equation (7.12), yield Equation (7.8).

7.3 Statistical Tools for the Identification of Stochastic Models of Uncertainties

In this section, we present

- the least-square method for estimating the hyperparameters.
- the maximum likelihood method for estimating the hyperparameters.
- the Bayes method for estimating the posterior probability distribution from a prior probability model.

7.3.1 Notation and Scheme for the Identification of Model-Parameter Uncertainties Using Experimental Data

We consider the identification of the model-parameter uncertainties of a computational model by using experimental data for which the scheme is displayed in Figure 7.2.

☐ *Real system.* The experimental configuration is defined in Figure 7.2-left. We have

$$\mathbf{U}^{\mathrm{exp}} = \mathbf{U}^r + \mathbf{B}\,, \qquad (7.13)$$

in which

- $\mathbf{U}^r$ is the $\mathbb{R}^m$-valued random observation of the real system (randomness is due to the variability of the real system).
- $\mathbf{B}$ is the $\mathbb{R}^m$-valued random noise which is assumed to be a second-order and centered random variable whose probability distribution is $P_{\mathbf{B}}(d\mathbf{b}) = p_{\mathbf{B}}(\mathbf{b})\,d\mathbf{b}$.
- $\mathbf{U}^{\mathrm{exp}}$ is the $\mathbb{R}^m$-valued random measured observation of the real system for which ν independent experimental observations $\mathbf{u}^{\mathrm{exp},1},\ldots,\mathbf{u}^{\mathrm{exp},\nu}$ are assumed to be available.

☐ *Computational model.* The computational configuration is defined in Figure 7.2-right. We have

$$\mathbf{U} = \mathbf{h}(\mathbf{X})\,, \qquad (7.14)$$

in which

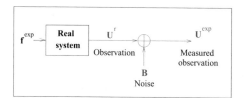
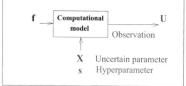

Fig. 7.2 Scheme describing the identification of model-parameter uncertainties using experimental data. Experimental configuration (left figure). Computational configuration (right figure).

- $\mathbf{X}$ is the $\mathbb{R}^n$-valued random variable (uncertain parameter) whose probability distribution is $P_{\mathbf{X}}(d\mathbf{x}\,;\mathbf{s}) = p_{\mathbf{X}}(\mathbf{x}\,;\mathbf{s})\,d\mathbf{x}$ that depends on the $\mathbb{R}^\mu$-valued hyperparameter $\mathbf{s} \in \mathcal{C}_{\mathrm{ad}} \subset \mathbb{R}^\mu$. The support of the pdf is such that $\mathrm{Supp}\,p_{\mathbf{X}} = \mathcal{S}_n \subset \mathbb{R}^n$, which is assumed independent of $\mathbf{s}$.
- $\mathbf{U}$ is the $\mathbb{R}^m$-valued random observation of the computational model and $\mathbf{h}$ is a given deterministic mapping from $\mathbb{R}^n$ into $\mathbb{R}^m$ such that $\mathbf{U} = \mathbf{h}(\mathbf{X})$ is a second-order random variable.

7.3.2 Least-Square Method for Estimating the Hyperparameter

☐ *Formulation.* The objective of the least-square method is to calculate an optimal value $\mathbf{s}^{\mathrm{opt}}$ of hyperparameter $\mathbf{s}$ of the prior pdf $p_{\mathbf{X}}(\mathbf{x}\,;\mathbf{s})$ by solving the optimization problem,

$$\mathbf{s}^{\mathrm{opt}} = \arg \min_{\mathbf{s}\in\mathcal{C}_{\mathrm{ad}}} \mathcal{J}(\mathbf{s}), \tag{7.15}$$

in which the cost function is written as

$$\mathcal{J}(\mathbf{s}) = E\{\|\mathbf{U}^{\mathrm{exp}} - \mathbf{U}(\mathbf{s})\|^2\}. \tag{7.16}$$

We have introduced the notation $\mathbf{U}(\mathbf{s})$ for indicating that the random variable $\mathbf{U} = \mathbf{h}(\mathbf{X})$ depends on $\mathbf{s}$ because the pdf $p_{\mathbf{X}}(\mathbf{x}\,;\mathbf{s})$ of $\mathbf{X}$ depends on $\mathbf{s}$. By developing the square of the norm that appears in the mathematical expectation in the right-hand side of Equation (7.16), it can be seen that

$$\mathcal{J}(\mathbf{s}) = E\{\|\mathbf{U}^{\mathrm{exp}} - \overline{\mathbf{u}^{\mathrm{exp}}}\|^2\} + E\{\|\mathbf{m}_{\mathbf{U}}(\mathbf{s}) - \mathbf{U}(\mathbf{s})\|^2\}$$
$$+ \|\overline{\mathbf{u}^{\mathrm{exp}}} - \mathbf{m}_{\mathbf{U}}(\mathbf{s})\|^2, \tag{7.17}$$

in which $\mathbf{m}_{\mathbf{U}}(\mathbf{s}) = E\{\mathbf{U}(\mathbf{s})\}$ is the mean function of the random variable $\mathbf{U}(\mathbf{s})$, and where $\overline{\mathbf{u}^{\mathrm{exp}}} = \frac{1}{\nu}\sum_{\ell=1}^{\nu} \mathbf{u}^{\mathrm{exp},\ell}$ is the empirical mean of the experimental data.

☐ *Interpretation et comments.* The three terms in the right-hand side member of Equation (7.17) can be interpreted as follows:

- the term $E\{\|\mathbf{U}^{\mathrm{exp}} - \overline{\mathbf{u}^{\mathrm{exp}}}\|^2\}$ is the variance of the real system, which cannot be reduced.
- the term $E\{\|\mathbf{m}_{\mathbf{U}}(\mathbf{s}) - \mathbf{U}(\mathbf{s})\|^2\}$ is the variance of the model that increases with the level of the statistical fluctuations.
- the term $\|\overline{\mathbf{u}^{\mathrm{exp}}} - \mathbf{m}_{\mathbf{U}}(\mathbf{s})\|^2$ is the bias between the real system and the model, which has to be reduced.

Consequently, the minimization of the bias increases the variance of the model and, the minimization of the variance of the model increases the bias. The least-square method allows for constructing a compromise for minimizing the bias.

However, with such a cost function, the least-square method does not allow for identifying the level of the statistical fluctuations of the model with the level of the statistical fluctuations of experiments (see the next paragraph).

□ *Introduction of a cost function that allows for identifying the level of the statistical fluctuations of the model with the experiments.* Since the cost function defined by Equation (7.16) does not allow for correctly taking into account the level of the statistical fluctuations between the model and the experiments, the following one is introduced:

$$J(\mathbf{s}) = \alpha \, \frac{\|\mathbf{m}_{\mathbf{U}}(\mathbf{s}) - \overline{\mathbf{u}^{\exp}}\|^2}{\|\overline{\mathbf{u}^{\exp}}\|^2} + (1 - \alpha) \, \frac{(\Delta_{\mathbf{U}}(\mathbf{s}) - \Delta^{\exp})^2}{(\Delta^{\exp})^2}, \tag{7.18}$$

$$\Delta_{\mathbf{U}}(\mathbf{s}) = \sqrt{\frac{E\{\|\mathbf{U}(\mathbf{s})\|^2\}}{\|\mathbf{m}_{\mathbf{U}}(\mathbf{s})\|^2} - 1} \quad, \quad \Delta^{\exp} = \sqrt{\frac{\overline{\|\mathbf{u}^{\exp}\|^2}}{\|\overline{\mathbf{u}^{\exp}}\|^2} - 1}, \tag{7.19}$$

with $\alpha \in]0, 1[$, and where

$$\overline{\mathbf{u}^{\exp}} = \frac{1}{\nu} \sum_{\ell=1}^{\nu} \mathbf{u}^{\exp,\ell} \in \mathbb{R}^m \quad, \quad \overline{\|\mathbf{u}^{\exp}\|^2} = \frac{1}{\nu} \sum_{\ell=1}^{\nu} \|\mathbf{u}^{\exp,\ell}\|^2 \in \mathbb{R}^+. \tag{7.20}$$

7.3.3 Maximum Likelihood Method for Estimating the Hyperparameters

In order to simplify the presentation, it is assumed that there is no noise **B** on the data. The extension of the presented results is straightforward if such a noise must be taken into account.

□ *Formulation.* The objective of the maximum likelihood method is to calculate an optimal value $\mathbf{s}_\nu^{\mathrm{opt}}$ of the hyperparameter $\mathbf{s} \in \mathcal{C}_{\mathrm{ad}} \subset \mathbb{R}^\mu$ of the prior pdf $\mathbf{x} \mapsto p_{\mathbf{X}}(\mathbf{x}; \mathbf{s})$ by solving the optimization problem,

$$\mathbf{s}_\nu^{\mathrm{opt}} = \arg \max_{\mathbf{s} \in \mathcal{C}_{\mathrm{ad}} \subset \mathbb{R}^\mu} L(\mathbf{s}; \mathbf{u}^{\exp,1}, \ldots, \mathbf{u}^{\exp,\nu}), \tag{7.21}$$

in which $L(\mathbf{s}; \mathbf{u}^{\exp,1}, \ldots, \mathbf{u}^{\exp,\nu})$ is the likelihood function defined by

$$L(\mathbf{s}; \mathbf{u}^{\exp,1}, \ldots, \mathbf{u}^{\exp,\nu}) = p_{\mathbf{U}}(\mathbf{u}^{\exp,1}; \mathbf{s}) \times \ldots \times p_{\mathbf{U}}(\mathbf{u}^{\exp,\nu}; \mathbf{s}), \tag{7.22}$$

where $\mathbf{u} \mapsto p_{\mathbf{U}}(\mathbf{u}; \mathbf{s})$ is the pdf of the $\mathbb{R}^m$-valued random variable $\mathbf{U} = \mathbf{h}(\mathbf{X})$. The optimization problem defined by Equation (7.21) can be rewritten as follows:

$$\mathbf{s}_\nu^{\mathrm{opt}} = \arg \max_{\mathbf{s} \in \mathcal{C}_{\mathrm{ad}} \subset \mathbb{R}^\mu} \mathcal{L}(\mathbf{s}; \mathbf{u}^{\exp,1}, \ldots, \mathbf{u}^{\exp,\nu}), \tag{7.23}$$

in which $\mathcal{L}(\mathbf{s}; \mathbf{u}^{exp,1}, \ldots, \mathbf{u}^{exp,\nu}) = \ln\{L(\mathbf{s}; \mathbf{u}^{exp,1}, \ldots, \mathbf{u}^{exp,\nu})\}$ is the log-likelihood function that can be written as

$$\mathcal{L}(\mathbf{s}; \mathbf{u}^{exp,1}, \ldots, \mathbf{u}^{exp,\nu}) = \ln\{p_\mathbf{U}(\mathbf{u}^{exp,1}; \mathbf{s})\} + \ldots + \ln\{p_\mathbf{U}(\mathbf{u}^{exp,\nu}; \mathbf{s})\}, \quad (7.24)$$

which has the same maximum as the likelihood function. For fixed $\mathbf{s}$ in $\mathcal{C}_{ad} \subset \mathbb{R}^\mu$ and for ℓ in $\{1, \ldots, \nu\}$, the value $\ln\{p_\mathbf{U}(\mathbf{u}^{exp,\ell}; \mathbf{s})\}$ of the log-pdf is estimated with the computational model and the stochastic model of $\mathbf{X}$ defined by $p_\mathbf{X}(\mathbf{x}; \mathbf{s})$, using a stochastic solver, for instance, a Monte Carlo solver (see Section 6.4), the multivariate kernel density estimation (see Section 7.2), and generally, a MCMC method for generating the realizations of $\mathbf{X}$ (see Chapter 4).

☐ *A more advanced mathematical result concerning the asymptotic probability distribution of the maximum likelihood estimator.* Let $\mathbf{U}^{(1)}, \ldots, \mathbf{U}^{(\nu)}$ be ν independent copies of $\mathbb{R}^m$-valued random variable $\mathbf{U}$,

$$p_{\mathbf{U}^{(1)}, \ldots, \mathbf{U}^{(\nu)}}(\mathbf{u}^1, \ldots \mathbf{u}^\nu; \mathbf{s}) = p_\mathbf{U}(\mathbf{u}^1; \mathbf{s}) \times \ldots \times p_\mathbf{U}(\mathbf{u}^\nu; \mathbf{s}). \quad (7.25)$$

For all $\mathbf{s}$ in $\mathcal{C}_{ad} \subset \mathbb{R}^\mu$, the random log-likelihood function is defined as

$$\mathcal{L}(\mathbf{s}; \mathbf{U}^{(1)}, \ldots, \mathbf{U}^{(\nu)}) = \ln\{p_\mathbf{U}(\mathbf{U}^{(1)}; \mathbf{s})\} + \ldots + \ln\{p_\mathbf{U}(\mathbf{U}^{(\nu)}; \mathbf{s})\}. \quad (7.26)$$

Let $\widehat{\mathbf{S}}_\nu$ be the $\mathbb{R}^\mu$-valued random variable whose probability distribution $P_{\widehat{\mathbf{S}}_\nu}(d\mathbf{s})$ is unknown, but for which the support is $\mathcal{C}_{ad} \subset \mathbb{R}^\mu$, which is defined by

$$\widehat{\mathbf{S}}_\nu = \arg \max_{\mathbf{s} \in \mathcal{C}_{ad} \subset \mathbb{R}^\mu} \mathcal{L}(\mathbf{s}; \mathbf{U}^{(1)}, \ldots, \mathbf{U}^{(\nu)}). \quad (7.27)$$

The random vector $\widehat{\mathbf{S}}_\nu$ is the maximum likelihood estimator of the unknown true value $\mathbf{s}^t$ in $\mathcal{C}_{ad}$, which can formally be written as

$$\widehat{\mathbf{S}}_\nu = \widehat{\mathbf{s}}(\mathbf{U}^{(1)}, \ldots, \mathbf{U}^{(\nu)}), \quad (7.28)$$

in which $\widehat{\mathbf{s}}$ is a deterministic measurable mapping from $\mathbb{R}^m \times \ldots \times \mathbb{R}^m$ onto $\mathcal{C}_{ad}$. The maximum likelihood estimation, which corresponds to $(\mathbf{u}^{exp,1}, \ldots, \mathbf{u}^{exp,\nu})$ that is considered as a realization of $(\mathbf{U}^{(1)}, \ldots, \mathbf{U}^{(\nu)})$, is written as

$$\mathbf{s}_\nu^{opt} = \widehat{\mathbf{s}}(\mathbf{u}^{exp,1}, \ldots, \mathbf{u}^{exp,\nu}). \quad (7.29)$$

The following result that is related to the asymptotic probability distribution of the maximum likelihood estimator [178] is important for evaluating the quality of the convergence of the solution constructed by the maximum likelihood as a function of the number ν of data. Under regularity conditions for the function $\mathbf{s} \mapsto p_\mathbf{U}(\cdot, \mathbf{s})$, the Fisher information matrix $[J_F(\mathbf{s})] \in \mathbb{M}_\mu(\mathbb{R})$, which is defined, for all $\mathbf{s}$ in $\mathcal{C}_{ad}$, by

$$[J_F(\mathbf{s})]_{ij} = E\left\{\frac{\partial \mathcal{L}(\mathbf{s}; \mathbf{U}^{(1)}, \dots, \mathbf{U}^{(\nu)})}{\partial s_i} \frac{\partial \mathcal{L}(\mathbf{s}; \mathbf{U}^{(1)}, \dots, \mathbf{U}^{(\nu)})}{\partial s_j}\right\}, \quad (7.30)$$

is positive definite,

$$\forall \mathbf{s} \in \mathcal{C}_{\text{ad}} \quad , \quad [J_F(\mathbf{s})] \in \mathbb{M}_\mu^+(\mathbb{R}). \quad (7.31)$$

▷ *For $\nu \to +\infty$, the sequence of random vectors $\widehat{\mathbf{S}}_\nu$ converges in probability distribution to a Gaussian random vector with the mean vector $\mathbf{s}^t$ and with the covariance matrix $[C(\nu)] = [J_F(\mathbf{s}^t)]^{-1}$.*

In practice, since $\mathbf{s}^t$ is unknown, the approximation $[C(\nu)] \simeq [J_F(\mathbf{s}_\nu^{\text{opt}})]^{-1}$ is used, and the error of the estimate, $\mathbf{s}_\nu^{\text{opt}}$, for each one of its component $s_{\nu,i}^{\text{opt}}$ can be calculated by using the standard deviation $\sigma_i(\nu) = \{[C(\nu)]_{ii}\}^{1/2}$ that is approximated by

$$\sigma_i(\nu) \simeq \sqrt{\{[J_F(\mathbf{s}_\nu^{\text{opt}})]^{-1}\}_{ii}} \quad , \quad i = 1, \dots, \mu. \quad (7.32)$$

7.3.4 Bayes' Method for Estimating the Posterior Probability Distribution from a Prior Probability Model

□ *Objective.* The objective of the Bayes method is to calculate a posterior pdf $p_{\mathbf{X}}^{\text{post}}(\mathbf{x})$ of the $\mathbb{R}^n$-valued random variable $\mathbf{X}$ that is a modeling of the uncertain parameter of the computational model for which

- a prior pdf $p_{\mathbf{X}}^{\text{prior}}(\mathbf{x})$ is given.
- ν experimental data $\mathbf{u}^{\text{exp},1}, \dots, \mathbf{u}^{\text{exp},\nu}$, considered as ν independent realizations of the $\mathbb{R}^m$-valued random vector $\mathbf{U}^{\text{exp}}$, are available.
- there is a noise $\mathbf{B}$ that is assumed to be a second-order $\mathbb{R}^{m_b}$-valued random variable with probability distribution $P_{\mathbf{B}}(d\mathbf{b}) = p_{\mathbf{B}}(\mathbf{b}) \, d\mathbf{b}$, possibly with $m_b = m$, and which is independent of $\mathbf{X}$.
- in order to correctly compare the random observation $\mathbf{U} = \mathbf{h}(\mathbf{X})$ of the computational model with the corresponding measured quantity $\mathbf{U}^{\text{exp}}$ (in presence of noise $\mathbf{B}$), the random vector $\mathbf{U}^{\text{out}}$ is introduced such that

$$\mathbf{U}^{\text{out}} = \mathbf{g}(\mathbf{X}, \mathbf{B}), \quad (7.33)$$

in which $(\mathbf{x}, \mathbf{b}) \mapsto \mathbf{g}(\mathbf{x}, \mathbf{b})$ is a mapping from $\mathbb{R}^n \times \mathbb{R}^{m_b}$ into $\mathbb{R}^m$. For instance, for an additive noise on the observation, we have

$$\mathbf{U}^{\text{out}} = \mathbf{h}(\mathbf{X}) + \mathbf{B} \quad \text{with} \quad m_b = m. \quad (7.34)$$

□ *Formulation.* The Bayes method is based on the conditional probability of events (see Section 2.1.2),

$$P(A \cap B) = P(A|B) \times P(B) = P(B|A) \times P(A).$$

Taking $P(A) = \mathcal{P}^{\text{prior}}(A)$, $P(A|B^{\text{exp}}) = \mathcal{P}^{\text{post}}(A)$ and $1/P(B) = c$ yield

$$\mathcal{P}^{\text{post}}(A) = c \times P(B^{\text{exp}}|A) \times \mathcal{P}^{\text{prior}}(A). \tag{7.35}$$

By applying Equation (7.35) to the problem defined before yields

$$p_{\mathbf{X}}^{\text{post}}(\mathbf{x}) = c_n L(\mathbf{x}) p_{\mathbf{X}}^{\text{prior}}(\mathbf{x}), \tag{7.36}$$

in which $L(\mathbf{x})$ is the likelihood function (without including the constant of normalization) defined on $\mathcal{S}_n \subset \mathbb{R}^n$, with values in $\mathbb{R}^+$, such that

$$L(\mathbf{x}) = \prod_{\ell=1}^{\nu} p_{\mathbf{U}^{\text{out}}|\mathbf{X}}(\mathbf{u}^{\text{exp},\ell}|\mathbf{x}), \tag{7.37}$$

where $p_{\mathbf{U}^{\text{out}}|\mathbf{X}}(\mathbf{u}|\mathbf{x})$ is the conditional pdf of the observation $\mathbf{U}^{\text{out}}$ of the computational model given $\mathbf{X} = \mathbf{x}$, and where c_n is the constant of normalization such that

$$\frac{1}{c_n} = \int_{\mathcal{S}_n} L(\mathbf{x}) p_{\mathbf{X}}^{\text{prior}}(\mathbf{x}) \, d\mathbf{x}. \tag{7.38}$$

□ *Computational aspects for the high dimensions.* In general, we are interested in computing the mathematical expectation of a quantity of interest, $q(\mathbf{U})$,

$$E\{q(\mathbf{U})\} = E\{q(\mathbf{h}(\mathbf{X}))\},$$
$$= \int_{\mathbb{R}^n} q(\mathbf{h}(\mathbf{x})) p_{\mathbf{X}}^{\text{post}}(\mathbf{x}) \, d\mathbf{x}, \tag{7.39}$$

in which $\mathbf{u} \mapsto q(\mathbf{u})$ is a given mapping from $\mathbb{R}^m$ into $\mathbb{R}$. Such a calculation cannot directly be done by a direct computation for the following reasons:

- In general, the integral is in high dimension (n is big).
- The pdf $p_{\mathbf{X}}^{\text{post}}(\mathbf{x})$ is a function of the normalization constant, c_n, that is very difficult to directly compute, and even impossible to compute for the high dimensions.

Consequently, the MCMC methods presented in Chapter 4 must be used because, as we have explained, the MCMC methods

- allow for computing the integral for high dimension n and the speed of convergence is independent of the value of n.
- do not require the calculation of the normalization constant c_n of the pdf $p_{\mathbf{X}}^{\text{post}}$,
- are not intrusive with respect to the computational model.

7.4 Application of the Bayes Method to the Output-Prediction-Error Method

The output-prediction-error method is often used to take into account model uncertainties induced by modeling errors in a computational model (see the analysis given in Section 8.1 concerning such a method). In this paragraph, we present the application of the Bayes method for the parametric probabilistic approach of uncertainties in presence of modeling errors that are taken into account by using the output-prediction-error method. The pdf p_B is assumed to be Gaussian.

□ *Description.* The $\mathbb{R}^m$-valued random variable $\mathbf{U}^{out}$ is written (assumed to be additive) as

$$\mathbf{U}^{out} = \mathbf{h}(\mathbf{X}) + \mathbf{B}, \qquad (7.40)$$

in which $\mathbf{B}$ is a Gaussian, second-order, centered, $\mathbb{R}^m$-valued random variable (the noise), for which the covariance matrix $[C_\mathbf{B}] = E\{\mathbf{B}\mathbf{B}^T\}$ is given in $\mathbb{M}_m^+(\mathbb{R})$. Its pdf is thus written as

$$p_\mathbf{B}(\mathbf{b}) = \frac{1}{(2\pi)^{m/2}(\det[C_\mathbf{B}])^{1/2}} \exp(-\frac{1}{2} < [C_\mathbf{B}]^{-1}\mathbf{b}, \mathbf{b} >). \qquad (7.41)$$

Such a Gaussian noise model corresponds to the use of the MaxEnt principle for which the available information is the mean vector $\mathbf{m}_\mathbf{B} = \mathbf{0}$ and the covariance matrix $[C_\mathbf{B}]$ that must be known (see Section 5.3.7).

□ *Calculation of the conditional pdf of the observation of the computational model given $\mathbf{X} = \mathbf{x}$.* For given $\mathbf{X} = \mathbf{x}$, the conditional random variable $\mathbf{U}^{out}|\mathbf{X} = \mathbf{h}(\mathbf{x}) + \mathbf{B}$ is thus a Gaussian random variable with mean vector $\mathbf{h}(\mathbf{x})$ and with a covariance matrix $[C_\mathbf{B}]$,

$$p_{\mathbf{U}^{out}|\mathbf{X}}(\mathbf{u}|\mathbf{x}) = p_\mathbf{B}(\mathbf{u} - \mathbf{h}(\mathbf{x})). \qquad (7.42)$$

The developments presented in Section 7.3 can directly be used.

Chapter 8
Uncertainty Quantification in Computational Structural Dynamics and Vibroacoustics

Abstract *This chapter deals with an ensemble of methodologies for stochastic modeling of both the model-parameter uncertainties and the model uncertainties induced by modeling errors, in computational structural dynamics and in computational vibroacoustics, for which basic illustrations and experimental validations are given. The developments use all the previous chapters and in particular, the random matrix theory (Chapter 5) for constructing the nonparametric probabilistic approach of uncertainties, the stochastic solvers (Chapter 6), and the tools for statistical inverse problems (Chapter 7 with Chapter 4) in order to perform the identification of the stochastic models.*

8.1 Parametric Probabilistic Approach of Uncertainties in Computational Structural Dynamics

In this section, the following items are presented:

1. construction of the computational model.

© Springer International Publishing AG 2017

C. Soize, *Uncertainty Quantification*, Interdisciplinary Applied Mathematics 47,
DOI 10.1007/978-3-319-54339-0_8

2. construction of a reduced-order model, and convergence analysis. The introduction of a reduced-order model is required for decreasing the numerical cost (such as for design optimization, performance optimization, etc.), but also for implementing the nonparametric probabilistic approach of model uncertainties induced by the modeling errors. As soon as a reduced-order model is introduced, a convergence analysis must be carried out with respect to the reduction order.
3. Presentation of the parametric probabilistic approach for the model-parameter uncertainties.
4. Estimation of a posterior stochastic model of uncertainties using the output-prediction-error method based on the use of the Bayesian method.

8.1.1 Computational Model

The finite element model of the nonlinear (or linear) dynamical system is written as

$$[\mathbb{M}(\mathbf{x})]\,\ddot{\mathbf{y}}(t) + [\mathbb{D}(\mathbf{x})]\,\dot{\mathbf{y}}(t) + [\mathbb{K}(\mathbf{x})]\,\mathbf{y}(t) + \mathbf{f}_{NL}(\mathbf{y}(t), \dot{\mathbf{y}}(t); \mathbf{x}) = \mathbf{f}(t; \mathbf{x}),\qquad (8.1)$$

in which t is the time and where

- $\mathbf{x} = (x_1, \ldots, x_n)$ is the uncertain model parameter with values in an admissible set $\mathcal{S}_n$, which is assumed to be a subset of $\mathbb{R}^n$.
- $\mathbf{y}(t)$ is the unknown time-dependent displacement vector of the m_{DOF} degrees of freedom (DOFs). For simplifying the presentation, it is assumed that the boundary conditions are such that there are no rigid body displacements.
- $\dot{\mathbf{y}}(t) = d\mathbf{y}(t)/dt$ is the time-dependent velocity vector and $\ddot{\mathbf{y}}(t) = d^2\mathbf{y}(t)/dt^2$ is the time-dependent acceleration vector.
- $[\mathbb{M}(\mathbf{x})]$, $[\mathbb{D}(\mathbf{x})]$, and $[\mathbb{K}(\mathbf{x})]$ are the mass, the damping, and the stiffness positive-definite symmetric real $(m_{DOF} \times m_{DOF})$ matrices of the linear part of the computational model, which depend on $\mathbf{x}$. The DOFs that correspond to the zero Dirichlet conditions are eliminated from the equation.
- $(\mathbf{y}, \mathbf{z}) \mapsto \mathbf{f}_{NL}(\mathbf{y}, \mathbf{z}; \mathbf{x})$ is a nonlinear mapping that models the local nonlinear forces that are assumed independent of time, and which depends on $\mathbf{x}$.
- $\mathbf{f}(t; \mathbf{x})$ is the given time-dependent external load vector of the m_{DOF} inputs, which depends on $\mathbf{x}$.

Let $\mathcal{C}_{ad,y}$ be the admissible set for the displacement vector, which is assumed to be a vector space. For simplifying the presentation, it is assumed that $\mathcal{C}_{ad,y}$ coincides with $\mathbb{R}^{m_{DOF}}$. Therefore, for all fixed time t, $\mathbf{y}(t)$ belongs to $\mathcal{C}_{ad,y}$. The time-evolution problem consists in constructing the solution $\{\mathbf{y}(t), t \geq 0\}$ of Equation (8.1) for $t > 0$ with the initial conditions $\mathbf{y}(0) = \mathbf{y}_0$ and $\dot{\mathbf{y}}(0) = \mathbf{v}_0$. The forced response problem consists, for $\mathbf{f}(t; \mathbf{x})$ defined for all t in $\mathbb{R}$, in constructing a function $\{\mathbf{y}(t), t \in \mathbb{R}\}$, which verifies Equation (8.1) for all $t \in \mathbb{R}$ (no initial conditions are given).

8.1.2 Reduced-Order Model and Convergence Analysis

□ *Preambles concerning the use of the elastic modes of the underlying linear computational model for constructing a reduced-order basis.*

For constructing the reduced-order model (ROM), a reduced-order basis (ROB) must be constructed. Such an ROB is either a vector basis $\{\phi^\alpha\}_\alpha$ (independent of $\mathbf{x}$) of $\mathcal{C}_{\text{ad,y}}$ or is a family $\{\{\phi^\alpha(\mathbf{x})\}_\alpha, \mathbf{x} \in \mathcal{S}_n\}$ of vector bases of $\mathcal{C}_{\text{ad,y}}$ (in practice, it is a discrete family associated with a finite sampling in $\mathbf{x}$ of $\mathcal{S}_n$). These two cases can be commented as follows:

Single vector basis.

1. A possible choice consists in using the elastic modes of the underlying linear computational model, $\{\underline{\phi}^\alpha\}_\alpha$, in which the vector bases $\underline{\phi}^\alpha = \phi^\alpha(\underline{\mathbf{x}})$ are computed for $\mathbf{x}$ fixed to a value $\underline{\mathbf{x}}$ in $\mathcal{S}_n$ (for instance, in choosing for $\mathbf{x}$, its nominal value $\underline{\mathbf{x}} = \mathbf{x}_0$). This approach does not require to solve many generalized eigenvalue problems. Nevertheless, the projection that is required for constructing the ROM must be done for each $\mathbf{x}$, which is intrusive for the usual commercial software.

2. Another choice consists in constructing a vector basis $\{\phi^\alpha\}_\alpha$ (independent of $\mathbf{x}$) that is obtained by computing, for a set of sampling points $\mathbf{x}$ in $\mathcal{S}_n$, the proper orthogonal decomposition (POD) from snapshots, and then by performing a fusion by using a singular value decomposition (SVD) [2, 4, 39, 104, 116, 176, 221]. This type of methods is introduced in Section 8.6.2 for constructing an $\mathbf{x}$-parametric nonlinear ROM.

3. Some specific methods for constructing nonlinear ROM in the framework of geometrical nonlinearity in three-dimensional elasticity are presented in Section 8.5.4.

Family of vector bases. The elastic modes are computed for a set of sampling points $\mathbf{x}$ in $\mathcal{S}_n$. This approach allows for obtaining a fast convergence and is not intrusive, but in counter part, requires to solve many generalized eigenvalue problems that can be reduced using an interpolation procedure of the elastic modes with respect to $\mathbf{x}$ (see, for instance, [3]).

In Section 8.1, the family of elastic modes of the underlying linear computational model, indexed by $\mathbf{x}$, is chosen, but any another choice could be done without modifying the probabilistic approach that is presented.

□ *Construction of the ROM by using a family of vector bases.*

Definition of the ROB. For all $\mathbf{x}$ fixed in $\mathcal{S}_n$, let $\{\phi^1(\mathbf{x}), \ldots, \phi^{m_{\text{DOF}}}(\mathbf{x})\}$ be a vector basis of $\mathcal{C}_{\text{ad,y}}$. At fixed $\mathbf{x}$, the ROM is obtained by the projection of Equation (8.1) on the subspace $V_N(\mathbf{x})$ of $\mathcal{C}_{\text{ad,y}}$ spanned by $\{\phi^1(\mathbf{x}), \ldots, \phi^N(\mathbf{x})\}$, with $N \ll m_{\text{DOF}}$. Let $[\phi(\mathbf{x})] = [\phi^1(\mathbf{x}) \ldots \phi^N(\mathbf{x})]$ be the matrix in $\mathbb{M}_{m_{\text{DOF}}, N}(\mathbb{R})$ of the ROB.

Construction of the reduced matrices. We introduce the reduced matrices

$$[M(\mathbf{x})] = [\phi(\mathbf{x})]^T [\mathbb{M}(\mathbf{x})] [\phi(\mathbf{x})], \tag{8.2}$$
$$[D(\mathbf{x})] = [\phi(\mathbf{x})]^T [\mathbb{D}(\mathbf{x})] [\phi(\mathbf{x})], \tag{8.3}$$
$$[K(\mathbf{x})] = [\phi(\mathbf{x})]^T [\mathbb{K}(\mathbf{x})] [\phi(\mathbf{x})], \tag{8.4}$$

that are positive definite and that belong to $\mathbb{M}_N^+(\mathbb{R})$.

Definition of the vector basis. The vector basis is chosen as the elastic modes of the underlying linear computational model. In such a case, for all fixed $\mathbf{x}$, the ROB $[\phi(\mathbf{x})] \in \mathbb{M}_{m_{\mathrm{DOF}}, N}(\mathbb{R})$ is computed by solving the following generalized eigenvalue problem:

$$[\mathbb{K}(\mathbf{x})] [\phi(\mathbf{x})] = [\mathbb{M}(\mathbf{x})] [\phi(\mathbf{x})] [\Lambda(\mathbf{x})], \tag{8.5}$$

in which $[\Lambda(\mathbf{x})]$ is the $(N \times N)$ diagonal matrix of the N smallest positive eigenvalues $0 < \lambda_1(\mathbf{x}) \leq \dots \leq \lambda_N(\mathbf{x})$, where $[M(\mathbf{x})]$ is the $(N \times N)$ diagonal positive-definite matrix (often the normalization of the eigenvectors are chosen in order that $[M(\mathbf{x})] = [I_N]$), and where $[K(\mathbf{x})] = [M(\mathbf{x})] [\Lambda(\mathbf{x})]$ is an $(N \times N)$ diagonal positive-definite matrix. The eigenfrequencies of the underlying linear system are $0 < \omega_1(\mathbf{x}) \leq \dots \leq \omega_N(\mathbf{x})$ with $\omega_\alpha(\mathbf{x}) = \lambda_\alpha(\mathbf{x})^{1/2}$. In general, the positive-definite matrix $[D(\mathbf{x})]$ is full.

Construction of the ROM. For any $\mathbf{x}$ fixed in $\mathcal{S}_n$, the projection of Equation (8.1) on the subspace $V_N(\mathbf{x})$ yields

$$\mathbf{y}^N(t) = [\phi(\mathbf{x})] \mathbf{q}(t), \tag{8.6}$$

$$[M(\mathbf{x})] \ddot{\mathbf{q}}(t) + [D(\mathbf{x})] \dot{\mathbf{q}}(t) + [K(\mathbf{x})] \mathbf{q}(t) + \mathbf{F}_{\mathrm{NL}}(\mathbf{q}(t), \dot{\mathbf{q}}(t); \mathbf{x}) = \mathbf{f}(t; \mathbf{x}), \tag{8.7}$$

in which the projection of the nonlinear forces is a vector in $\mathbb{R}^N$, which is written as

$$\mathbf{F}_{\mathrm{NL}}(\mathbf{q}(t), \dot{\mathbf{q}}(t); \mathbf{x}) = [\phi(\mathbf{x})]^T \mathbf{f}_{\mathrm{NL}}([\phi(\mathbf{x})] \mathbf{q}(t), [\phi(\mathbf{x})] \dot{\mathbf{q}}(t); \mathbf{x}), \tag{8.8}$$

and where the projection of the reduced time-dependent external load vector is a vector in $\mathbb{R}^N$, which is such that

$$\mathbf{f}(t; \mathbf{x}) = [\phi(\mathbf{x})]^T \mathbb{f}(t; \mathbf{x}). \tag{8.9}$$

☐ *Uniform convergence analysis with respect to* $\mathbf{x} \in \mathcal{S}_n \subset \mathbb{R}^n$.

It is assumed that the order N of the ROM is fixed to the value N_0 (independent of $\mathbf{x}$) such that, for all $\mathbf{x} \in \mathcal{S}_n$, the response $\mathbf{y}^N$ (for the time-evolution problem or for the forced response problem) is converged (for an adapted choice of a norm) towards $\mathbf{y}$, with respect to N, for a given accuracy that is assumed to be independent of $\mathbf{x}$.

8.1.3 Parametric Probabilistic Approach of Model-Parameter Uncertainties

8.1.3.1 Methodology

The methodology of the parametric probabilistic approach of model-parameter uncertainties consists in modeling the uncertain model parameter $\mathbf{x}$ by a random variable $\mathbf{X}$, defined on a probability space $(\Theta, \mathcal{T}, \mathcal{P})$, with values in $\mathcal{S}_n \subset \mathbb{R}^n$. As we have previously explained (see Chapter 5), the prior stochastic model of $\mathbf{X}$ must carefully be done. Consequently,

- the reduced matrices become random matrices, $[M(\mathbf{X})]$, $[D(\mathbf{X})]$, and $[K(\mathbf{X})]$, with values in $\mathbb{M}_N^+(\mathbb{R})$.
- for all $\mathbf{u}$ and $\mathbf{v}$ in $\mathbb{R}^N$, the reduced nonlinear force becomes a random vector $\mathbf{F}_{\mathrm{NL}}(\mathbf{u}, \mathbf{v}; \mathbf{X})$ with values in $\mathbb{R}^N$.
- the reduced time-dependent external force becomes a time-dependent random vector, $\mathbf{f}(t; \mathbf{X})$, that is to say is a stochastic process with values in $\mathbb{R}^N$.

It should be noted that, for a given construction of the probability distribution of random variable $\mathbf{X}$, which must be such that the random matrices are second-order random variables, the mean values of these random matrices are such that

$$[M_{\mathrm{mean}}] = E\{[M(\mathbf{X})]\} \;,\; [D_{\mathrm{mean}}] = E\{[D(\mathbf{X})]\} \;,\; [K_{\mathrm{mean}}] = E\{[K(\mathbf{X})]\} \;, \quad (8.10)$$

in which $[M_{\mathrm{mean}}]$, $[D_{\mathrm{mean}}]$, and $[K_{\mathrm{mean}}]$ are matrices in $\mathbb{M}_N^+(\mathbb{R})$, which are, in general, different from the nominal values $[\underline{M}] = [M(\underline{\mathbf{x}})]$, $[\underline{D}] = [D(\underline{\mathbf{x}})]$, and $[\underline{K}] = [K(\underline{\mathbf{x}})]$.

8.1.3.2 Prior Stochastic Model of Random Vector X

The prior stochastic model of random vector $\mathbf{X}$ is constructed as explained in Chapter 5, by using the direct approach (MaxEnt, Section 5.3) or the indirect approaches (PCE, Section 5.5 or PAR, Section 5.6). The support of the pdf, $p_{\mathbf{X}}$, is the set $\mathcal{S}_n$ that is given,

$$\mathrm{Supp}\, p_{\mathbf{X}} = \mathcal{S}_n \subset \mathbb{R}^n \,. \tag{8.11}$$

For simplifying the presentation with respect to the definition of the hyperparameter $\mathbf{s}$ of the prior stochastic model, we assume that the MaxEnt or the PAR is used. Therefore, the probability density function $\mathbf{x} \mapsto p_{\mathbf{X}}(\mathbf{x}; \mathbf{s})$ of $\mathbf{X}$ depends on hyperparameter $\mathbf{s}$ that is assumed to be written as

$$\mathbf{s} = (\underline{\mathbf{x}}, \boldsymbol{\delta}_X) \in \mathcal{C}_{\mathrm{ad}} = \mathcal{S}_n \times \mathcal{C}_X \subset \mathbb{R}^\mu \,, \tag{8.12}$$

in which $\mathcal{C}_{\mathrm{ad}}$ is the admissible set of $\mathbf{s}$, where $\underline{\mathbf{x}}$ is a value of $\mathbf{x}$ in $\mathcal{S}_n$ (for instance, the mean value of $\mathbf{X}$), and where $\mathcal{C}_X$ is the admissible set of an $\mathbb{R}^{\mu_X}$-valued hyperparameter $\boldsymbol{\delta}_X$ (which allows the control of the level of statistical fluctuations of $\mathbf{X}$, that is to say, the control of the level of uncertainties). We then have $\mu = n + \mu_X$.

8.1.3.3 Stochastic Reduced-Order Model Constructed with the Parametric Probabilistic Approach of Uncertainties

From Equations (8.6) and (8.7), it can be deduced the stochastic reduced-order model (SROM) that is written as

$$\mathbf{Y}(t) = [\,\phi(\mathbf{X})\,]\,\mathbf{Q}(t)\,, \tag{8.13}$$

$$[M(\mathbf{X})]\,\ddot{\mathbf{Q}}(t) + [D(\mathbf{X})]\,\dot{\mathbf{Q}}(t) + [K(\mathbf{X})]\,\mathbf{Q}(t) + \mathbf{F}_{\mathrm{NL}}(\mathbf{Q}(t),\dot{\mathbf{Q}}(t)\,;\mathbf{X}) = \mathbf{f}(t\,;\mathbf{X})\,. \tag{8.14}$$

As we have explained in Chapter 5, the prior stochastic model of $\mathbf{X}$ must be constructed in order that $E\{\|\mathbf{Y}(t)\|^2\} < +\infty$. The solution of the SROM is obtained by using the methods presented in Chapter 6.

8.1.3.4 Estimation of the Hyperparameter of the Prior Stochastic Model of the Uncertain Model Parameter

□ *First case: no data are available.* The pdf $p_{\mathbf{X}}(\mathbf{x}\,;\mathbf{s})$ of random variable $\mathbf{X}$, which is the prior stochastic model of the model-parameter uncertainties, depends on the hyperparameter $\mathbf{s} = (\underline{\mathbf{x}},\delta_X)$ that belongs to the admissible set $\mathcal{C}_{\mathrm{ad}} = \mathcal{S}_n \times \mathcal{C}_X$. Then, $\underline{\mathbf{x}}$ is fixed to the nominal value $\mathbf{x}_0$ of $\mathbf{x}$, and δ_X must be considered as a parameter for performing a sensitivity analysis of the stochastic solution with respect to the level of uncertainties. Such a prior stochastic model of model-parameter uncertainties then allows for analyzing the robustness of the stochastic solution $\mathbf{Y}$ with respect to the level of model-parameter uncertainties, which is controlled by δ_X.

□ *Second case: data are available.* Let $\mathbf{U}$ be the random observation with values in $\mathbb{R}^m$, independent of time t, but depending on $\{\mathbf{Y}(t), t \in \tau\}$ in which τ is a given part of $\mathbb{R}$ on which the stochastic solution has been constructed. In such a case, $\underline{\mathbf{x}}$ can be updated and δ_X can be estimated using data $\mathbf{u}^{\mathrm{exp},1}, \ldots, \mathbf{u}^{\mathrm{exp},\nu}$, relative to the observation vector $\mathbf{U}$. For all $\mathbf{s} = (\underline{\mathbf{x}},\delta_X) \in \mathcal{C}_{\mathrm{ad}} = \mathcal{S}_n \times \mathcal{C}_X$, the pdf of $\mathbf{U}$ is denoted by $p_{\mathbf{U}}(\mathbf{u};\mathbf{s})$. The optimal value $\mathbf{s}^{\mathrm{opt}}$ of $\mathbf{s}$ is estimated by using the maximum likelihood method (see Section 7.3.3),

$$\mathbf{s}^{\mathrm{opt}} = \arg\max_{\mathbf{s}\in\mathcal{C}_{\mathrm{ad}}} \sum_{\ell=1}^{\nu} \ln\{p_{\mathbf{U}}(\mathbf{u}^{\mathrm{exp},\ell};\mathbf{s})\}\,. \tag{8.15}$$

For all ℓ, the quantity $\ln\{p_{\mathbf{U}}(\mathbf{u}^{\mathrm{exp},\ell};\mathbf{s})\}$ can be estimated by using the multivariate kernel density estimation method (see Section 7.2) for which ν_s independent realizations $\mathbf{u}^{r,1}, \ldots, \mathbf{u}^{r,\nu_s}$ of $\mathbf{U}$ are computed with the SROM using a stochastic solver (see Chapter 6).

8.1.4 Estimation of a Posterior Stochastic Model of Uncertainties Using the Output-Prediction-Error Method Based on the Use of the Bayesian Method

Let $p_{\mathbf{X}}^{\text{prior}}(\mathbf{x}) = p_{\mathbf{X}}(\mathbf{x}; \mathbf{s}^{\text{opt}})$ be the optimal prior pdf of $\mathbf{X}$. The Bayesian method, coupled with the output-prediction-error method (see Section 7.4) is used for estimating the posterior pdf $p_{\mathbf{X}}^{\text{post}}(\mathbf{x})$ of $\mathbf{X}$. The random output $\mathbf{U}^{\text{out}}$ with values in $\mathbb{R}^m$, for which data are $\mathbf{u}^{\text{exp},1}, \ldots, \mathbf{u}^{\text{exp},\nu}$, is then written as

$$\mathbf{U}^{\text{out}} = \mathbf{U} + \mathbf{B} \quad , \quad \mathbf{U} = \mathbf{h}(\mathbf{X}), \tag{8.16}$$

in which $\mathbf{B}$ is the additive noise whose pdf has been defined in Section 7.4,

$$p_{\mathbf{B}}(\mathbf{b}) = \frac{1}{(2\pi)^{m/2}\,(\det[C_{\mathbf{B}}])^{1/2}}\,\exp(-\frac{1}{2} < [C_{\mathbf{B}}]^{-1}\,\mathbf{b}\,,\mathbf{b} >), \tag{8.17}$$

and where $\mathbf{U} = \mathbf{h}(\mathbf{X})$ is the formal expression that connects the uncertain parameter to the observation through the SROM. Using Bayes' method presented in Sections 7.3.4 and 7.4 yields the posterior pdf of $\mathbf{X}$,

$$p_{\mathbf{X}}^{\text{post}}(\mathbf{x}) = c_n\,L(\mathbf{x})\,p_{\mathbf{X}}^{\text{prior}}(\mathbf{x}), \tag{8.18}$$

in which $L(\mathbf{x})$ is the likelihood function defined on $\mathcal{S}_n \subset \mathbb{R}^n$, with values in $\mathbb{R}^+$, such that

$$L(\mathbf{x}) = \prod_{\ell=1}^{\nu} p_{\mathbf{U}^{\text{out}}|\mathbf{X}}(\mathbf{u}^{\text{exp},\ell}|\mathbf{x}), \tag{8.19}$$

in which the conditional probability density function of $\mathbf{U}^{\text{out}}$ given $\mathbf{X} = \mathbf{x}$ is written as

$$p_{\mathbf{U}^{\text{out}}|\mathbf{X}}(\mathbf{u}^{\text{exp},\ell}|\mathbf{x}) = p_{\mathbf{B}}(\mathbf{u}^{\text{exp},\ell} - \mathbf{h}(\mathbf{x})). \tag{8.20}$$

Important remarks about the use of the output-prediction-error method in structural dynamics.

- The posterior pdf of $\mathbf{X}$ strongly depends on the choice of the pdf of output additive noise $\mathbf{B}$.
- If the output-prediction-error method is used to take into account the model uncertainties induced by the modeling errors, data are absolutely required (the method cannot be used if there are no data).
- If data are available, the computational model cannot learn from data because the stochastic model concerns the output (this can be a serious problem for the robust optimization and for the robust design).
- The available information relative to $\mathbf{B}$ are very poor for constructing a prior pdf of $\mathbf{B}$ with the MaxEnt principle. When giving $\mathbf{m}_{\mathbf{B}} = 0$ and the covariance matrix $[C_{\mathbf{B}}]$, the MaxEnt principle yields a Gaussian pdf. So, the question is: why the

model uncertainties $\mathbf{U}^{out} - \mathbf{h}(\mathbf{X})$ induced by the modeling errors would be a Gaussian random vector. Such a stochastic modeling is surely a rough approximation in structural dynamics.

8.2 Nonparametric Probabilistic Approach of Uncertainties in Computational Structural Dynamics

In this section, the following items are presented.

1. In a first paragraph, we explain what is the nonparametric probabilistic approach of uncertainties.
2. Then, the mean computational model is introduced.
3. A reduced-order model is then constructed by using the elastic modes of the underlying linear mean computational model and the convergence analysis is mentioned.
4. The methodology of the nonparametric probabilistic approach of both the modeling errors and the model-parameter uncertainties is detailed.
5. The estimation of the hyperparameters of the prior stochastic model of the uncertain model parameter is presented.
6. We present a simple example that shows the capability of the nonparametric probabilistic approach to take into account model uncertainties induced by modeling errors in structural dynamics.
7. We then give an experimental validation that is related to the vibration of a composite sandwich panel.
8. In complex mechanical systems, the level of modeling errors can be different from a part to another part of the mechanical system. In such a case, the model uncertainties are not homogeneous in the computational model. We thus show how the use of the substructuring techniques and the use of the nonparametric probabilistic approach of uncertainties allow for analyzing this case. An experimental validation is presented.
9. In a real system, some boundary conditions can be uncertain due to the manufacturing process that does not exactly restitute those indicated in the design. We thus show how such uncertain boundary conditions can be taken into account in the framework of the nonparametric probabilistic approach of model uncertainties.

8.2.1 What Is the Nonparametric Probabilistic Approach of Uncertainties

8.2.1.1 Why the Model Uncertainties Induced by the Modeling Errors Must Be Taken into Account in the Computational Models

The poor predictions given by the predictive computational models of the complex systems are mainly due to both the model-parameter uncertainties and the model uncertainties induced by the modeling errors. Today, it is well understood that the parametric probabilistic approach is an efficient and a powerful method to address the model-parameter uncertainties for the predictive computational models (see Section 8.1), but cannot address the modeling errors.

8.2.1.2 Why the Parametric Probabilistic Approach Cannot Take into Account the Model Uncertainties Induced by the Modeling Errors

Let us introduce a simple explanation in the framework of a reduced-order model of dimension N for which the parametric probabilistic approach of uncertainties is summarized in Figure 8.1. Let us consider a reduced matrix $[A(\mathbf{x})]$ of the ROM (for instance, the reduced stiffness matrix that has been denoted by $[K(\mathbf{x})]$), which belongs to $\mathbb{M}_N^+(\mathbb{R})$, and which depends on the uncertain parameter $\mathbf{x}$). In Figure 8.1 (left and right), the set $\mathbb{M}_N^+(\mathbb{R})$ is represented by the biggest ellipsoid in blue color (or in dark gray). The smallest ellipsoid in yellow color (or in light gray) represents the admissible set $\mathcal{S}_n$ of the uncertain parameter $\mathbf{x}$. When $\mathbf{x}$ travels through $\mathcal{S}_n$, $[A(\mathbf{x})]$ spans a subset $[A(\mathcal{S}_n)]$ of $\mathbb{M}_N^+(\mathbb{R})$, represented by the ellipsoid contained in the biggest one and in red color (or in medium grey). In general, subset $[A(\mathcal{S}_n)]$ does not coincide with $\mathbb{M}_N^+(\mathbb{R})$. As we have explained, the parametric probabilistic approach consists in modeling $\mathbf{x}$ by a random variable $\mathbf{X}$ with values in $\mathcal{S}_n$. Consequently, the random matrix $[\mathbf{A}_{par}] = [A(\mathbf{X})]$ is with values in the subset $[A(\mathcal{S}_n)]$ of $\mathbb{M}_N^+(\mathbb{R})$. If the "experimental reference" $[A_{exp}]$ does not belong to the set $[A(\mathcal{S}_n)]$ (see Figure 8.1 right), but belongs to the subset $\mathbb{M}_N^+(\mathbb{R}) \backslash [A(\mathcal{S}_n)]$, then the distance of $[A_{exp}]$ to subset $[A(\mathcal{S}_n)]$ cannot be reduced and therefore, there exist modeling errors that cannot be taken into account by the parametric probabilistic approach.

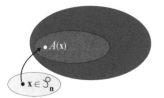 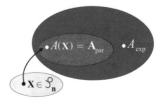

Fig. 8.1 Scheme summarizing the parametric probabilistic approach of uncertainties concerning a reduced matrix $[A(\mathbf{x})]$ of the reduced-order model of dimension N.

8.2.1.3 The Nonparametric Probabilistic Approach as a Way to Take into Account the Model Uncertainties Induced by the Modeling Errors

In this paragraph, we consider the case that we have analyzed in Section 8.2.1.2 and we reuse the same notations. The ROM is constructed for the value $\mathbf{x} = \underline{\mathbf{x}}$ fixed in $\mathcal{S}_n$, which is generally chosen as the nominal value $\mathbf{x}_0$ of $\mathbf{x}$. Consequently, the reduced matrix is denoted by $[A(\underline{\mathbf{x}})]$. The nonparametric probabilistic approach of both the model-parameter uncertainties and the model uncertainties has been introduced in 1999 [184, 185] and consists in directly constructing with the MaxEnt principle, the prior probability distribution of random matrix $[\mathbf{A}_{\text{nonpar}}]$ for which the support is the entire set $\mathbb{M}_N^+(\mathbb{R})$. Such a construction requires the use of the random matrix theory that has been presented in Section 5.4, in particular, requires the use of ensembles $\text{SG}_0^+, \text{SG}_\varepsilon^+, \text{SE}_0^+, \text{SE}_\varepsilon^+$ of random matrices. Consequently, see Figure 8.2, if the "experimental reference" $[A_{\text{exp}}]$ does not belong to the set $[A(\mathcal{S}_n)]$, but belongs to the subset $\mathbb{M}_N^+(\mathbb{R}) \backslash [A(\mathcal{S}_n)]$, since the support of the probability distribution of $[\mathbf{A}_{\text{nonpar}}]$ is the entire set $\mathbb{M}_N^+(\mathbb{R})$, then the distance of $[A_{\text{exp}}]$ to $[\mathbf{A}_{\text{nonpar}}]$ can be reduced and therefore, the existing modeling errors can be taken into account by the proposed nonparametric probabilistic approach.

8.2.1.4 Remarks About the Nonparametric Probabilistic Approach of the Model Uncertainties

The nonparametric probabilistic approach,

1. can be used even if there are no available data.
2. is an alternative way to the output-prediction-error method, if data are available, but in contrary to the output-prediction-error method, the computational model can learn from data because the stochastic model concerns the model itself and not the output.
3. does not introduce an arbitrary Gaussian hypothesis for the output contribution of the modeling errors.

Fig. 8.2 Scheme summarizing the nonparametric probabilistic approach of uncertainties concerning a reduced matrix $[A(\mathbf{x})]$ of the reduced-order model of dimension N.

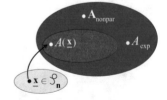

8.2.2 Mean Computational Model

The mean computational model is the computational model introduced in Section 8.1.1 for which $\mathbf{x}$ is fixed to a nominal value $\underline{\mathbf{x}}$. This mean computational model is thus written as

$$[\underline{M}]\,\ddot{\mathbf{y}}(t) + [\underline{D}]\,\dot{\mathbf{y}}(t) + [\underline{K}]\,\mathbf{y}(t) + \mathbf{f}_{\mathrm{NL}}(\mathbf{y}(t), \dot{\mathbf{y}}(t)) = \underline{\mathbf{f}}(t)\,, \qquad (8.21)$$

in which

- $\mathbf{y}(t)$ is the unknown time-dependent displacement vector of the m_{DOF} DOFs (again, for simplifying the presentation, it is assumed that the boundary conditions are such that there are no rigid body displacements).
- $\dot{\mathbf{y}}(t) = d\mathbf{y}(t)/dt$ is the time-dependent velocity vector and $\ddot{\mathbf{y}}(t) = d^2\mathbf{y}(t)/dt^2$ is the time-dependent acceleration vector.
- $[\underline{M}]$, $[\underline{D}]$, and $[\underline{K}]$ are the nominal values of the mass, the damping, and the stiffness positive-definite symmetric real $(m_{\mathrm{DOF}} \times m_{\mathrm{DOF}})$ matrices of the linear part of the computational model (the DOFs that correspond to the zero Dirichlet conditions are eliminated from the equation).
- $(\mathbf{y}, \mathbf{z}) \mapsto \mathbf{f}_{\mathrm{NL}}(\mathbf{y}, \mathbf{z})$ is a nonlinear mapping that models the nominal local nonlinear forces that are assumed independent of time.
- $\underline{\mathbf{f}}(t)$ is the given nominal time-dependent external load vector of the m_{DOF} inputs.

8.2.3 Reduced-Order Model and Convergence Analysis

The approach based on the use of a "single vector basis," which has been introduced in Section 8.1.2, is reused. The ROM is constructed by using an ROB, $\{\boldsymbol{\phi}^\alpha\}_\alpha$, which is a vector basis of the admissible space $\mathcal{C}_{\mathrm{ad},\mathbf{y}}$ related to the mean computational model defined by Equation (8.21).

□ *ROM constructed with a single vector basis.*

Definition of the ROB. The ROM is obtained by the projection of Equation (8.21) on the subspace V_N of $\mathcal{C}_{\mathrm{ad},\mathbf{y}}$ spanned by $\{\underline{\boldsymbol{\phi}}^1, \ldots, \underline{\boldsymbol{\phi}}^N\}$, with $N \ll m_{\mathrm{DOF}}$. Let $[\underline{\phi}] = [\underline{\boldsymbol{\phi}}^1 \ldots \underline{\boldsymbol{\phi}}^N]$ be the matrix in $\mathbb{M}_{m_{\mathrm{DOF}}, N}(\mathbb{R})$ of the ROB.

Construction of the reduced matrices. We introduce the reduced matrices

$$[\underline{M}] = [\underline{\phi}]^T [\underline{M}] [\underline{\phi}]\,, \quad [\underline{D}] = [\underline{\phi}]^T [\underline{D}] [\underline{\phi}]\,, \quad [\underline{K}] = [\underline{\phi}]^T [\underline{K}] [\underline{\phi}]\,, \qquad (8.22)$$

that are positive definite and that belong to $\mathbb{M}_N^+(\mathbb{R})$.

Definition of the vector basis. The vector basis is defined as the elastic modes of the underlying linear mean computational model. The ROB $[\underline{\phi}] \in \mathbb{M}_{m_{\mathrm{DOF}}, N}(\mathbb{R})$ is computed by solving the following generalized eigenvalue problem:

$$[\underline{K}] [\underline{\phi}] = [\underline{M}] [\underline{\phi}] [\underline{\varLambda}]\,, \qquad (8.23)$$

in which $[\Lambda]$ is the $(N \times N)$ diagonal matrix of the N smallest positive eigenvalues $0 < \lambda_1 \leq \ldots \leq \lambda_N$, where $[\underline{M}]$ is the $(N \times N)$ diagonal positive-definite matrix (often the normalization of the eigenvectors are chosen in order that $[\underline{M}] = [I_N]$), and where $[\underline{K}] = [\underline{M}][\Lambda]$ is an $(N \times N)$ diagonal positive-definite matrix. The eigenfrequencies of the underlying linear mean computational model are $0 < \underline{\omega}_1 \leq \ldots \leq \underline{\omega}_N$ with $\underline{\omega}_\alpha = \lambda_\alpha^{1/2}$. In general, the positive-definite matrix $[\underline{D}]$ is full.

Construction of the ROM. The projection of Equation (8.21) on the subspace V_N yields

$$\mathbf{y}^N(t) = [\underline{\phi}]\, \mathbf{q}(t)\,, \tag{8.24}$$

$$[\underline{M}]\, \ddot{\mathbf{q}}(t) + [\underline{D}]\, \dot{\mathbf{q}}(t) + [\underline{K}]\, \mathbf{q}(t) + \mathbf{F}_{\mathrm{NL}}(\mathbf{q}(t), \dot{\mathbf{q}}(t)) = \mathbf{f}(t)\,, \tag{8.25}$$

in which the projection of the nonlinear forces is the following vector in $\mathbb{R}^N$,

$$\mathbf{F}_{\mathrm{NL}}(\mathbf{q}(t), \dot{\mathbf{q}}(t)) = [\underline{\phi}]^T\, \mathbf{f}_{\mathrm{NL}}([\underline{\phi}]\, \mathbf{q}(t), [\underline{\phi}]\, \dot{\mathbf{q}}(t))\,, \tag{8.26}$$

and where the projection of the reduced time-dependent external load vector is a vector in $\mathbb{R}^N$ such that

$$\mathbf{f}(t) = [\underline{\phi}]^T\, \mathbb{f}(t)\,. \tag{8.27}$$

☐ *Convergence analysis.* It is assumed that, for a given accuracy, the order N of the ROM is fixed to the value N_0 such that the response $\mathbf{y}^N$ (for the time-evolution problem or for the forced response problem) is converged (for an adapted choice of the norm) towards y.

8.2.4 Methodology for the Nonparametric Probabilistic Approach of Both the Modeling Errors and the Model-Parameter Uncertainties

8.2.4.1 Introduction of Random Matrices

For fixed $N = N_0$, the nonparametric probabilistic approach of uncertainties consists in replacing matrices $[\underline{M}]$, $[\underline{D}]$, and $[\underline{K}]$ in Equation (8.25) by the random matrices,

$$[\mathbf{M}] = \{\theta' \mapsto [\mathbf{M}(\theta')]\}\,, \quad [\mathbf{D}] = \{\theta' \mapsto [\mathbf{D}(\theta')]\}\,, \quad [\mathbf{K}] = \{\theta' \mapsto [\mathbf{K}(\theta')]\}\,, \tag{8.28}$$

defined on another probability space $(\Theta', \mathcal{T}', \mathcal{P}')$. The ROM defined by Equations (8.24) and (8.25) yields the stochastic reduced-order computational model (SROM),

$$\mathbf{Y}(t) = [\underline{\phi}]\, \mathbf{Q}(t)\,, \tag{8.29}$$

$$[\mathbf{M}]\, \ddot{\mathbf{Q}}(t) + [\mathbf{D}]\, \dot{\mathbf{Q}}(t) + [\mathbf{K}]\, \mathbf{Q}(t) + \mathbf{F}_{\mathrm{NL}}(\mathbf{Q}(t), \dot{\mathbf{Q}}(t)) = \mathbf{f}(t)\,. \tag{8.30}$$

8.2.4.2 Construction of the Available Information for the Three Random Matrices in the SROM

Since there are no rigid body displacements, the three random matrices have the same algebraic properties. Let A representing M, D or K. The available information are defined as follows:

1. The random matrix $[\mathbf{A}]$ is with values in $\mathbb{M}_N^+(\mathbb{R})$. This means that the support of the pdf of $[\mathbf{A}]$ is $\mathbb{M}_N^+(\mathbb{R})$.
2. The mean value is chosen such that

$$E\{[\mathbf{A}]\} = [\underline{A}] \in \mathbb{M}_N^+(\mathbb{R}). \tag{8.31}$$

3. The underlying linear SROM ($\mathbf{F}_{\mathrm{NL}}$ equal to zero) must have a second-order solution, and therefore, for any arbitrary lower bound $[A_\ell]$ in $\mathbb{M}_N^+(\mathbb{R})$, one must have (see [185, 203, 207]) the constraint,

$$0 < [A_\ell] < [\mathbf{A}] \text{ a.s. }, \quad E\{\log(\det([\mathbf{A}] - [A_\ell]))\} = \nu_A , \quad |\nu_A| < +\infty. \tag{8.32}$$

4. There is no available information concerning statistical dependencies between the three random matrices $[\mathbf{M}]$, $[\mathbf{D}]$, and $[\mathbf{K}]$.

8.2.4.3 Probability Distributions and Generators for the Random Matrices

Taking into account the available information defined in Section 8.2.4.2, from Sections 5.4.7.2 and 5.4.8.2, it can be deduced the following stochastic model.

1. The matrices $[\mathbf{M}]$, $[\mathbf{D}]$, and $[\mathbf{K}]$ are statistically independent.
2. Each random matrix $[\mathbf{A}]$, with $A = M, D$ or K, is written as

$$[\mathbf{A}] = [L_A]^T [\mathbf{G}_A] [L_A] \quad , \quad [\mathbf{G}_A] = \frac{1}{1+\varepsilon}([\mathbf{G}_{0,A}] + \varepsilon [I_N]). \tag{8.33}$$

in which the random matrix $[\mathbf{G}_{0,A}]$ belongs to ensemble SG_0^+ defined in Section 5.4.7.1, where $[L_A]$ is the upper triangular matrix such that

$$[\underline{A}] = [L_A]^T [L_A] \in \mathbb{M}_n^+(\mathbb{R}), \tag{8.34}$$

and where $\varepsilon > 0$.
3. The level of uncertainties for random matrix $[\mathbf{A}]$ is controlled by the coefficient of variation of the random matrix $[\mathbf{G}_{0,A}]$, denoted as δ_A (see Equation (5.59)).
4. The generator of realizations of random matrix $[\mathbf{G}_{0,A}]$ is defined by Equations (5.67) and (5.68).

5. The level of uncertainties of random matrices $\{[\mathbf{M}], [\mathbf{D}], [\mathbf{K}]\}$ is controlled by the vector-valued hyperparameter,

$$\boldsymbol{\delta}_G = (\delta_M, \delta_D, \delta_K) \in \mathcal{C}_G = [0, \delta_{\max}]^3 \subset \mathbb{R}^3, \tag{8.35}$$

in which $\delta_{\max} = \{(N+1)/(N+5)\}^{1/2}$.

8.2.5 Estimation of the Hyperparameters of the Prior Stochastic Model of Uncertainties with the Nonparametric Probabilistic Approach

□ *First case: no data are available.* The prior stochastic model depends on the hyperparameter $\boldsymbol{\delta}_G = (\delta_M, \delta_D, \delta_K) \in \mathcal{C}_G$ that allows for controlling the level of both the model-parameter uncertainties and the model uncertainties induced by the modeling errors. Consequently, hyperparameter $\boldsymbol{\delta}_G$ must be considered as a parameter to perform a sensitivity analysis of the stochastic solution $\mathbf{Y}$ constructed with the SROM. Such a prior stochastic model of both the model-parameter uncertainties and the modeling errors then allows for analyzing the robustness of the stochastic solution as a function of the level of uncertainties.

□ *Second case: data are available.* This statistical inverse problem is solved as explained in Chapter 7. The hyperparameter $\boldsymbol{\delta}_G$ can be estimated by using the data, $\mathbf{u}^{\exp,1}, \ldots, bf u^{\exp,\nu}$, relative to an observation vector $\mathbf{U}$, with values in $\mathbb{R}^m$, independent of t but depending on $\{\mathbf{Y}(t), t \in \tau\}$ in which τ is any part of $\mathbb{R}$ for which the stochastic solution is constructed with the SROM. For all $\mathbf{s} = \boldsymbol{\delta}_G \in \mathcal{C}_{\mathrm{ad}} = \mathcal{C}_G$, the pdf of $\mathbf{U}$ is denoted by $p_{\mathbf{U}}(\mathbf{u}; \mathbf{s})$. The optimal value $\mathbf{s}^{\mathrm{opt}}$ of $\mathbf{s}$ is estimated by using the maximum likelihood method (see Section 7.3.3),

$$\mathbf{s}^{\mathrm{opt}} = \arg \max_{\mathbf{s} \in \mathcal{C}_{\mathrm{ad}}} \sum_{\ell=1}^{\nu} \ln\{p_{\mathbf{U}}(\mathbf{u}^{\exp,\ell}; \mathbf{s})\}. \tag{8.36}$$

For all ℓ, the quantity $\ln\{p_{\mathbf{U}}(\mathbf{u}^{\exp,\ell}; \mathbf{s})\}$ is estimated with the multivariate kernel density estimation method (see Section 7.2) for which ν_s independent realizations $\mathbf{u}^{r,1}, \ldots, \mathbf{u}^{r,\nu_s}$ of $\mathbf{U}$ are computed with the SROM by using a stochastic solver (see Chapter 6).

8.2.6 Simple Example Showing the Capability of the Nonparametric Probabilistic Approach to Take into Account the Model Uncertainties Induced by the Modeling Errors in Structural Dynamics

The following simple example demonstrates the capability of the nonparametric probabilistic approach to take into account model uncertainties in structural dynamics [189].

□ *Designed system.* The geometry of the designed mechanical system is displayed in Figure 8.3. It is made up of a simply supported slender cylindrical body for which the dimensions are $10\,m \times 1\,m \times 1.5\,m$. This system is analyzed in the frequency band of analysis $]0\,, 1000]\,Hz$. This body is made up of an elastic composite material.

□ *Real system, simulated experiments, and mean computational model.*

– The real system and the sources of variabilities are shown in Figure 8.4 (left). For the real system, the main sources of variabilities are (1) the geometry of the real system that is uncertain due to the manufacturing tolerances, (2) the boundary conditions that are uncertain due to the manufacturing and differ from those of the designed system, (3) the manufactured composite material has mechanical properties that differ from those of the designed composite material.

– A simulated experiment of the real system is generated with a finite element model made up of $28\,275$ DOFs corresponding to a mesh done with $80 \times 8 \times 12$ 3D 8-nodes solid elements.

– The mean computational model is defined in Figure 8.4 (right). It is made up of a simply supported Euler beam. Such a mean computational model has been chosen willingly simple in order to generate modeling errors. The point force is applied at $\zeta_1 = 4.2\,m$ and the two observations are at point Obs1 ($\zeta_1 = 5.0\,m$) and at point Obs2 ($\zeta_1 = 6.4\,m$).

Figure 8.5 displays the frequency responses of the simulated experiments that are compared with the frequency responses calculated with the mean computational model. It can be seen that the modeling errors increase with the frequency. The

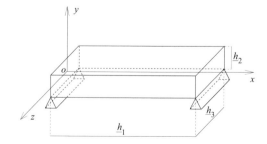

Fig. 8.3 Geometry of the designed mechanical system analyzed in the frequency domain. [Figure from [189]].

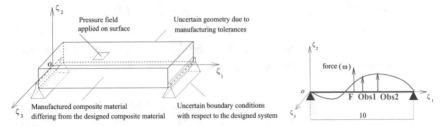

Fig. 8.4 Manufactured system from the designed system and definition of the sources of uncertainties (left figure) [Figure from [189]]. Mean computational model (right figure).

mean computational model prediction is good for frequencies lower than 120 Hz and not really good for frequencies greater than 120 Hz.

□ *Estimation of the dispersion parameters for the nonparametric and for the parametric probabilistic approaches of uncertainties.* The SROM is constructed with $N = 80$ elastic modes and a Monte Carlo stochastic solver is used with $\nu_s = 3\,000$ realizations. The estimation of the optimal value $\mathbf{s}^{\text{opt}} = (\delta_M^{\text{opt}}, \delta_D^{\text{opt}}, \delta_K^{\text{opt}})$ of the hyperparameter $\mathbf{s} = (\delta_M, \delta_D, \delta_K)$ of the SROM, constructed with the nonparametric probabilistic approach, is performed by using the least-square method (see Section 7.3.2) and yields $\delta_M^{\text{opt}} = 0.29$, $\delta_D^{\text{opt}} = 0.30$, and $\delta_K^{\text{opt}} = 0.68$. In order to compare the gain given by the nonparametric probabilistic approach with respect to the parametric probabilistic approach of model-parameter uncertainties, the following SROM is constructed by using the parametric probabilistic approach for which the uncertain parameters are the mass density, the geometric parameters, the Young modulus, and the damping ratio. The coefficients of variations of these uncertain parameters are calculated to obtain the same level of dispersion that the nonparametric probabilistic approach for the first random eigenfrequency.

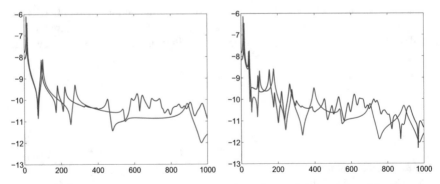

Fig. 8.5 Horizontal axis: frequency (Hz). Vertical axis: $\log_{10}$ of the modulus of the frequency response in displacement (m). Frequency response of the simulated experiment (blue oscillating line) compared with the frequency response calculated with the mean computational model (red smooth line) for Obs1 (left figure) and for Obs2 (right figure). [Figure from [189]].

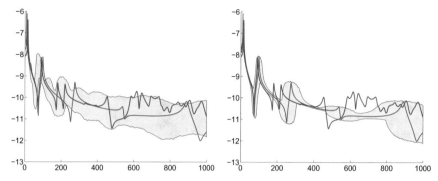

Fig. 8.6 Horizontal axis: frequency (Hz). Vertical axis: $\log_{10}$ of the modulus of the frequency response in displacement (m). Frequency response for Obs1: simulated experiment (blue oscillating line), prediction with the mean computational model (red smooth line), confidence region corresponding to the probability level 0.98 (region in yellow color or in gray), predicted with the nonparametric probabilistic approach (left figure) and predicted with the parametric probabilistic approach (right figure). [Figure from [189]].

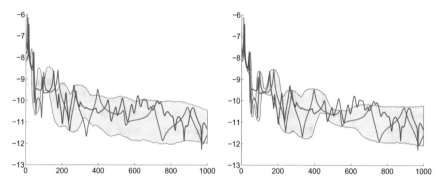

Fig. 8.7 Horizontal axis: frequency (Hz). Vertical axis: $\log_{10}$ of the modulus of the frequency response in displacement (m). Frequency response for Obs2: simulated experiments (blue oscillating line), prediction with the mean computational model (red smooth line), confidence region corresponding to the probability level 0.98 (region in yellow color or in gray), predicted with the nonparametric probabilistic approach (left figure) and predicted with the parametric probabilistic approach (right figure). [Figure from [189]].

□ *Comparisons of the nonparametric approach with the parametric one with the same level of dispersion.* Figure 8.6 (observation Obs1) and Figure 8.7 (observation Obs2) show the frequency responses computed with the nonparametric probabilistic approach, which are compared with the frequency responses computed with the parametric probabilistic approach (with the same level of uncertainties for the first eigenfrequency). It can be seen that the nonparametric probabilistic approach of model uncertainties induced by the modeling errors give good predictions while the parametric probabilistic approach of the model-parameter uncertainties does not have the capability to take into account the modeling errors.

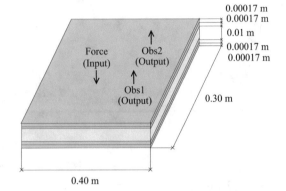

Fig. 8.8 Designed system made up of 2 thin carbon-resin layers and 1 layer made of a stiff closed-cell foam core. Location of the driven point and location of the two observation points, Obs1 and Obs2.

8.2.7 An Experimental Validation: Nonparametric Probabilistic Model of Uncertainties for the Vibration of a Composite Sandwich Panel

The following application presents an experimental validation of the nonparametric probabilistic model of uncertainties for the vibration of a composite sandwich panel [47].

□ *Designed system.* The mechanical system is a composite sandwich panel for which the dimensions are $0.4\,m \times 0.3\,m \times 0.01068\,m$, which is constituted of 2 thin carbon-resin layers made up of 2 unidirectional plies whose orientations are $[60^\circ/-60^\circ]$ and each ply has a thickness of $0.00017\,m$, and 1 layer made up of a stiff closed-cell core for which the thickness is $0.01\,m$ (see Figure 8.8). The location of the point force that is applied (input) and the location of the two observation points (Obs1 and Obs2) in which the out-plane acceleration is observed (output) are shown in Figure 8.8.

□ *Real system and experimental data.* Eight panels have been manufactured with the same process. For the experimental configuration, the plate is in free-free boundary conditions. The frequency response function (FRF) are measures in the frequency band $[10, 4500]\,Hz$ for each one of the 8 panels. A normal force is applied to the driven point. Some out-plane accelerations, distributed in the plate, are measured (one is the observation point). An experimental modal analysis has been performed and the first eleven elastic modes have been identified for each panel. Figure 8.9 shows the experimental frequency response functions of the 8 experimental panels for Obs1 and for Obs2. It can be seen that the dispersion (the variability of the real system) increases with the frequency.

□ *Mean computational model, ROM, and predictions.* The mean computational model is made up of $25\,155$ DOFs with 128×64 finite elements (4-nodes

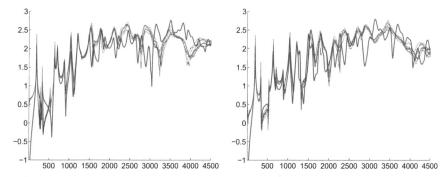

Fig. 8.9 Horizontal axis: frequency (Hz). Vertical axis: $\log_{10}$ of the modulus of the FRF for the out-plane acceleration (m/s^2). Graphs of the FRF for Obs1 (left figure) and for Obs2 (right figure). Experimental FRF corresponding to the eight panels (eight blue thin lines). Numerical FRF predicted with the ROM (red thick line). [Figure from [47]].

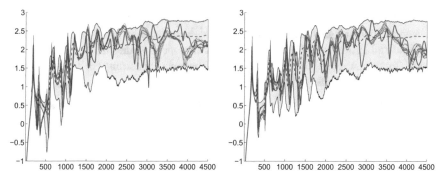

Fig. 8.10 Horizontal axis: frequency (Hz). Vertical axis: $\log_{10}$ of the modulus of the FRF for the out-plane acceleration (m/s^2). Graphs of the FRF for Obs1 (left figure) and for Obs2 (right figure). Experimental FRF corresponding to the eight panels (eight blue thin lines). Numerical FRF predicted with the ROM (red thick line). Confidence region corresponding to the probability level 0.98, predicted with the SROM (region in yellow color of in gray). [Figure from [47]].

Reissner-Mindlin multilayer composite plate element with orthotropic layers). The mean computational model is updated by using the first 4 average experimental eigenfrequencies (averaging on the 8 panels). The ROM is constructed with $N = 200$ elastic modes. Figure 8.9 displays the predictions of the FRFs performed with the ROM, which are compared with the experimental FRFs corresponding to the eight panels. It can be seen that the predictions are good in the frequency band $[0, 1500]\, Hz$ and that the effects of the modeling errors increase in the frequency band $[100, 4500]\, Hz$.

□ *Prediction of the confidence region with the SROM and comparison with the experiments.* The dimension of the SROM is $N = 200$. The Monte Carlo stochastic solver is used with $\nu_s = 2\,000$ realizations. The estimation of the optimal value $\mathbf{s}^{\mathrm{opt}} = (\delta_M^{\mathrm{opt}}, \delta_D^{\mathrm{opt}}, \delta_K^{\mathrm{opt}})$ of the hyperparameter of the SROM constructed with the nonparametric probabilistic approach is performed by using [47] and yields $\delta_M^{\mathrm{opt}} = 0.23, \delta_D^{\mathrm{opt}} = 0.43, \delta_K^{\mathrm{opt}} = 0.25$. Figure 8.10 displays the confidence region corresponding to the probability level 0.98, which is predicted with the SROM. It can be seen that the nonparametric probabilistic approach gives a good prediction of the modeling errors.

8.2.8 Case of Nonhomogeneous Uncertainties Taken into Account by Using Substructuring Techniques

The concept of substructures was first introduced by Argyris and Kelsey in 1959 [8] and was extended by Guyan and Irons in 1965 [102, 111]. In 1960 and 1965, Hurty [108, 109] considered the case of two substructures coupled through a geometrical interface. Finally, Craig and Bampton in 1968 [54] adapted the Hurty method in order to represent each substructure of the same manner consisting in using the elastic modes of the substructure with fixed geometrical interface and the static boundary functions on its geometrical interface. MacNeal in 1971 [128] introduced the concept of residual flexibility which has then been used by Rubin in 1975 [171]. Benfield and Hruda in 1971 [22] proposed a component mode substitution using the Craig-Bampton method for each appendage. Reviews concerning the substructuring techniques can be found in [55, 61, 125, 143, 154, 157].

The implementation of the nonparametric probabilistic approach of uncertainties in the framework of the substructuring techniques has been done in [187] with experimental validations in [46, 71] in order to introduce the capability of the nonparametric probabilistic approach to take into account nonhomogeneous uncertainties in the computational model. Such a case is encountered in the computational models of complex mechanical systems for which the level of modeling errors is different from a part to another part of the mechanical system. In such a case, the model uncertainties are not homogeneous in the computational model.

□ *Setting the problem.* The nonparametric probabilistic approach of uncertainties consists in modeling the reduced matrices of the ROM by random matrices, the level of uncertainties for each one being controlled by only one scalar hyperparameter δ. If the modeling errors are nonhomogeneous in the entire structure and if the structure can be described as an union of substructures for which the level of modeling errors is approximatively homogeneous for each substructure, then the nonparametric probabilistic approach of uncertainties can be used substructure by substructure, and the level of uncertainties can then be different for each substructure. Hereinafter, the Craig-Bampton method is used.

□ *Brief review of the Craig-Bampton substructuring technique for a linear substructure.* Let us consider the mean computational model of a substructure in linear dynamics,

$$\left\{ [\mathbb{M}] \frac{d^2}{dt^2} + [\mathbb{D}] \frac{d}{dt} + [\mathbb{K}] \right\} \begin{bmatrix} y^i(t) \\ y^c(t) \end{bmatrix} = \begin{bmatrix} \mathbb{f}^i(t) \\ \mathbb{f}^c(t) + \mathbb{f}^c_{\text{coup}}(t) \end{bmatrix} . \tag{8.37}$$

in which

- $y(t) \in \mathbb{R}^m$ is the displacement vector of the $m = m_i + m_c$ DOFs, which is written as $y(t) = (y^i(t), y^c(t))$ where $y^i(t) \in \mathbb{R}^{m_i}$ is the displacement vector of the m_i internal DOFs of the substructure and where $y^c(t) \in \mathbb{R}^{m_c}$ is the displacement vector of the m_c coupling DOFs.
- $[\underline{M}]$, $[\underline{D}]$, and $[\underline{K}]$ are the nominal values of the mass, the damping, and the stiffness positive-definite symmetric real $(m \times m)$ matrices of the mean computational model of the substructure.
- $\mathbb{f}(t) \in \mathbb{R}^m$ is the given time-dependent external load vector of the m inputs, which is decomposed as $\mathbb{f}(t) = (\mathbb{f}^i(t), \mathbb{f}^c(t))$ with respect to the internal and to the coupling DOFs.
- $\mathbb{f}^c_{\text{coup}}(t) \in \mathbb{R}^{m_c}$ is the vector of the coupling forces.

The Craig-Bampton transformation of coordinates is written as

$$\begin{bmatrix} y^i(t) \\ y^c(t) \end{bmatrix} = [H] \begin{bmatrix} \mathbf{q}(t) \\ y^c(t) \end{bmatrix} \quad , \quad [H] = \begin{bmatrix} \varPhi & \mathbb{S} \\ [0] & [I_{m_c}] \end{bmatrix} . \tag{8.38}$$

in which $[\varPhi]$ denotes the $(m_i \times N)$ modal matrix of the N selected elastic modes of the substructure with fixed coupling interface and where

$$[\mathbb{S}] = -[\underline{K}^{ii}]^{-1}[\underline{K}^{ic}] \in \mathbb{M}_{m_i, m_c}(\mathbb{R}) , \tag{8.39}$$

is related to the static boundary functions of the mean computational model of the substructure ($[\underline{K}^{ii}]$ and $[\underline{K}^{ic}]$ correspond to the block decomposition of $[\underline{K}]$). The projection defined by Equation (8.38) of Equation (8.37) yields the Craig-Bampton ROM,

$$\left\{ [\underline{M}]\frac{d^2}{dt^2} + [\underline{D}]\frac{d}{dt} + [\underline{K}] \right\} \begin{bmatrix} \mathbf{q}(t) \\ y^c(t) \end{bmatrix} = \begin{bmatrix} \mathcal{F}(t) \\ \mathbb{F}^c(t) + \mathbb{f}^c_{\text{coup}}(t) \end{bmatrix} . \tag{8.40}$$

in which

$$\mathcal{F}(t) = [\varPhi]^T \mathbb{f}^i(t) \in \mathbb{R}^N \quad , \quad \mathbb{F}^c(t) = [\mathbb{S}]^T \mathbb{f}^i(t) + \mathbb{f}^c(t) \in \mathbb{R}^{m_c} , \tag{8.41}$$

$$[\underline{M}] = [H]^T [\underline{M}] [H] \ , \ [\underline{D}] = [H]^T [\underline{D}] [H] \ , \ [\underline{K}] = [K]^T [\underline{K}] [H] . \tag{8.42}$$

□ *Algebraic properties of the reduced matrices in the Craig-Bampton ROM.* In order to implement the nonparametric probabilistic approach of uncertainties in the substructure, we need to analyze the algebraic properties of the reduced matrices in the Craig-Bampton ROM defined by Equation (8.40) and to construct decomposition of these matrices. Let be $\mu = N + m_c$.

- Concerning the reduced mass matrix, the Cholesky decomposition of $[\underline{M}]$ is written as

$$[\underline{M}] = [L_M]^T [L_M] \in \mathbb{M}^+_\mu(\mathbb{R}) , \tag{8.43}$$

in which $[L_M]$ is a triangular matrix in $\mathbb{M}_\mu(\mathbb{R})$.

− Concerning the reduced damping matrix and the reduced stiffness matrix, two cases have to be considered.

1. Case of a fixed substructure. The matrices $[\underline{D}]$ and $[\underline{K}]$ belong to $\mathbb{M}_\mu^+(\mathbb{R})$ and we have the Cholesky decomposition,

$$[\underline{D}] = [L_D]^T [L_D] \quad , \quad [\underline{K}] = [L_K]^T [L_K], \tag{8.44}$$

in which $[L_D]$ and $[L_K]$ are triangular matrices in $\mathbb{M}_\mu(\mathbb{R})$.

2. Case of a free substructure with μ_{rig} rigid body modes. The matrices $[\underline{D}]$ and $[\underline{K}]$ belong to $\mathbb{M}_\mu^{+0}(\mathbb{R})$ and the following decompositions can be constructed,

$$[\underline{D}] = [L_D]^T [L_D] \quad , \quad [\underline{K}] = [L_K]^T [L_K], \tag{8.45}$$

in which $[L_D]$ and $[L_K]$ are rectangular matrices in $\mathbb{M}_{\mu - \mu_{\text{rig}}, \mu}(\mathbb{R})$.

□ *Construction of the SROM with the nonparametric probabilistic approach of uncertainties.* Following the methodology presented in Section 8.2.4, the SROM, based on the ROM defined by Equations (8.38) and (8.40), is written as

$$\begin{bmatrix} \mathbf{Y}^i(t) \\ \mathbf{Y}^c(t) \end{bmatrix} = [H] \begin{bmatrix} \mathbf{Q}(t) \\ \mathbf{Y}^c(t) \end{bmatrix}, \tag{8.46}$$

$$\left\{ [\mathbf{M}] \frac{d^2}{dt^2} + [\mathbf{D}] \frac{d}{dt} + [\mathbf{K}] \right\} \begin{bmatrix} \mathbf{Q}(t) \\ \mathbf{Y}^c(t) \end{bmatrix} = \begin{bmatrix} \mathcal{F}(t) \\ \mathbf{F}^c(t) + \mathbf{F}^c_{\text{coup}}(t) \end{bmatrix}, \tag{8.47}$$

in which, for fixed time t, $\mathbf{F}^c_{\text{coup}}(t)$ is the random vector of the coupling forces. The random matrices $[\mathbf{M}]$, $[\mathbf{D}]$, and $[\mathbf{K}]$ are statistically independent, and for $A = M, D$ or K, each random matrix $[\mathbf{A}]$ is written as

$$[\mathbf{A}] = [L_A]^T [\mathbf{G}_A] [L_A] \quad , \quad [\mathbf{G}_A] = \frac{1}{1 + \varepsilon} ([\mathbf{G}_{0,A}] + \varepsilon [I_N]), \tag{8.48}$$

in which the random matrix $[\mathbf{G}_{0,A}]$ belongs to ensemble SG_0^+, with dimension μ (for a fixed substructure) or $\mu - \mu_{\text{rig}}$ (for a free substructure), and where $\varepsilon > 0$. The level of uncertainties for random matrix $[\mathbf{A}]$ is controlled by the coefficient of variation δ_A of the random matrix $[\mathbf{G}_{0,A}]$. The generator of realizations of random matrix $[\mathbf{G}_{0,A}]$ is detailed in Section 5.4.7.1. For the substructure, the levels of uncertainties of the random matrices $[\mathbf{M}]$, $[\mathbf{D}]$, and $[\mathbf{K}]$ are controlled by the hyperparameter,

$$\delta_G = (\delta_M, \delta_D, \delta_K) \in \mathcal{C}_G = [0, \delta_{\max}]^3 \subset \mathbb{R}^3, \tag{8.49}$$

in which $\delta_{\max} = \{(N + 1)/(N + 5)\}^{1/2}$.

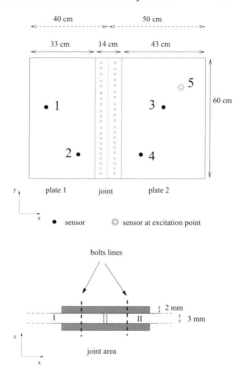

Fig. 8.11 The real system is made up of two plates connected together through a complex joint constituted of two smaller plates tightened by 2 lines of 20 bolts. Plate 1 (left) and plate 2 (right) are isotropic thin plates. The complex joint is modeled by an equivalent orthotropic plate. The excitation (the input) is at point 5 in plate 2 and the observation (the output) is at point 1 in plate 1.

□ *Experimental validation.* We now present an experimental validation [46] of this nonparametric probabilistic approach for nonhomogeneous uncertainties in a linear dynamical system.

– *Real system and experiments.* The real system is defined in Figure 8.11. It is made up of two plates (plate 1 and plate 2) connected together through a complex joint constituted of two smaller plates tightened by two lines of 20 bolts. The driven force (input) is applied to point 5 in plate 2 and the observation (output) is the out-plane displacement at point 1 in plate 1. The cross-FRF between the input and the output is analyzed over the frequency band $[0, 2000]\ Hz$. This cross-FRF allows for analyzing the propagation of the vibration through the complex joint for bending modes and consequently, for analyzing the role played by the modeling errors in the complex joint. In order to study the influence of the screw-bolt prestresses, 21 experimental configurations have been measured. Each one corresponds to a random screwing couple of the 40 screw bolts. The cross-FRF between the input and the output have been measured for the 21 experimental configurations.

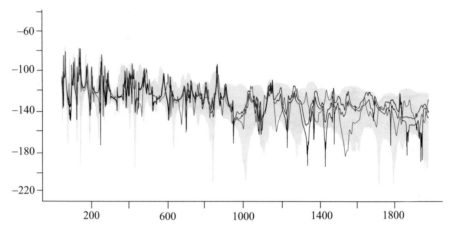

Fig. 8.12 Horizontal axis: frequency (*Hz*) over the frequency band [0 , 2 000]. Vertical axis: $20 \log_{10}$ of the modulus of the cross-FRF for the out-plane displacement (*m*) over the interval [−220 , −40]. Upper and the lower envelopes of the experimental cross-FRF corresponding to 21 random distributions of screw-bolt prestresses (black thick lines). Numerical cross-FRF computed with the mean computational model (blue tin line). Confidence region corresponding to the probability level 0.95, predicted with the SROM (region in yellow color or in gray). [Figure from [46]]

— *Mean computational model.* The mean computational model is constructed by using the Craig-Bampton substructuring method. Plates 1 and 2 are modeled by two isotropic thin plates. The complex joint is modeled by an equivalent orthotropic plate. Three substructures are introduced, plate 1, plate 2, and the complex joint. The finite element models of the three substructures are carried out with compatible meshes made up of four-nodes bending plate elements. The mesh size is $0.01 \, m \times 0.01 \, m$. The assembled structure has 16 653 DOFs, the number of internal DOFs of plate 1, plate 2, and the complex joint are respectively, 6 039, 2 379, and 7 869. The number of coupling DOFs for each interface is 183.

— *Stochastic reduced-order model (SROM) and comparison of the experiments with the predictions.* The model uncertainties induced by the modeling errors are larger in the complex joint than in plate 1 and in plate 2. The computation has been carried out for the following values of the hyperparameter. For plate 1 and plate 2, $\delta_M = 0$, $\delta_D = 0.1$, and $\delta_K = 0.15$. For the complex joint, $\delta_M = 0, \delta_D = 0.8$, and $\delta_K = 0.8$. Figure 8.12 is related to the cross-FRF in dB scale ($20 \log_{10}$ of the modulus of the FRF) and displays the upper and the lower envelopes of the experimental cross-FRF corresponding to 21 random distributions of screw-bolt prestresses, the numerical cross-FRF computed with the mean computational model, and the confidence region corresponding to the probability level 0.95, which is predicted with the SROM.

8.2.9 Stochastic Reduced-Order Computational Model in Linear Structural Dynamics for Structures with Uncertain Boundary Conditions

Often, some boundary conditions in a real system can be uncertain due to the manufacturing process that does not exactly restitute those indicated by the design. In addition, certain boundary conditions that are designed can have a complex behavior and cannot correctly be modeled and consequently, some modeling errors are systematically introduced in the mean computational model. For instance, a part of the boundary of a structure that is considered as clamped for the design can be not perfectly clamped for the structure that is manufactured. Instead to be clamped, the boundary corresponds to a flexible boundary. We thus present an approach [141] that allows for taking into account uncertain boundary conditions by using the nonparametric probabilistic approach of model uncertainties.

We consider the following framework.

— We consider the mean computational model in linear structural dynamics formulated in the frequency domain.
— A part of the boundary of the structure is not perfectly clamped, but corresponds to an uncertain flexible boundary with unknown and nonhomogeneous elastic properties.
— The objective is thus to construct an adapted ROM formulated in the frequency domain by using the substructuring technique and then to implement the nonparametric probabilistic approach of uncertainties in order to deduce a SROM. The method presented hereinafter can directly be extended to the uncertain coupling between substructures (see [141]).

□ *Construction of an adapted ROM and deducing the SROM.* For constructing an adapted ROM in the frequency domain, two reductions are successively used. The first one corresponds to the Craig-Bampton method presented in Section 8.2.8 and the second one is a reduction of the physical boundary DOFs introduced in the mean computational model. By reusing the notations introduced in Section 8.2.8 and by using Equation (8.46) that is rewritten in the frequency domain, the two transformations of the coordinates are written as

$$
\begin{bmatrix} \mathbf{Y}^i(\omega) \\ \mathbf{Y}^c(\omega) \end{bmatrix} = \begin{bmatrix} [\varPhi] & [\mathbb{S}] \\ [0] & [I_{m_c}] \end{bmatrix} \begin{bmatrix} \mathbf{Q}(\omega) \\ \mathbf{Y}^c(\omega) \end{bmatrix} , \tag{8.50}
$$

$$
\begin{bmatrix} \mathbf{Q}(\omega) \\ \mathbf{Y}^c(\omega) \end{bmatrix} = \begin{bmatrix} [I_N] & [0] \\ [0] & [\varPsi] \end{bmatrix} \begin{bmatrix} \mathbf{Q}(\omega) \\ \mathbf{R}(\omega) \end{bmatrix} , \tag{8.51}
$$

in which $[\varPsi]$ is the matrix in $\mathbb{M}_{m_c,N_c}(\mathbb{R})$ with $N_c < m_c$, which are made up, for instance, of some selected eigenvectors of the boundary stiffness resulting from the Craig-Bampton reduction. Equation (8.40) is rewritten in the frequency domain and is projected by using Equation (8.51). A flexible boundary is introduced and the nonparametric probabilistic approach is implemented for this flexible boundary that is considered as an uncertain boundary condition. We then obtain the following random equation, for all ω in $\mathbb{R}$,

$$\left\{ -\omega^2 \left[M \right] + i\omega \left[D \right] + \begin{bmatrix} [K^{qq}] & [K^{qr}] \\ [K^{rq}] & [K^{rr}] + k [\mathbf{K}^{rr}] \end{bmatrix} \begin{bmatrix} \mathbf{Q}(\omega) \\ \mathbf{R}(\omega) \end{bmatrix} \right\}$$
$$= \begin{bmatrix} \mathcal{F}(\omega) \\ \mathcal{F}^r(\omega) \end{bmatrix}, \quad (8.52)$$

in which the quantities in Equation (8.52) are defined as follows:

- The positive parameter k that is introduced in the model allows for controlling the level of flexibility of the boundary (if $k = +\infty$, then the boundary is perfectly clamped, while if $k = 0$, then the boundary is perfectly free).
- The vector $\mathcal{F}^r(\omega)$ in $\mathbb{C}^{N_c}$ is written as $[\Psi]^T \mathbb{F}^c(\omega)$.
- The real $(N + N_c) \times (N + N_c)$ matrices $[M]$, $[D]$ and $[K]$ are obtained by projecting the matrices $[\underline{M}]$, $[\underline{D}]$ and $[\underline{K}]$ (that are defined by Equation (8.42)) by using the transformation defined by Equation (8.51). The matrix $[K]$ is rewritten in a block form, related to $(\mathbf{q}, \mathbf{r})$ in $\mathbb{R}^N \times \mathbb{R}^{N_c}$, as

$$[K] = \begin{bmatrix} [K^{qq}] & [K^{qr}] \\ [K^{rq}] & [K^{rr}] \end{bmatrix}. \quad (8.53)$$

1. If $[K^{rr}]$ belongs to $\mathbb{M}^+_{N_c}(\mathbb{R})$, then its Cholesky decomposition is written as $[K^{rr}] = [L^{rr}]^T [L^{rr}]$ in which $[L^{rr}]$ is a triangular matrix that belongs to $\mathbb{M}_{N'_c, N_c}(\mathbb{R})$ with $N'_c = N_c$.

2. If $[K^{rr}]$ belongs to $\mathbb{M}^{+0}_{N_c}(\mathbb{R})$ and if its null space has dimension $1 \leq \mu_{\text{rig}} < N_c$, then the decomposition $[K^{rr}] = [L^{rr}]^T [L^{rr}]$ can be constructed in which $[L^{rr}]$ is a rectangular matrix that is in $\mathbb{M}_{N'_c, N_c}(\mathbb{R})$ with $N'_c = N_c - \mu_{\text{rig}}$.

- The full random matrix $[\mathbf{K}^{rr}]$ with values in $\mathbb{M}^+_{N_c}$ or in $\mathbb{M}^{+0}_{N_c}$ is written as

$$[\mathbf{K}^{rr}] = [L^{rr}]^T [\mathbf{G}][L^{rr}], \quad (8.54)$$

in which $[L^{rr}]$ belongs to $\mathbb{M}_{N'_c, N_c}(\mathbb{R})$. The random matrix $[\mathbf{G}]$ with values in $\mathbb{M}^+_{N'_c}(\mathbb{R})$ is written as

$$[\mathbf{G}] = \frac{1}{1+\varepsilon}([\mathbf{G}_0] + \varepsilon [I_{N'_c}]), \quad (8.55)$$

where $[\mathbf{G}_0]$ belongs to ensemble SG^+_0, with dimension N'_c and where $\varepsilon > 0$. The level of uncertainties is controlled by the hyperparameter δ defined by Equation (5.59) (which allows for controlling the level of statistical fluctuations of $[\mathbf{G}_0]$). The generator of realizations of random matrix $[\mathbf{G}_0]$ is detailed in Section 5.4.7.1.

□ *Experimental identification of the flexibility k and of the level of uncertainties δ.* For the experimental identification, we introduce a random "energy" related to the flexible boundary, which is sensitive to the deformations of the boundary, and which thus depends on k and δ,

$$\mathcal{E}_{bc}(k, \delta) = \overline{\mathbf{Y}^{c,\mathrm{obs}}(k, \delta)}^{T} [A_{bc}] \mathbf{Y}^{c,\mathrm{obs}}(k, \delta), \qquad (8.56)$$

in which,

- the matrix $[A_{bc}]$ is given in $\mathbb{M}_{m_c}^+(\mathbb{R})$. For instance, if there are translation and rotation DOFs on the flexible boundary, the matrix $[A_{bc}]$ can be chosen as a diagonal matrix for which the entries are equal to 1 for the translations and are equal to a given weight larger than 1 for the rotations.
- the random vector $\mathbf{Y}^{c,\mathrm{obs}}(k, \delta)$ is a particular observation constructed from the random response $\mathbf{Y}^c(\omega)$ that is the stochastic solution of Equations (8.50) to (8.52), which is sensitive to the deformation of the flexible boundary, and which then depends on parameters k and δ. For instance, we can take $\mathbf{Y}^{c,\mathrm{obs}}(k, \delta) = \mathbf{Y}^c(\omega_\alpha)$ in which ω_α is the α-th resonance associated with the α-th elastic mode of the structure.
- the experimental identification of k and δ can be performed by using either the least-square method with the control of the standard deviation (Section 7.3.2) or the maximum likelihood method (Section 7.3.3) for the random variable $\mathcal{E}_{bc}(k, \delta)$, by using the experimental measurements $\mathcal{E}_{bc}^{\mathrm{exp},1}, \ldots, \mathcal{E}_{bc}^{\mathrm{exp},\nu}$.

□ *Numerical illustration for clarifying the effects of the parameters k and δ.* The designed system is made up of an aluminum plate of dimensions $0.3556\,m \times 0.254\,m \times 0.001m$. The mean computational model is developed by using plate finite elements. The ROM is constructed with $N = 10$ clamped elastic modes and $N_c = 120$ boundary elastic modes. The matrix $[A_{bc}]$ is diagonal with 1 on translations and 10 on rotations. The observation is constructed at frequency $\omega = \omega_1$ corresponding to the first elastic modes. Figure 8.13 displays the mean value and the coefficient of variation of the random variable $\mathcal{E}_{bc}(k, \delta)$ as a function of k and δ. These two figures clearly show that the mean of $\mathcal{E}_{bc}$ provides an unambiguous estimation of k, while the coefficient of variation of $\mathcal{E}_{bc}$ strongly depends on δ. Consequently, these results show that the statistical inverse problem based on the least-square method with the control of the standard deviation is a well-posed problem for the experimental identification of k and δ.

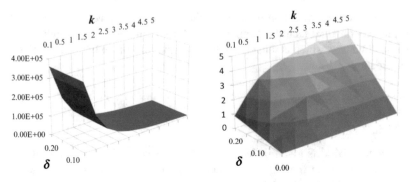

Fig. 8.13 Mean (left figure) and coefficient of variation in percent (right figure) of $\mathcal{E}_{bc}(k, \delta)$ corresponding to the response at the first resonance associated with the first elastic modes, as a function of k and δ [Figure from [141]].

8.3 Nonparametric Probabilistic Approach of Uncertainties in Computational Vibroacoustics

We will consider the linear computational vibroacoustics (or structural acoustics) in the frequency domain and formulated with the structural displacement and with the acoustic pressure (see, for instance, [154, 156]).

– Concerning the deterministic modeling in computational vibroacoustics, two ingredients should be taken into account, but for limiting the presentation, we will not introduce them, and we will focus on the presentation of the implementation of the nonparametric probabilistic approach of uncertainties for the computational interior vibroacoustic problem for which the structure is dissipative but is not viscoelastic. The first ingredient concerns the coupling between the structure and an exterior unbounded acoustic medium, which requires the construction of a coupling operator obtained by solving the exterior Neumann problem related to the Helmholtz equation (see [154, 156]). The second one is related to vibroacoustics involving a viscoelastic structure, and we refer the reader to [154, 155] for the deterministic aspects involving the Hilbert transform and to [37, 156] for the nonparametric probabilistic approach of model uncertainties, which requires the use of the ensemble SE^{HT} of random functions detailed in Section 5.4.8.5, which is related to the use of the Hilbert transform for the stochastic formulation.

– For the nonparametric probabilistic approach of uncertainties in computational vibroacoustics, the approach that we have presented in Section 8.2, for taking into account the uncertainties in computational structural dynamics, is reused for implementing the uncertainties, on the one hand, in the reduced dynamic structural stiffness matrix and, on the other hand, in the reduced dynamic acoustical "stiffness" matrix. Nevertheless, an additional term appears in the computational model, which is the rectangular matrix that describes the acoustic-structure coupling for which the model uncertainties can be significant. Consequently, the ensemble SE^{rect} of rectangular random matrices presented in Section 5.4.8.4 will be used for a nonparametric stochastic modeling of the acoustic-structure coupling.

– For the nonparametric probabilistic approach of uncertainties in computational interior vibroacoustics that we will present in this section, the experimental validation published in [72] will be summarized. Some additional developments can be found in [9, 77, 80, 113].

– The details concerning the proposed formulation can be found in [156].

8.3.1 Mean Boundary Value Problem for the Vibroacoustic System

We consider a linear vibroacoustic system made up of a dissipative structure coupled with a dissipative internal acoustic cavity. The acoustic cavity is assumed to be an almost closed (nonsealed wall) acoustic cavity (see [156], page 80). The computational model is developed in the three-dimensional space $\mathbb{R}^3$ whose generic point is $\zeta = (\zeta_1, \zeta_2, \zeta_3)$ and is formulated in the frequency domain. The geometry, in particular the unit normal to the boundaries, and the notations are defined in Figure 8.14.

– *Structure.* The structure occupies the bounded domain $\Omega_s \subset \mathbb{R}^3$ with the boundary $\partial\Omega_s = \Gamma_s \cup \Gamma_0 \cup \Gamma_a$. The $\mathbb{C}^3$-valued displacement field is $\{\mathbf{u}(\zeta, \omega) = (u_1(\zeta, \omega), u_2(\zeta, \omega), u_3(\zeta, \omega)), \zeta \in \Omega_s\}$. The structure is fixed on the boundary Γ_0 is made up of a dissipative anisotropic nonhomogeneous elastic medium with mass density $\rho_s(\zeta)$, and is subjected to given external loads that are a body force field $\mathbf{g}^{vol}(\omega, \zeta) \in \mathbb{C}^3$ in Ω_s and a surface force field $\mathbf{g}^{sur}(\omega, \zeta) \in \mathbb{C}^3$ on Γ_s.

– *Acoustic cavity.* The cavity is a bounded domain $\Omega_a \subset \mathbb{R}^3$ with the boundary $\partial\Omega_a = \Gamma_a$ that is the acoustic-structure coupling interface. The acoustic cavity is filled with a dissipative acoustic fluid with a constant mass density ρ_a and a constant speed of sound c_a at equilibrium, with a constant dissipative coefficient τ due to the viscosity of the fluid (thermal conduction is neglected). The $\mathbb{C}$-valued acoustic pressure field is $\{p(\zeta, \omega), \zeta \in \Omega_a\}$. There is a given $\mathbb{C}$-valued acoustic source density $\{s(\zeta, \omega), \zeta \in \Omega_a\}$ inside Ω_a.

– *Boundary value problem.* Using the summations over repeated Latin indices, for fixed frequency ω, and for j in $\{1, 2, 3\}$, the boundary value problem is written [156] as

$$-\omega^2 \rho_s u_j - \frac{\partial \sigma_{jk}}{\partial \zeta_k} = g_j^{vol} \quad \text{in} \quad \Omega_s,$$

$$\sigma_{jk}(\mathbf{u}) n_{s,k} = g_j^{sur} \quad \text{on} \quad \Gamma_s,$$

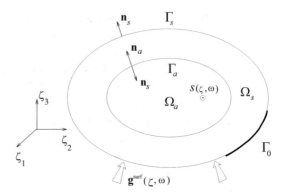

Fig. 8.14 Scheme defining
the vibroacoustic system.

$$\sigma_{jk}(\mathbf{u})n_{s,k} = -p\,n_{s,j} \quad \text{on} \quad \Gamma_a \,,$$

$$u_j = 0 \quad \text{on} \quad \Gamma_0$$

$$-\frac{\omega^2}{\rho_a c_a^2}p - i\omega\frac{\tau}{\rho_a}\nabla^2 p - \frac{1}{\rho_a}\nabla^2 p = -\frac{\tau}{\rho_a}c_a^2\nabla^2 s + i\frac{\omega}{\rho_a}s \quad \text{in} \quad \Omega_a \,,$$

$$\frac{1}{\rho_a}(1+i\omega\tau)\frac{\partial p}{\partial\mathbf{n}_a} = \omega^2\mathbf{u}\cdot\mathbf{n}_a + \tau\frac{c_a^2}{\rho_a}\frac{\partial s}{\partial\mathbf{n}_a} \quad \text{on} \quad \Gamma_a \,,$$

in which the second-order stress tensor σ_{jk} is related to the second-order strain
tensor $\varepsilon_{h\ell}(\mathbf{u}) = (\partial u_h/\partial\zeta_\ell + \partial u_\ell/\partial\zeta_h)/2$ by the constitutive equation $\sigma_{jk} = \mathbb{C}_{jkh\ell}(\boldsymbol{\zeta})\epsilon_{h\ell}(\mathbf{u}) + i\omega\,\widetilde{\mathbb{C}}_{jkh\ell}(\boldsymbol{\zeta})\varepsilon_{h\ell}(\mathbf{u})$, in which $\mathbb{C}_{jkh\ell}(\boldsymbol{\zeta})$ is the fourth-order real
elastic tensor and where $\widetilde{\mathbb{C}}_{jkh\ell}(\boldsymbol{\zeta})$ is the fourth-order real dissipative tensor,
which depend on $\boldsymbol{\zeta}$ but which are assumed to be independent of frequency ω.

8.3.2 Reduced-Order Model of the Vibroacoustic System

The analysis is performed in the frequency domain over the frequency band of anal-
ysis $\mathcal{B} = [\omega_{\min}\,,\omega_{\max}]$ (angular frequency in rad/s) with $\omega_{\min} > 0$.

□ *Mean computational model.* The mean computational model is obtained by using a
finite element discretization of the boundary value problem defined in Section 8.3.1.
It involves the vector $\mathbf{y}(\omega)$ in $\mathbb{C}^{m_{\mathrm{DOF}}}$ of the displacements in which m_{DOF} is the number of structural DOFs and the vector $\mathbb{p}(\omega)$ in $\mathbb{C}^{m_f}$ of the acoustical pressures in
which m_f is the number of acoustical DOFs.

□ *Reduced-order model (ROM).* The reduced-order basis (ROB) is constituted of the
structural ROB and of the acoustical ROB. The structural ROB $[\phi] = [\phi^1 \dots \phi^N]$
is the matrix in $\mathbb{M}_{m_{\mathrm{DOF}},N}(\mathbb{R})$ of the first N elastic modes of the structure in vacuo
with free wall Γ_a (see Section 8.2.3). The acoustical ROB $[\psi] = [\psi^1 \dots \psi^{N_f}]$ is
the matrix in $\mathbb{M}_{m_f,N_f}(\mathbb{R})$ of the first N_f acoustic modes of the acoustic cavity with

rigid wall Γ_a. The ROM is constructed by projecting the mean computational model by using the ROB and is written [156] as

$$\mathbf{y}^{(N,N_f)}(\omega) = [\underline{\phi}]\,\mathbf{q}(\omega) \quad , \quad \mathbf{p}^{(N,N_f)}(\omega) = [\underline{\psi}]\,\mathbf{q}^f(\omega)\,, \tag{8.57}$$

$$\begin{bmatrix} [\underline{Z}(\omega)] & [\underline{C}] \\ \omega^2[\underline{C}]^T & [\underline{Z}_f(\omega)] \end{bmatrix} \begin{bmatrix} \mathbf{q}(\omega) \\ \mathbf{q}^f(\omega) \end{bmatrix} = \begin{bmatrix} \mathbf{f}(\omega) \\ \mathbf{f}^f(\omega) \end{bmatrix}\,, \tag{8.58}$$

in which the reduced dynamic structural stiffness matrix $[\underline{Z}(\omega)] \in \mathbb{M}_N(\mathbb{C})$ is written as

$$[\underline{Z}(\omega)] = -\omega^2\,[\underline{M}] + i\omega\,[\underline{D}] + [\underline{K}]\,, \tag{8.59}$$

where the reduced dynamic acoustical stiffness matrix $[\underline{Z}_f(\omega)] \in \mathbb{M}_{N_f}(\mathbb{C})$ is written as

$$[\underline{Z}_f(\omega)] = -\omega^2\,[\underline{M}_f] + i\omega\,[\underline{D}_f] + [\underline{K}_f]\,, \tag{8.60}$$

with $[\underline{D}_f] = \tau\,[\underline{K}_f]$, and where $[\underline{C}] \in \mathbb{M}_{N,N_f}(\mathbb{R})$ is the rectangular matrix that represents the acoustic-structure coupling.

□ *Uniform convergence analysis with respect to $\omega \in \mathcal{B}$.* The order reduction N (resp. N_f) is fixed to the value N_0 (resp. $N_{f,0}$), independent of ω such that, for all $\omega \in \mathcal{B}$, the response $(\mathbf{y}^{(N,N_f)}(\omega), \mathbf{p}^{(N,N_f)}(\omega))$ is converged to $(\mathbb{y}(\omega), \mathbb{p}(\omega))$ (with respect to (N, N_f)) for a given accuracy that is independent of ω.

□ *Decomposition of the matrices of the ROM.* The Cholesky decomposition of the matrices $[\underline{M}]$, $[\underline{D}]$, and $[\underline{K}]$ that belong to $\mathbb{M}_N^+(\mathbb{R})$, are written as,

$$[\underline{M}] = [L_M]^T[L_M]\,, \; [\underline{D}] = [L_D]^T[L_D]\,, \; [\underline{K}] = [L_K]^T[L_K]\,. \tag{8.61}$$

As the acoustic cavity is almost closed (nonsealed wall), the matrices $[\underline{M}_f]$, $[\underline{D}_f]$, and $[\underline{K}_f]$ belong to $\mathbb{M}_N^+(\mathbb{R})$, and consequently, their Cholesky decompositions yield

$$[\underline{M}_f] = [L_{M_f}]^T[L_{M_f}]\,, \; [\underline{D}_f] = [L_{D_f}]^T[L_{D_f}]\,, \; [\underline{K}_f] = [L_{K_f}]^T[L_{K_f}]\,. \tag{8.62}$$

As $[\underline{D}_f] = \tau\,[\underline{K}_f]$, then $[L_{D_f}] = \sqrt{\tau}\,[L_{K_f}]$. The reduced coupling matrix can be written (see Equations (5.105) and (5.106)) as

$$[\underline{C}] = [U]\,[T]\,, \tag{8.63}$$

in which the rectangular matrix $[U] \in \mathbb{M}_{N,N_f}(\mathbb{R})$ is such that $[U]^T\,[U] = [I_{N_f}]$ and where the square matrix $[T] \in \mathbb{M}_{N_f}^+(\mathbb{R})$ is written (Cholesky decomposition) as

$$[T] = [L_T]^T\,[L_T] \in \mathbb{M}_{N_f}^+(\mathbb{R})\,. \tag{8.64}$$

8.3.3 Stochastic Reduced-Order Model of the Structural-Acoustic System Constructed with the Nonparametric Probabilistic Approach of Uncertainties

□ *Construction of the stochastic reduced-order model (SROM).* For all ω fixed in the frequency band of analysis $\mathcal{B}$, the construction of the SROM is performed by implementing the nonparametric probabilistic approach of uncertainties into Equations (8.57) to (8.64) and yields

$$\mathbf{Y}(\omega) = [\phi]\,\mathbf{Q}(\omega) \quad , \quad \mathbf{P}(\omega) = [\psi]\,\mathbf{Q}^f(\omega)\,, \tag{8.65}$$

$$\begin{bmatrix} [\mathbf{Z}(\omega)] & [\mathbf{C}] \\ \omega^2[\mathbf{C}]^T & [\mathbf{Z}_f(\omega)] \end{bmatrix} \begin{bmatrix} \mathbf{Q}(\omega) \\ \mathbf{Q}^f(\omega) \end{bmatrix} = \begin{bmatrix} \mathbf{f}(\omega) \\ \mathbf{f}^f(\omega) \end{bmatrix}\,, \tag{8.66}$$

where $[\mathbf{Z}(\omega)]$, and $[\mathbf{Z}_f(\omega)]$ are written as

$$[\mathbf{Z}(\omega)] = -\omega^2[\mathbf{M}] + i\omega[\mathbf{D}] + [\mathbf{K}]\,, \tag{8.67}$$

$$[\mathbf{Z}_f(\omega)] = -\omega^2[\mathbf{M}_f] + i\omega[\mathbf{D}_f] + [\mathbf{K}_f]\,, \tag{8.68}$$

and where $[\mathbf{M}]$, $[\mathbf{D}]$, $[\mathbf{K}]$, $[\mathbf{M}_f]$, $[\mathbf{D}_f]$ and $[\mathbf{K}_f]$ are random matrices that are assumed to be mutually independent, each one belonging to the ensemble SE_0^+ defined in Section 5.4.8.1, and where $[\mathbf{C}]$ is independent of the previous ones, and belongs to the ensemble $\mathrm{SE}^{\mathrm{rect}}$ of random matrices defined in Section 5.4.8.4.

□ *Probability distributions and generators of the random matrices.* Let A representing M, D, K, M_f, D_f, or K_f. Therefore, the random matrix $[\mathbf{A}]$ that belongs to SE_0^+ is written as

$$[\mathbf{A}] = [L_A]^T[\mathbf{G}_A][L_A] \quad , \quad [\mathbf{G}_A] = \frac{1}{1+\varepsilon}([\mathbf{G}_{0,A}] + \varepsilon\,[I])\,, \tag{8.69}$$

in which the random matrix $[\mathbf{G}_{0,A}]$ belongs to ensemble SG_0^+, with dimension N or N_f, where $\varepsilon > 0$, and where $[I]$ is either $[I_N]$ or $[I_{N_f}]$. The level of uncertainties for each random matrix $[\mathbf{A}]$ is controlled by the coefficient of variation δ_A of the random matrix $[\mathbf{G}_{0,A}]$. The random matrix $[\mathbf{C}]$ belongs to the ensemble $\mathrm{SE}^{\mathrm{rect}}$ of random matrices, which is written as

$$[\mathbf{C}] = [U]\,[L_T]^T[\mathbf{G}_C][L_T] \quad , \quad [\mathbf{G}_C] = \frac{1}{1+\varepsilon}([\mathbf{G}_{0,C}] + \varepsilon\,[I_{N_f}])\,, \tag{8.70}$$

in which the random matrix $[\mathbf{G}_{0,C}]$ belongs to ensemble SG_0^+, with dimension N_f, and where $\varepsilon > 0$. The level of uncertainties for random matrix $[\mathbf{C}]$ is controlled by the coefficient of variation δ_C of random matrix $[\mathbf{G}_{0,C}]$. The level of uncertainties of random matrices $[\mathbf{M}]$, $[\mathbf{D}]$, $[\mathbf{K}]$, $[\mathbf{M}_f]$, $[\mathbf{D}_f]$, $[\mathbf{K}_f]$, and $[\mathbf{C}]$ is controlled by the vector-valued hyperparameter,

$$\boldsymbol{\delta}_G = (\delta_M, \delta_D, \delta_K, \delta_{M_f}, \delta_{D_f}, \delta_{K_f}, \delta_C) \in \mathcal{C}_G = [0, \delta_{\max}]^7 \subset \mathbb{R}^7, \qquad (8.71)$$

in which $\delta_{\max} = \{(N+1)/(N+5)\}^{1/2}$. The generator of realizations of random matrices $[\mathbf{G}_{0,A}]$ and $[\mathbf{G}_{0,C}]$ is detailed in Section 5.4.7.1.

8.3.4 Experimental Validation with a Complex Computational Vibroacoustic Model of an Automobile

We present an experimental validation of the nonparametric probabilistic approach of uncertainties for a complex computational vibroacoustic model of an automobile [72].

□ *Description and methodology.* The vibroacoustic system is a car of a given type with several optional extra, for which a single mean computational model is developed. The experimental variabilities are due to the manufacturing process and to the optional extra. The objective is to predict the booming noise for which the engine rotation is $[1500, 4800]$ rpm (rotation per minute), which corresponds to the frequency band $[50, 160]$ Hz, for which the input forces are applied to the engine supports, and for which the output observation is the acoustic pressure at a given point localized in the acoustic cavity.

□ *Mean computational model and SROM.* The mean computational model is a finite element model of the structure and of the acoustic cavity shown in Figure 8.15. The structure is modeled with $978\,733$ structural DOFs of displacement and the acoustic cavity is modeled with $8\,139$ acoustical DOFs of pressure. The structural ROB is such that $N = 1\,722$ and the acoustical ROB is such that $N_f = 57$. The hyperparameters of the SROM are $\boldsymbol{\delta} = (\delta_M, \delta_D, \delta_K)$ for the structure, $\delta_f = \delta_{M_f} = \delta_{D_f} = \delta_{K_f}$ for the acoustic cavity, and δ_C for the vibroacoustic coupling.

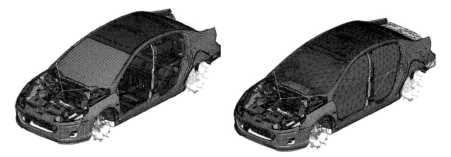

Fig. 8.15 Finite element model of the structure (left figure). Finite element mesh of the computational structural-acoustic model (right figure). [Figure from [72]].

□ *Experimental identification of hyperparameter δ_f.* The acoustical input is an acoustic source placed inside the acoustic cavity. The acoustical measurements have been performed for $\nu = 30$ cars of the same type with different configurations corresponding to different seat positions, different internal temperatures, and different number of passengers. The acoustical pressures have been measured with $\nu_m = 32$ microphones distributed inside the acoustic cavity. For the statistical inverse problem, the observation is the real-valued random variable U defined by

$$U = \int_B dB(\omega)\, d\omega \quad , \quad dB(\omega) = 10\, \log_{10}\left\{\frac{1}{p_{\text{ref}}^2}\frac{1}{\nu_m}\sum_{j=1}^{\nu_m} |P_{k_j}(\omega)|^2\right\}, \quad (8.72)$$

in which $P_{k_1}(\omega), \ldots, P_{k_{\nu_m}}(\omega)$ are the components of $\mathbf{P}(\omega)$, which correspond to the observed DOFs, and which are computed with the SROM defined by Equations (8.65) and (8.66). Let $u^{\text{exp},1}, \ldots, u^{\text{exp},\nu}$ be the corresponding measurements for the $\nu = 30$ cars. The identification of the hyperparameter δ_f is performed by using the maximum likelihood method (Section 7.3.3),

$$\delta_f^{\text{opt}} = \arg\max_{\delta_f} \mathcal{L}(\delta_f) \quad , \quad \mathcal{L}(\delta_f) = \sum_{\ell=1}^{\nu} \log_{10}(p_U(u^{\text{exp},\ell};\delta_f)). \quad (8.73)$$

For $\ell = 1, \ldots, \nu$, the value $p_U(u^{\text{exp},\ell};\delta_f)$ of the pdf of the random variable U is estimated with the kernel density estimation method (Section 7.2) by using the SROM for which the stochastic solver has been chosen as the Monte Carlo method (Section 6.4) with $\nu_s = 2\,000$ realizations.

□ *Experimental identification of hyperparameter $\boldsymbol{\delta} = (\delta_{M_s}, \delta_{D_s}, \delta_{K_s})$.* The structural inputs are some forces applied to the engine supports. The measurements of the displacements (via the accelerations) in the structure have been performed for $\nu = 20$ cars of the same type. The random vector-valued observation is $\mathbf{U}(\omega) = (U_1(\omega), \ldots, U_6(\omega))$ such that

$$U_j(\omega) = \log_{10}(\omega^2\, |Y_{k_j}(\omega)|) \quad , \quad j = 1, \ldots, 6, \quad (8.74)$$

in which $Y_{k_1}(\omega), \ldots, Y_{k_6}(\omega)$ are the six components of $\mathbf{Y}(\omega)$, which correspond to the observed structural DOFs, and which are computed with the SROM defined by Equations (8.65) and (8.66). Let $\mathbf{u}^{\text{exp},1}, \ldots, \mathbf{u}^{\text{exp},\nu}$ be the corresponding measurements for the ν cars. The identification of the hyperparameter $\boldsymbol{\delta} = (\delta_{M_s}, \delta_{D_s}, \delta_{K_s})$ has been performed by using the least-square method (Section 7.3.2). The Monte Carlo method (Section 6.4) has been used as the stochastic solver with $\nu_s = 1\,000$ realizations.

□ *Experimental validation.* The hyperparameters are fixed to their identified values, $\delta_f = \delta_f^{\text{opt}}$ and $\boldsymbol{\delta} = \boldsymbol{\delta}^{\text{opt}}$, while δ_C is fixed to a given value. The SROM defined by Equations (8.65) and (8.66) is solved by using the Monte Carlo method (Section 6.4) with $\nu_s = 600$ realizations. The prediction, with the identified SROM, of

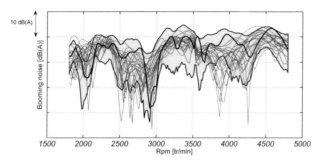

Fig. 8.16 The horizontal axis is the engine rotation expressed in rotation per minute corresponding to the frequency band $[50, 160]$ Hz. The vertical axis is the level of the acoustic pressure in dB(A). Experimental measurements (20 blue thin lines). Prediction with the ROM (black thick line). Prediction with the identified SROM of the confidence region corresponding to the probability level 0.95 (region in yellow color of in gray). [Figure from [72]].

the confidence region of the internal noise at a given point of observation due to the engine excitation is displayed in Figure 8.16. It can be seen that this prediction is good for representing the big variabilities of the measurements, while the response given by the ROM (the mean computational model) gives only a rough idea of the real system.

8.4 Generalized Probabilistic Approach of Uncertainties in a Computational Model

The generalized probabilistic approach to construct a prior stochastic model of both the model-parameter uncertainties and the model uncertainties induced by the modeling errors has been introduced in [197]. It consists in combining the parametric probabilistic approach of the model-parameter uncertainties with the nonparametric probabilistic approach of the model uncertainties induced by the modeling errors. Such an approach is particularly interesting when a computational model is very sensitive to an uncertain model parameter **x** that has a small dimension (n small), in presence of model uncertainties induced by the modeling errors in the computational model. If dimension n is big, then the nonparametric probabilistic approach presented in Sections 8.2 and 8.3 must be used instead of the generalized approach.

In such an approach, the prior stochastic models of the model-parameter uncertainties and of the model uncertainties induced by the modeling errors are separately constructed.

- The model-parameter uncertainties are taken into account with the parametric probabilistic approach presented in Section 8.1.
- The model uncertainties are taken into account with the nonparametric probabilistic approach presented in Section 8.2.

Framework of the presentation. In order to simplify the presentation, the generalized approach is presented in linear/nonlinear computational structural dynamics (for a fixed structure), but this approach can be used without additional difficulties for the linear or the nonlinear computational models, in static or in dynamics, in computational solid mechanics, in computational fluid mechanics, or for coupled mechanical systems such as those encountered in computational vibroacoustics (for a fixed or for a free structure). All the previous notations, concepts, and methods, which have been introduced for presenting the parametric probabilistic approach (Section 8.1) and the nonparametric probabilistic approach (Section 8.2) are directly reused, in particular the construction of the ROM.

8.4.1 Principle for the Construction of the Generalized Probabilistic Approach

In this section, we give a simple explanation of the generalized probabilistic approach in the framework of a reduced-order model of dimension N, which is summarized in Figure 8.17. We simultaneously consider the two cases that we have analyzed in Sections 8.2.1.2 and 8.2.1.3, and we reuse the same notations.

Consequently, $[A(\mathbf{x})]$ is assumed to be one of the reduced matrices of the ROM (for instance, the reduced stiffness matrix), which belongs to $\mathbb{M}_N^+(\mathbb{R})$, and which depends on the uncertain parameter $\mathbf{x}$). In Figure 8.17, the set $\mathbb{M}_N^+(\mathbb{R})$ is represented by the biggest ellipsoid in blue color (or in dark gray). The smallest ellipsoid in yellow color (or in light gray) represents the admissible set $\mathcal{S}_n$ of the uncertain parameter $\mathbf{x}$. When $\mathbf{x}$ travels through $\mathcal{S}_n$, $[A(\mathbf{x})]$ spans a subset $[A(\mathcal{S}_n)]$ of $\mathbb{M}_N^+(\mathbb{R})$, represented by the ellipsoid contained in the biggest one and in red color (or in medium grey).

Within the framework of the scheme displayed in Figure 8.17, the generalized probabilistic approach is characterized by the following items.

1. The uncertain parameter $\mathbf{x}$ is with values in $\mathcal{S}_n \subset \mathbb{R}^n$.
2. The family $\{[A(\mathbf{x})], \mathbf{x} \in \mathcal{S}_n\}$ of deterministic matrices that belong to $\mathbb{M}_N^+(\mathbb{R})$ is replaced by a family $\{[\mathbf{A}(\mathbf{x})], \mathbf{x} \in \mathcal{S}_n\}$ of random matrices with values in

Fig. 8.17 Scheme summarizing the generalized probabilistic approach of uncertainties concerning a reduced matrix $[A(\mathbf{x})]$ of the reduced-order model of dimension N

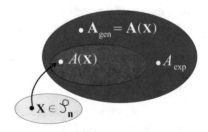

$\mathbb{M}_N^+(\mathbb{R})$. For all $\mathbf{x}$ fixed in $\mathcal{S}_n$, each random matrix $[\mathbf{A}(\mathbf{x})]$ is constructed by using the nonparametric probabilistic approach of the modeling errors.

3. The uncertain parameter $\mathbf{x}$ with values in $\mathcal{S}_n$ is represented by a random variable $\mathbf{X}$ with values in $\mathcal{S}_n \subset \mathbb{R}^n$, which is constructed with the parametric probabilistic approach of the model-parameter uncertainties.

4. The family $\{[\mathbf{A}(\mathbf{x})], \mathbf{x} \in \mathcal{S}_n\}$ of the random matrices with values in $\mathbb{M}_N^+(\mathbb{R})$ is replaced by the random matrix $[\mathbf{A}_{\text{gen}}] = [\mathbf{A}(\mathbf{X})]$ with values in $\mathbb{M}_N^+(\mathbb{R})$.

8.4.2 ROM Associated with the Computational Model and Convergence Analysis

□ *ROM associated with the computational model.* The time-dependent response vector $y(t)$ of the m_{DOF} DOFs is approximated, for $N < m_{\text{DOF}}$, by the $\mathbb{R}^{m_{\text{DOF}}}$-valued function $\mathbf{y}^N$,

$$\mathbf{y}^N(t) = [\phi(\mathbf{x})]\,\mathbf{q}(t)\,, \tag{8.75}$$

$$[M(\mathbf{x})]\,\ddot{\mathbf{q}}(t) + [D(\mathbf{x})]\,\dot{\mathbf{q}}(t) + [K(\mathbf{x})]\,\mathbf{q}(t) + \mathbf{F}_{\text{NL}}(\mathbf{q}(t), \dot{\mathbf{q}}(t)\,;\mathbf{x}) = \mathbf{f}(t\,;\mathbf{x})\,, \tag{8.76}$$

in which $\mathbf{x}$ belongs to $\mathcal{S}_n \subset \mathbb{R}^n$ and is an uncertain model parameter. For all $\mathbf{x}$ in $\mathcal{S}_n$, the reduced matrices $[M(\mathbf{x})]$, $[D(\mathbf{x})]$, and $[K(\mathbf{x})]$ belong to $\mathbb{M}_N^+(\mathbb{R})$ and, for A representing M, D or K, are written as

$$[A(\mathbf{x})] = [\phi(\mathbf{x})]^T\,[\mathbb{A}(\mathbf{x})]\,[\phi(\mathbf{x})]\,. \tag{8.77}$$

□ *Uniform convergence analysis with respect to* $\mathbf{x} \in \mathcal{S}_n \subset \mathbb{R}^n$. The reduction order N is fixed to the value N_0 (independent of $\mathbf{x}$) such that, for all $\mathbf{x} \in \mathcal{S}_n$, the response $\mathbf{y}^N$ is converged to $\mathbf{y}$ (with respect to N) for a given accuracy that is assumed to be independent of $\mathbf{x}$.

8.4.3 Methodology for the Generalized Probabilistic Approach

□ *Construction of the prior stochastic model of uncertainties.*

1. The parametric probabilistic approach is used for modeling the uncertain parameter $\mathbf{x}$ by a random variable $\mathbf{X}$ that is defined on a probability space $(\Theta, \mathcal{T}, \mathcal{P})$, with values in $\mathcal{S}_n \subset \mathbb{R}^n$.

2. The nonparametric probabilistic approach is used for taking into account the model uncertainties. For all $\mathbf{x}$ fixed in $\mathcal{S}_n$, the matrices $[M(\mathbf{x})]$, $[D(\mathbf{x})]$, and $[K(\mathbf{x})]$ are modeled by independent random matrices $[\mathbf{M}(\mathbf{x})] = \{\theta' \mapsto [\mathbf{M}(\theta';\mathbf{x})]\}$, $[\mathbf{D}(\mathbf{x})] = \{\theta' \mapsto [\mathbf{D}(\theta';\mathbf{x})]\}$, and $[\mathbf{K}(\mathbf{x})] = \{\theta' \mapsto [\mathbf{K}(\theta';\mathbf{x})]\}$, that are defined on a probability space $(\Theta', \mathcal{T}', \mathcal{P}')$, and which belong to the ensemble SE_ε^+ of random matrices defined in Section 5.4.8.2,

□ *Stochastic reduced-order model (SROM).* The SROM of order N is written as

$$\mathbf{Y}(t) = [\phi(\mathbf{X})]\,\mathbf{Q}(t)\,, \tag{8.78}$$

$$[\mathbf{M}(\mathbf{X})]\,\ddot{\mathbf{Q}}(t) + [\mathbf{D}(\mathbf{X})]\,\dot{\mathbf{Q}}(t) + [\mathbf{K}(\mathbf{X})]\,\mathbf{Q}(t)$$
$$+ \mathbf{F}_{\mathrm{NL}}(\mathbf{Q}(t), \dot{\mathbf{Q}}(t)\,;\mathbf{X}) = \mathbf{f}(t\,;\mathbf{X})\,, \quad (8.79)$$

in which $\mathbf{Y}(t) = \{(\theta,\theta') \mapsto \mathbf{Y}(\theta,\theta';t)\}$ is an $\mathbb{R}^{m_{\mathrm{DOF}}}$-valued random vector and $\mathbf{Q}(t) = \{(\theta,\theta') \mapsto \mathbf{Q}(\theta,\theta';t)\}$ is an $\mathbb{R}^N$-valued random vector defined on $(\Theta \times \Theta', \mathcal{T} \otimes \mathcal{T}', \mathcal{P} \otimes \mathcal{P}')$.

□ *Statistical realization of the SROM.* For any realization $\mathbf{X}(\theta)$ of random variable $\mathbf{X}$ with θ in Θ, for any realization $[\mathbf{M}(\theta';\mathbf{x})]$, $[\mathbf{D}(\theta';\mathbf{x})]$, and $[\mathbf{K}(\theta';\mathbf{x})]$ of the independent random matrices $[\mathbf{M}(\mathbf{x})]$, $[\mathbf{D}(\mathbf{x})]$, and $[\mathbf{K}(\mathbf{x})]$ with θ' in Θ' and with $\mathbf{x}$ fixed in $\mathcal{S}_n$, and for all fixed time t, the realization $\mathbf{Y}(\theta,\theta';t)$ of random variable $\mathbf{Y}(t)$ satisfies the following nonlinear differential equation:

$$\mathbf{Y}(\theta,\theta';t) = [\phi(\mathbf{X}(\theta))]\,\mathbf{Q}(\theta,\theta';t)\,, \tag{8.80}$$

$$[\mathbf{M}(\theta';\mathbf{X}(\theta))]\,\ddot{\mathbf{Q}}(\theta,\theta';t) + [\mathbf{D}(\theta';\mathbf{X}(\theta))]\,\dot{\mathbf{Q}}(\theta,\theta';t) + [\mathbf{K}(\theta';\mathbf{X}(\theta))]\,\mathbf{Q}(\theta,\theta';t)$$
$$+ \mathbf{F}_{\mathrm{NL}}(\mathbf{Q}(\theta,\theta';t), \dot{\mathbf{Q}}(\theta,\theta';t);\mathbf{X}(\theta)) = \mathbf{f}(t;\mathbf{X}(\theta))\,. \tag{8.81}$$

□ *Construction of the prior stochastic model of the model-parameter uncertainties.* The prior stochastic model of random vector $\mathbf{X}$ is constructed by using the Max-Ent principle (Section 5.3) or the PCE approach (Section 5.5). For simplifying the presentation, let us assume that the MaxEnt is used. The pdf $\mathbf{x} \mapsto p_{\mathbf{X}}(\mathbf{x}\,;\mathbf{s}_{\mathrm{par}})$ of $\mathbf{X}$ has a support that is the subset $\mathcal{S}_n$ of $\mathbb{R}^n$, and depends on a hyperparameter $\mathbf{s}_{\mathrm{par}} = (\underline{\mathbf{x}}, \boldsymbol{\delta}_X)$ that belongs to an admissible set $\mathcal{C}_{\mathrm{ad}} = \mathcal{S}_n \times \mathcal{C}_X$, which is a subset of $\mathbb{R}^\mu = \mathbb{R}^n \times \mathbb{R}^{\mu_X}$. The mean values of the random matrices $[M(\mathbf{X})]$, $[D(\mathbf{X})]$, and $[K(\mathbf{X})]$ are such that

$$[M_{\mathrm{mean}}] = E\{[M(\mathbf{X})]\}\,, \quad [D_{\mathrm{mean}}] = E\{[D(\mathbf{X})]\}\,, \quad [K_{\mathrm{mean}}] = E\{[K(\mathbf{X})]\}\,, \tag{8.82}$$

in which $[M_{\mathrm{mean}}]$, $[D_{\mathrm{mean}}]$, and $[K_{\mathrm{mean}}]$ are matrices in $\mathbb{M}_N^+(\mathbb{R})$, which are, in general, different from the nominal values $[\underline{M}] = [M(\underline{\mathbf{x}})]$, $[\underline{D}] = [D(\underline{\mathbf{x}})]$, and $[\underline{K}] = [K(\underline{\mathbf{x}})]$.

□ *Construction of the prior stochastic model of model uncertainties.* For all $\mathbf{x}$ fixed in $\mathcal{S}_n \subset \mathbb{R}^n$, each random matrix $[\mathbf{A}]$, in which A represents M, D or K, is taken in ensemble $\mathrm{SE}_\varepsilon^+$ (see Section 5.4.8.2), and is then written as

$$[\mathbf{A}(\mathbf{x})] = [L_A(\mathbf{x})]^T [\mathbf{G}_A]\,[L_A(\mathbf{x})] \quad , \quad [\mathbf{G}_A] = \frac{1}{1+\varepsilon}([\mathbf{G}_{0,A}] + \varepsilon\,[I_N])\,, \tag{8.83}$$

in which the random matrix $[\mathbf{G}_{0,A}]$ defined on $(\Theta', \mathcal{T}', \mathcal{P}')$ belongs to ensemble SG_0^+ (see Section 5.4.7.1), where $[L_A(\mathbf{x})]$ is the upper triangular matrix such that

$$[A(\mathbf{x})] = [L_A(\mathbf{x})]^T [L_A(\mathbf{x})] \in \mathbb{M}_N^+(\mathbb{R}), \tag{8.84}$$

and where $\varepsilon > 0$. The level of uncertainties for random matrix $[\mathbf{A}(\mathbf{x})]$ is controlled by the coefficient of variation of the random matrix $[\mathbf{G}_{0,A}]$, denoted by δ_A that is assumed to be independent of $\mathbf{x}$. The generator of realizations of random matrix $[\mathbf{G}_{0,A}]$ is detailed in Section 5.4.7.1.

□ *Prior stochastic model of uncertainties constructed with the generalized approach.* It consists of the mutually dependent random matrices $[\mathbf{M}(\mathbf{X})]$, $[\mathbf{D}(\mathbf{X})]$, and $[\mathbf{K}(\mathbf{X})]$, defined on the probability space $(\Theta \times \Theta', \mathcal{T} \otimes \mathcal{T}', \mathcal{P} \otimes \mathcal{P}')$, and which are constructed as

$$[\mathbf{M}(\mathbf{X})] = [L_M(\mathbf{X})]^T [\mathbf{G}_M] [L_M(\mathbf{X})], \tag{8.85}$$
$$[\mathbf{D}(\mathbf{X})] = [L_D(\mathbf{X})]^T [\mathbf{G}_D] [L_D(\mathbf{X})], \tag{8.86}$$
$$[\mathbf{K}(\mathbf{X})] = [L_K(\mathbf{X})]^T [\mathbf{G}_K] [L_K(\mathbf{X})]. \tag{8.87}$$

The level of uncertainties is controlled by the hyperparameter $\boldsymbol{\delta}_X \in \mathcal{C}_X \subset \mathbb{R}^{\mu_X}$ for the model-parameter uncertainties and by the hyperparameter $\boldsymbol{\delta}_G = (\delta_M, \delta_D, \delta_K) \in \mathcal{C}_G \subset \mathbb{R}^3$ for the model uncertainties.

8.4.4 Estimation of the Parameters of the Prior Stochastic Model of Uncertainties

□ *First case: no data are available.* The prior stochastic model of uncertainties depends on $\underline{\mathbf{x}}$, $\boldsymbol{\delta}_X$, and $\boldsymbol{\delta}_G$ that belong to the admissible sets $\mathcal{S}_n$, $\mathcal{C}_X$, and $\mathcal{C}_G$. If no data are available, then $\underline{\mathbf{x}}$ is fixed to a nominal value $\mathbf{x}_0$. The hyperparameters $\boldsymbol{\delta}_X$ and $\boldsymbol{\delta}_G$ must be used for performing a sensitivity analysis of the stochastic solution in order to analyze the robustness of the uncertain computational model with respect to the level of uncertainties that is controlled by $\boldsymbol{\delta}_X$ and $\boldsymbol{\delta}_G$.

□ *Second case: data are available.* If data are available, $\underline{\mathbf{x}}$ can be updated. The hyperparameters $\boldsymbol{\delta}_X$ and $\boldsymbol{\delta}_G$ can be estimated using the data $\mathbf{u}^{\exp,1}, \ldots, \mathbf{u}^{\exp,\nu}$ related to the observation vector $\mathbf{U}$, that is with values in $\mathbb{R}^m$, that is independent of t, and that depends on $\{\mathbf{Y}(t), t \in \tau\}$, which is constructed with the SROM. For all $\mathbf{s} = (\underline{\mathbf{x}}, \boldsymbol{\delta}_X, \boldsymbol{\delta}_G) \in \mathcal{C}_{\mathrm{ad}} = \mathcal{S}_n \times \mathcal{C}_X \times \mathcal{C}_G$, the pdf of $\mathbf{U}$ is denoted by $p_{\mathbf{U}}(\mathbf{u}; \mathbf{s})$. The optimal value $\mathbf{s}^{\mathrm{opt}}$ of $\mathbf{s}$ can be estimated by using the maximum likelihood method (see Section 7.3.3),

$$\mathbf{s}^{\mathrm{opt}} = \arg \max_{\mathbf{s} \in \mathcal{C}_{\mathrm{ad}}} \sum_{\ell=1}^{\nu} \ln\{p_{\mathbf{U}}(\mathbf{u}^{\exp,\ell}; \mathbf{s})\}. \tag{8.88}$$

For each ℓ, the value $p_{\mathbf{U}}(\mathbf{u}^{\text{exp},\ell}\,;\mathbf{s})$ of the pdf is estimated by using the multivariate kernel density estimation method (see Section 7.2) for which ν_s independent realizations $\mathbf{u}^{r,1},\ldots,\mathbf{u}^{r,\nu_s}$ of $\mathbf{U}$ are computed with the SROM by using a stochastic solver (see Chapter 6).

8.4.5 Posterior Stochastic Model of Model-Parameter Uncertainties in Presence of the Prior Stochastic Model of the Modeling Errors, Using the Bayesian Method

Let $p_{\mathbf{X}}^{\text{prior}}(\mathbf{x}) = p_{\mathbf{X}}(\mathbf{x}\,;\underline{\mathbf{x}}^{\text{opt}},\delta_X^{\text{opt}})$ and $p_{\mathbf{G}}([G_M],[G_D],[G_K]\,;\delta_G^{\text{opt}})$ be the optimal prior probability density functions of $\mathbf{X}$, and of $[\mathbf{G}_M],[\mathbf{G}_D],[\mathbf{G}_K]$.

□ *Objective of the estimation.* For given $p_{\mathbf{G}}([G_M],[G_D],[G_K]\,;\delta_G^{\text{opt}})$ (the optimal prior stochastic model of model uncertainties induced by the modeling errors), the posterior probability density function $p_{\mathbf{X}}^{\text{post}}(\mathbf{x})$ of the uncertain model parameter $\mathbf{X}$ is estimated by using the Bayesian method with the available data (see Section 7.3.4). This means that we construct a posterior stochastic model of the model-parameter uncertainties in presence of the prior stochastic model of modeling errors.

□ *Posterior probability density function of* $\mathbf{X}$. The notations introduced in Section 7.3.4 are used. Let $\mathbf{u} \mapsto p_{\mathbf{U}^{\text{out}}|\mathbf{X}}(\mathbf{u}|\mathbf{x})$ be the conditional probability density function of the random output $\mathbf{U}^{\text{out}}$ given $\mathbf{X} = \mathbf{x}$ in $\mathcal{S}_n$, which depends on noise $\mathbf{B}$ and on the random observation $\mathbf{U}$ computed with the SROM. Then

$$p_{\mathbf{X}}^{\text{post}}(\mathbf{x}) = c_n\,L(\mathbf{x})\,p_{\mathbf{X}}^{\text{prior}}(\mathbf{x})\,, \tag{8.89}$$

in which $L(\mathbf{x})$ is the likelihood function (without including the constant of normalization) defined on $\mathcal{S}_n \subset \mathbb{R}^n$, with values in $\mathbb{R}^+$, such that

$$L(\mathbf{x}) = \prod_{\ell=1}^{\nu} p_{\mathbf{U}^{\text{out}}|\mathbf{X}}(\mathbf{u}^{\text{exp},\ell}|\mathbf{x})\,, \tag{8.90}$$

and where c_n is the constant of normalization. The calculation must be done as explained in Section 7.3.4. The generator of realizations associated with $p_{\mathbf{X}}^{\text{post}}$ must be constructed using the MCMC methods (see Chapter 4). The conditional pdf $p_{\mathbf{U}^{\text{out}}|\mathbf{X}}$ can be estimated with the multivariate kernel density method (see Section 7.2) by using the realizations that are constructed with the stochastic computational model and using a Monte Carlo stochastic solver (see Section 6.4) .

8.4.6 Simple Example Showing the Capability of the Generalized Probabilistic Approach to Take into Account the Model Uncertainties in Structural Dynamics

This section deals with a simple example that shows the capability of the generalized probabilistic approach to take into account model uncertainties in structural dynamics [197]. The mechanical system presented in Section 8.2.6 is revisited.

The mean computational model is based on the Euler beam theory for which the first eigenvalue is written as $\lambda_1 = a_1 \, x$ in which x is the uncertain parameter and a_1 is a given positive constant. The prior stochastic model of X is constructed by using the MaxEnt principle (see Section 5.3) that yields a Gamma probability distribution. The identification of the hyperparameter δ_X is performed by using the maximum likelihood method (See Section 7.3.3) for which the observation is the lowest random eigenvalue Λ_1 and yields $\delta_X^{\text{opt}} = 0.093$. For the stochastic model of model uncertainties, the hyperparameters δ_M and δ_K are estimated using the maximum

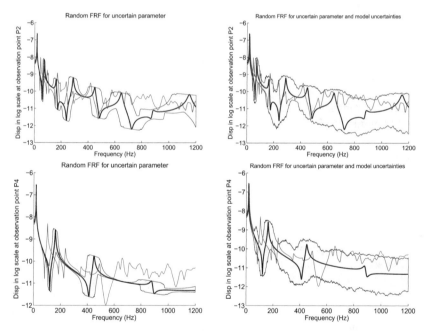

Fig. 8.18 Horizontal axis: frequency (Hz). Vertical axis: $\log_{10}$ of the modulus of the frequency response in displacement (m). Results for the observation point located at $\zeta_1 = 3.125 \, m$ (top figures) and at $\zeta_1 = 5.000 \, m$ (down figures). Frequency response of the simulated experiment (blue thin line). Frequency response calculated with the mean computational model (black thick line). Confidence region corresponding to a probability level $P_c = 0.98$ computed with the SROM (region in yellow color or in gray) and, with the model-parameter uncertainties but without the modeling errors (left figures), with both the model-parameter uncertainties and the model uncertainties (right figures). [Figure from [197]].

likelihood method for which the observation is the frequency response functions and yields $\delta_M^{\mathrm{opt}} = 0.9$ and $\delta_K^{\mathrm{opt}} = 0.15$. Figure 8.18 displays the frequency response function of the transversal displacement corresponding to the simulated experiment (the reference) and calculated with the mean computational model. Figure 8.18 also shows the confidence region corresponding to a probability level $P_c = 0.98$, which is estimated by using the SROM with two different stochastic modelings. The first one corresponds to the model-parameter uncertainties but without the modeling errors (left figures). The second one corresponds to both the model-parameter uncertainties and the model uncertainties (right figures). These results demonstrate the capability of the generalized probabilistic approach for correctly representing the uncertainties.

8.4.7 Simple Example Showing the Capability of the Generalized Probabilistic Approach Using Bayes' Method for Updating the Stochastic Model of an Uncertain Model Parameter in Presence of Model Uncertainties

The following simple example [205] shows the capability of the generalized probabilistic approach to take into account uncertainties that occur at two different scales (for instance, for vibrations in the low- and in the medium- frequency ranges) in a computational model by using the Bayes method. This method allows for updating the stochastic model of an uncertain model parameter whose statistical fluctuations induce significant effects for the first scale (for instance, for the low-frequency band) and in presence of model uncertainties induced by the modeling errors for which the statistical fluctuations induce significant effects for the second scale (for instance, for the medium-frequency band). This example is chosen in the field of the linear structural dynamics for a frequency analysis performed in the low- and in the medium-frequency ranges.

□ *Designed system and simulated experiments.* The designed system is a bounded 3D medium whose geometry is a slender cylinder with length $0.01\ m$, with a rectangular section $0.001\ m \times 0.002\ m$. The medium is made of a homogeneous and isotropic elastic material for which the Young modulus is $10^{10}\ N/m^2$, the Poisson coefficient is 0.15, and the mass density is $1\,500\ kg/m^3$. A damping term is added and is described by a critical damping rate of 0.01 for each elastic mode used for constructing the ROB. The frequency band of analysis is $\mathcal{B} =]0\,, 1.2 \times 10^6]\ Hz$. The external load is a point load whose Fourier transform is flat over $\mathcal{B}$.

The available data are made up of 5 simulated experiments that have been constructed using a stochastic 3D computational model made up of 8-nodes solid elements yielding $69\,867$ DOFs, for which random parameters have been introduced for generating a variability in the simulated experiments. Figure 8.19 displays the modulus of the "experimental" frequency response function (FRF) in $\log_{10}$ scale

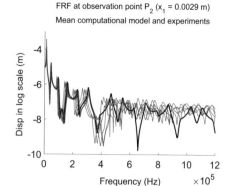

Fig. 8.19 Modulus of the FRF in $\log_{10}$ scale at a given observation point: simulated experiments (5 blue thin lines) and calculated with the mean computational model (black thick line). [Figure from [205]].

over band $\mathcal{B}$ for the transversal displacement at a given observation point. In analyzing the FRF shown in this figure, two regimes of vibration can be identified. On the low-frequency band LFB $=]0 , 3.6 \times 10^5]$ Hz, there are isolated resonances that have a small variability. On the medium-frequency band MFB $=]3.6 \times 10^5 , 1.2 \times 10^6]$ Hz, there are many resonances that have an important variability.

☐ *Mean computational model prediction and experimental comparison.* The mean computational model is made up of a damped homogeneous Timoshenko elastic beam with Young modulus $\underline{y} = 10^{10}$ N/m^2 and shear deformation factor $\underline{f}_s = 3$. Figure 8.19 compares the simulated experiments with the prediction obtained with the mean computational model for the FRF of the transversal displacement at a given observation point over the frequency band $\mathcal{B} = $ LFB $\cup$ MFB. The prediction is acceptable for the first resonances in the low-frequency band LFB, but there are significant differences due to the uncertainties induced by the modeling errors in the medium-frequency band MFB. The deterministic mean computational model does not have the capability to represent the variabilities of the experiments in the medium-frequency band MFB.

☐ *Identification of the optimal prior stochastic models of uncertainties.* A SROM is developed from the ROM that is constructed by the projection of the mean computational model with the ROB constituted of the elastic modes of the mean computational model. The uncertain model parameter is $\mathbf{x} = (y, f_s)$ in which y is the Young modulus and where f_s is the shear deformation factor of the Timoshenko beam, for which the nominal value is $\underline{\mathbf{x}} = (\underline{y}, \underline{f}_s)$, is modeled by the random variable $\mathbf{X} = (Y, F_s)$ whose prior probability density functions are constructed using the MaxEnt principle, which yields independent Gamma random variables Y and F_s with pdf p_Y and p_{F_s}. The identification of the hyperparameters of the SROM is performed as follows by using the simulated experiments.

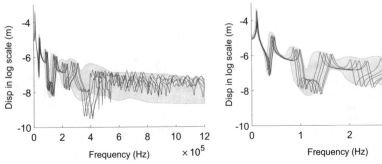

Fig. 8.20 Modulus of the FRF in $\log_{10}$ scale over band $\mathcal{B} = \text{LFB} \cup \text{MFB}$ (left figure) and over band LFB (right figure) for the transversal displacement at a given observation point: simulated experiments (5 blue lines) and confidence region with a probability level 0.95 (region in yellow color or in gray) predicted with the SROM for which the optimal prior stochastic model of uncertainties is used. [Figure from [205]]

Step 1: identification of the hyperparameters of the prior nonparametric stochastic model of the model uncertainties by using, as observations, the FRFs in the medium-frequency band MFB for the transversal displacements at 6 given observation points.

Step 2: in presence of the optimal prior nonparametric stochastic model of the model uncertainties identified by Step 1, identification of the hyperparameters of the prior parametric stochastic model of $\mathbf{X}$ using, as observations, the first six eigenfrequencies in the low-frequency band LFB.

□ *Response predicted with the optimal prior stochastic models of uncertainties.* Figure 8.20 displays the modulus of the FRF in $\log_{10}$ scale over band $\mathcal{B} = \text{LFB} \cup \text{MFB}$ (left figure) and over band LFB (right figure) for the transversal displacement at a given observation point. It can be seen the simulated experiments and the confidence region (corresponding to the probability level 0.95) that is predicted by using the SROM for which the optimal prior stochastic model of uncertainties (parametric and nonparametric) is used. The confidence region that is predicted is already not so bad, but such a prediction can be improved yet, as shown hereinafter.

□ *Identification of the posterior stochastic model of the model-parameter uncertainties using Bayes' method.* The Bayes method is used (without noise $\mathbf{B}$) for estimating the posterior probability density functions p_Y^{post} and $p_{F_s}^{\text{post}}$ of random variables Y and F_s, in presence of the optimal prior nonparametric stochastic model of model uncertainties. Figure 8.21 displays the prior pdf and the posterior pdf for the random Young modulus Y and for the random shear deformation factor F_s.

□ *Response predicted with the posterior stochastic model of model-parameter uncertainties and the optimal prior stochastic model of model uncertainties.* For the

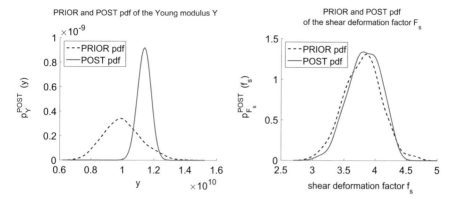

Fig. 8.21 Prior pdf (dashed line) and posterior pdf (solid line). Random Young modulus Y (left figure) and random shear deformation factor F_s (right figure). [Figure from [205]].

modulus of the FRF in $\log_{10}$ scale over band $\mathcal{B} = $ LFB $\cup$ MFB (left figure) and over band LFB (right figure) for the transversal displacement at a given observation point, Figure 8.22 displays the simulated experiments and the confidence region with a probability level 0.95 predicted by using the SROM for which the posterior stochastic model of $\mathbf{X}$ and the optimal prior stochastic model of the model uncertainties are used. The right figure shows that the confidence region in the low-frequency

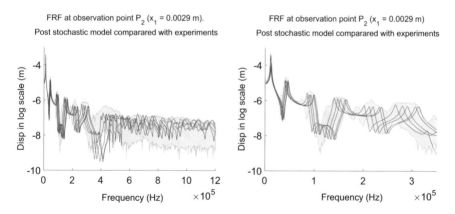

Fig. 8.22 Modulus of the FRF in $\log_{10}$ scale over band $\mathcal{B} = $ LFB $\cup$ MFB (left figure) and over band LFB (right figure) for the transversal displacement at a given observation point: simulated experiments (5 blue lines) and confidence region with a probability level 0.95 (region in yellow color or in gray) predicted with the SROM for which the posterior stochastic model of $\mathbf{X}$ and the optimal prior stochastic model of model uncertainties are used. [Figure from [205]].

band LFB has been improved while preserving the quality of the prediction in the medium-frequency band MFB.

8.5 Nonparametric Probabilistic Approach of Uncertainties in Computational Nonlinear Elastodynamics

The nonparametric probabilistic approach of uncertainties can be used to analyze the robustness of uncertain computational models in elastodynamics with geometric nonlinearities. Such an approach is particularly computationally attractive as it applies to reduced-order models of the nonlinear dynamical system. The nonparametric probabilistic approach of uncertainties, originally developed for linear operators, has been extended to the nonlinear operator of the 3D geometric nonlinear elasticity [139]. In this section, we present such a formulation and two applications that validate the theory proposed.

8.5.1 Boundary Value Problem for Elastodynamics with Geometric Nonlinearities

The elastic medium occupies a bounded domain Ω_0 in $\mathbb{R}^3$, corresponding to the undeformed configuration (natural state) for which the generic point is $\zeta = (\zeta_1, \zeta_2, \zeta_3)$. Its boundary is $\partial\Omega_0 = \partial\Omega_0^0 \cup \partial\Omega_0^1$ with external unit normal $\mathbf{n}^0$. The material is made up of a linear elastic material and is assumed to undergo large deformations induced by the geometric nonlinearities. In the reference configuration, the density is $\rho_0(\zeta) > 0$. A body-forces field $\mathbf{b}^0(\zeta)$ is applied in Ω_0 and a surface-forces field $\mathbf{g}^0(\zeta)$ is applied on $\partial\Omega_0^1$. Vectors $\mathbf{b}^0$ and $\mathbf{g}^0$ correspond to the transport to the undeformed (reference) configuration, of the body forces $\mathbf{b}$ and surface forces $\mathbf{g}$ applied to the deformed configuration. The convention of the summation over repeated Latin indices is used. For all time t, the $\mathbb{R}^3$-valued displacement field $\{\mathbf{u}(\zeta, t), \zeta \in \Omega_0\}$ is expressed in Ω_0 (total Lagrangian formulation) and satisfies the following boundary value problem,

$$\rho_0\,\ddot{u}_i - \frac{\partial}{\partial\zeta_k}(F_{ij}S_{jk}) = \rho_0\,b_i^0 \quad , \quad i = 1, 2, 3\,, \tag{8.91}$$

$$\mathbf{u} = 0 \quad \text{on} \quad \partial\Omega_0^0\,, \tag{8.92}$$

$$F_{ij}S_{jk}n_k^0 = g_i^0 \quad , \quad i = 1, 2, 3 \quad \text{on} \quad \partial\Omega_0^1\,, \tag{8.93}$$

on which F_{ij} is the deformation gradient tensor that is written as

$$F_{ij} = \partial(u_i + \zeta_i)/\partial\zeta_j = \delta_{ij} + \partial u_i/\partial\zeta_j\,, \tag{8.94}$$

and where S_{jk} is the second Piola-Kirchhoff stress tensor. The constitutive equation is written as

$$S_{jk} = C_{jki\ell} \, E_{i\ell} \, , \qquad (8.95)$$

in which $E_{i\ell}$ is the Green strain tensor such that

$$E_{i\ell} = \frac{1}{2}(F_{ki} \, F_{k\ell} - \delta_{i\ell}) \, , \qquad (8.96)$$

and where $C_{jki\ell}$ is the fourth order elasticity tensor satisfying the classical symmetry and positive definiteness properties.

8.5.2 Weak Formulation of the Boundary Value Problem

Let $\mathcal{C}_{\text{ad}}$ be the space of the admissible displacements defined by

$$\mathcal{C}_{\text{ad}} = \{\mathbf{v} \in \Omega_0 \, , \, \mathbf{v} \text{ sufficiently regular}, \, \mathbf{v} = 0 \text{ on } \partial\Omega_0^0\} \, , \qquad (8.97)$$

The weak formulation of the boundary value problem consists in finding the unknown displacement field $\boldsymbol{\zeta} \mapsto \mathbf{u}(\boldsymbol{\zeta}, t)$ in $\mathcal{C}_{\text{ad}}$ such that, for all time-independent field $\boldsymbol{\zeta} \mapsto \mathbf{v}(\boldsymbol{\zeta})$ in $\mathcal{C}_{\text{ad}}$,

$$\int_{\Omega_0} \rho_0 \, \ddot{u}_i v_i \, d\boldsymbol{\zeta} + \int_{\Omega_0} F_{ij} S_{jk} \frac{\partial v_i}{\partial \zeta_k} \, d\boldsymbol{\zeta} = \int_{\Omega_0} \rho_0 \, b_i^0 \, v_i \, d\boldsymbol{\zeta} + \int_{\partial\Omega_0^1} g_i^0 \, v_i \, ds \, . \quad (8.98)$$

8.5.3 Nonlinear Reduced-Order Model

□ *Equations of the nonlinear ROM.* The nonlinear ROM is deduced from Equation (8.98) in introducing an ROB $\{\boldsymbol{\phi}^1, \ldots, \boldsymbol{\phi}^N\}$ in $\mathcal{C}_{\text{ad}}$ (for instance, made up of the elastic modes of the underlying linear dynamical system). The approximation $\{\mathbf{u}^N(\boldsymbol{\zeta}, t), \, \boldsymbol{\zeta} \in \Omega_0\}$ of order N of the $\mathbb{R}^3$-valued displacement field $\{\mathbf{u}(\boldsymbol{\zeta}, t), \boldsymbol{\zeta} \in \Omega_0\}$ is then written as

$$\mathbf{u}^N(\boldsymbol{\zeta}, t) = \sum_{\alpha=1}^{N} q_\alpha(t) \, \boldsymbol{\phi}^\alpha(\boldsymbol{\zeta}) \, , \qquad (8.99)$$

in which the vector $\mathbf{q}(t) = (q_1(t), \ldots, q_N(t))$ of the generalized coordinates satisfies the nonlinear ROM,

$$[M] \, \ddot{\mathbf{q}}(t) + [D] \, \dot{\mathbf{q}}(t) + \mathbf{k}_{\text{NL}}(\mathbf{q}(t)) = \mathbf{f}(t) \, , \qquad (8.100)$$

$$\{\mathbf{k}_{\text{NL}}(\mathbf{q})\}_\alpha = \sum_{\beta=1}^{N} K_{\alpha\beta}^{(1)} \, q_\beta + \sum_{\beta,\gamma=1}^{N} K_{\alpha\beta\gamma}^{(2)} \, q_\beta q_\gamma + \sum_{\beta,\gamma,\delta=1}^{N} K_{\alpha\beta\gamma\delta}^{(3)} \, q_\beta q_\gamma q_\delta \, , \quad (8.101)$$

where $[D]\,\dot{\mathbf{q}}(t)$ is an additional damping term. The stress vector $\mathbf{s}(t)$ is written as

$$\mathbf{s}(t) = \mathbf{s}^0 + \sum_{\alpha=1}^{N} \mathbf{s}^{(\alpha)} q_\alpha(t) + \sum_{\alpha,\beta=1}^{N} \mathbf{s}^{(\alpha,\beta)} q_\alpha(t)\, q_\beta(t) . \tag{8.102}$$

The coefficients of the ROM are the second-order tensor $K_{\alpha\beta}^{(1)}$, the third-order tensor $K_{\alpha\beta\gamma}^{(2)}$, and the fourth-order tensor $K_{\alpha\beta\gamma\delta}^{(3)}$, which are defined by

$$K_{\alpha\beta}^{(1)} = \int_{\Omega_0} C_{jk\ell m}(\boldsymbol{\zeta}) \frac{\partial \varphi_j^\alpha(\boldsymbol{\zeta})}{\partial \zeta_k} \frac{\partial \varphi_\ell^\beta(\boldsymbol{\zeta})}{\partial \zeta_m}\, d\boldsymbol{\zeta} , \tag{8.103}$$

$$K_{\alpha\beta\gamma}^{(2)} = \frac{1}{2}\left(\widehat{K}_{\alpha\beta\gamma}^{(2)} + \widehat{K}_{\beta\gamma\alpha}^{(2)} + \widehat{K}_{\gamma\alpha\beta}^{(2)} \right) , \tag{8.104}$$

$$\widehat{K}_{\alpha\beta\gamma}^{(2)} = \int_{\Omega_0} C_{jk\ell m}(\boldsymbol{\zeta}) \frac{\partial \varphi_j^\alpha(\boldsymbol{\zeta})}{\partial \zeta_k} \frac{\partial \varphi_s^\beta(\boldsymbol{\zeta})}{\partial \zeta_\ell} \frac{\partial \varphi_s^\gamma(\boldsymbol{\zeta})}{\partial \zeta_m}\, d\boldsymbol{\zeta} , \tag{8.105}$$

$$K_{\alpha\beta\gamma\delta}^{(3)} = \int_{\Omega_0} C_{jk\ell m}(\boldsymbol{\zeta}) \frac{\partial \varphi_r^\alpha(\boldsymbol{\zeta})}{\partial \zeta_j} \frac{\partial \varphi_r^\beta(\boldsymbol{\zeta})}{\partial \zeta_k} \frac{\partial \varphi_s^\gamma(\boldsymbol{\zeta})}{\partial \zeta_\ell} \frac{\partial \varphi_s^\delta(\boldsymbol{\zeta})}{\partial \zeta_m}\, d\boldsymbol{\zeta} . \tag{8.106}$$

The following properties of the coefficients of the nonlinear ROM can be deduced from the symmetry properties of the fourth-order elasticity tensor $C_{jk\ell m}(\boldsymbol{\zeta})$,

$$K_{\alpha\beta}^{(1)} = K_{\beta\alpha}^{(1)} , \tag{8.107}$$

$$\widehat{K}_{\alpha\beta\gamma}^{(2)} = \widehat{K}_{\alpha\gamma\beta}^{(2)} , \tag{8.108}$$

$$K_{\alpha\beta\gamma}^{(2)} = K_{\beta\gamma\alpha}^{(2)} = K_{\gamma\alpha\beta}^{(2)} , \tag{8.109}$$

$$K_{\alpha\beta\gamma\delta}^{(3)} = K_{\alpha\beta\delta\gamma}^{(3)} = K_{\beta\alpha\gamma\delta}^{(3)} = K_{\gamma\delta\alpha\beta}^{(3)} . \tag{8.110}$$

In addition, the positiveness property of the fourth-order elasticity tensor $C_{jk\ell m}(\boldsymbol{\zeta})$ implies that $K_{\alpha\beta}^{(1)}$ and $K_{\alpha\beta\gamma\delta}^{(3)}$ are positive definite.

□ *Computation of the coefficients of the nonlinear ROM.*

1. *Direct computation.* The coefficients of the nonlinear ROM are directly computed by using a finite element code. Such a method requires the development of a "home software," but, in counter part, is very efficient in terms of CPU time (see, for instance, [34, 35]).
2. *Indirect computation.* The coefficients of the nonlinear ROM are computed using a commercial software that is assumed to be a black box. Consequently, such a method does not require the development of a "home software," but, in counter part, can be expensive in CPU time and in core memory (see, for instance, [117, 139]). A procedure has been proposed for identifying, with any commercial finite element code devoted to structural dynamics, the coefficients of the nonlinear ROM for any complex dynamical system using the finite ele-

ment model in nonlinear elastostatic with large displacements. Such a procedure consists

- in imposing an adapted series of static deflections and in determining with the finite element model the forces required and the stresses.
- in identifying the coefficients of the reduced-order model by solving a linear system of equations.

One such technique is the STEP method (STiffness Evaluation Procedure) that was initially conceived by [144] and later improved by M.P. Mignolet [117, 142].

□ *Algebraic properties of the nonlinear stiffness.* In order to implement the nonparametric probabilistic approach of uncertainties in the nonlinear ROM, we need to introduce algebraic properties related to the nonlinear stiffness. The tensor $\widehat{K}^{(2)}_{\alpha\beta\gamma}$ is reshaped into an $N \times N^2$ matrix and the tensor $K^{(3)}_{\alpha\beta\gamma\delta}$ is reshaped into an $N^2 \times N^2$ matrix such that

$$[\widetilde{K}^{(2)}]_{\alpha J} = \widehat{K}^{(2)}_{\alpha\beta\gamma} \quad , \quad J = N(\beta - 1) + \gamma \,,$$

$$[\widetilde{K}^{(3)}]_{IJ} = K^{(3)}_{\alpha\beta\gamma\delta} \quad , \quad I = N(\alpha - 1) + \beta \quad , \quad J = N(\gamma - 1) + \delta \,.$$

Let $[K_B]$ be the $\nu \times \nu$ matrix with $\nu = N + N^2$ such that

$$[K_B] = \begin{bmatrix} [K^{(1)}] & [\widetilde{K}^{(2)}] \\ [\widetilde{K}^{(2)}]^T & 2[\widetilde{K}^{(3)}] \end{bmatrix} . \tag{8.111}$$

It can then be proven the following fundamental algebraic properties: the matrix $[K_B]$ is symmetric positive definite. This property is strongly used for constructing the SROM that we present hereinafter.

8.5.4 Stochastic Reduced-Order Model of the Nonlinear Dynamical System Using the Nonparametric Probabilistic Approach of Uncertainties

The SROM is deduced from Equations (8.99) to (8.102) by substituting the deterministic matrices by random matrices chosen in the adapted ensembles of random matrices and is written as

$$\mathbf{U}(\zeta, t) = \sum_{\alpha=1}^{N} Q_\alpha(t) \, \phi^\alpha(\zeta) \,, \tag{8.112}$$

$$[\mathbf{M}] \, \ddot{\mathbf{Q}}(t) + [\mathbf{D}] \, \dot{\mathbf{Q}}(t) + \mathbf{K}_{\mathrm{NL}}(\mathbf{Q}(t)) = \mathbf{f}(t) \,, \tag{8.113}$$

$$\{\mathbf{K}_{\mathrm{NL}}(\mathbf{q})\}_{\alpha} = \sum_{\beta=1}^{N}[\mathbf{K}^{(1)}]_{\alpha\beta}\,q_{\beta} + \sum_{\beta,\gamma=1}^{N}\mathbf{K}^{(2)}_{\alpha\beta\gamma}\,q_{\beta}q_{\gamma} + \sum_{\beta,\gamma,\delta=1}^{N}\mathbf{K}^{(3)}_{\alpha\beta\gamma\delta}\,q_{\beta}q_{\gamma}q_{\delta}\,, \quad (8.114)$$

$$\mathbf{S}(t) = \mathbf{s}^{0} + \Sigma_{\alpha=1}^{N}\mathbf{s}^{(\alpha)}Q_{\alpha}(t) + \Sigma_{\alpha,\beta=1}^{N}\mathbf{s}^{(\alpha,\beta)}\,Q_{\alpha}(t)\,Q_{\beta}(t)\,. \quad (8.115)$$

$$\begin{bmatrix} [\mathbf{K}^{(1)}] & [\widetilde{\mathbf{K}}^{(2)}] \\ [\widetilde{\mathbf{K}}^{(2)}]^{T} & 2[\widetilde{\mathbf{K}}^{(3)}] \end{bmatrix} = [\mathbf{K}_{B}]\,. \quad (8.116)$$

1. Each symmetric positive-definite random matrix $[\mathbf{M}]$, $[\mathbf{D}]$, and $[\mathbf{K}_{B}]$ is constructed as independent random matrices in the ensemble SE_{0}^{+} or $\mathrm{SE}_{\varepsilon}^{+}$ (see Sections 5.4.8.1 and 5.4.8.2).
2. Dependent random matrices $[\mathbf{K}^{(1)}]$, $[\widetilde{\mathbf{K}}^{(2)}]$ and $[\widetilde{\mathbf{K}}^{(3)}]$ are deduced from $[\mathbf{K}_{B}]$.
3. Dependent random tensors $\mathbf{K}^{(2)}_{\alpha\beta\gamma}$ and $\mathbf{K}^{(3)}_{\alpha\beta\gamma\delta}$ are deduced from $[\widetilde{\mathbf{K}}^{(2)}]$ and $[\widetilde{\mathbf{K}}^{(3)}]$.
4. The hyperparameters δ_{M}, δ_{D}, and δ_{K} of random matrices $[\mathbf{M}]$, $[\mathbf{D}]$, and $[\mathbf{K}_{B}]$ allow for controlling the level of uncertainties.

8.5.5 Simple Example Illustrating the Capability of the Approach Proposed

The following example [139] allows for illustrating the use of the theory proposed. We consider a dynamical system that is made up of a steel beam with $0.2286\,m$ length, $0.0127\,m$ width, and $0.000775\,m$ thick, with a clamped/clamped boundary condition. The computational model is performed with 40 CBEAM elements in MSC-NASTRAN software. The deterministic excitation is a single time-dependent concentrated force applied to the center of the beam in the direction perpendicular to the beam axis in the undeformed configuration, for which its Fourier transform

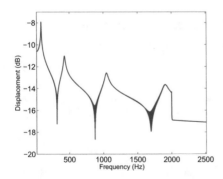 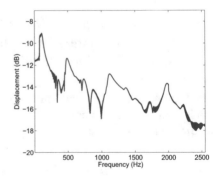

Fig. 8.23 Transverse displacement of the middle of the beam as a function of frequency computed with the linear mean computational model (left figure) and computed with the nonlinear mean computational model (right figure). [Figure from [139]].

is flat over the frequency band of analysis, $[0, 2000]$ Hz. Figure 8.23 displays the frequency response for the transverse displacement of the middle of the beam, which is computed with the linear mean computational model and with the nonlinear one. It should be noted a significant reduction in the peak response (by approximately 40 %) for the nonlinear response as compared to the response of the linear system and also some important differences of the response in the frequency domain.

The first eigenfrequencies of the transverse linear modes and of the in-plane linear modes for the dynamical system in the linear regime, and the modal damping rate in percent that are used for constructing the SROM, are given in Table 8.1. The ROB is made up of 12 in-plane linear modes and of 10 transverse linear modes for which the deterministic response of the nonlinear ROM is obtained. For constructing the nonlinear SROM, the level of uncertainties has been identified in order to obtain a 4 % mean-square variation of the first random eigenfrequency of the uncertain system around its corresponding value for the mean computational model. The trans-

Table 8.1 First eigenfrequencies in Hz for the transverse modes and for the in-plane modes for the dynamical system in the linear regime and modal damping rate in percent used in the ROM

Transverse eigenfreq	In-plane eigenfreq	Damping rate
79	11 168	2.69
218	22 353	1.22
427	33 573	1.03
706	44 844	1.20
1055	56 184	1.56
1473	67 611	2.05
1961	79 143	2.64
2518	90 795	2.64

verse displacement of the middle of the beam has been computed with the nonlinear SROM as a function of frequency. Figure 8.24 displays the response computed with the nonlinear mean computational model and computed with the nonlinear SROM. The confidence region is calculated with a probability level 0.95 and is shown for two values of the amplitudes in the nonlinear regime of the responses. The 5th-95th percentile band of the confidence region increases with increasing frequency and is broader near the resonances and the antiresonances of the mean computational model.

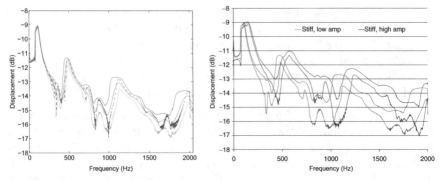

Fig. 8.24 Transverse displacement of the middle of the beam computed with the nonlinear model as a function of frequency. Left figure: response computed with the nonlinear mean computational model (dashed line). Response computed with the nonlinear SROM: mean value (green middle line close to the dashed line), lower envelope (light blue lower line), and upper envelope (red upper line) of the confidence region for a probability level 0.95. Right figure: confidence region for a probability level 0.95 computed with the nonlinear SROM, for the low amplitude (light blue lines or gray lines, the same as the one in the left figure) and for a high amplitude (dark blue lines or dark gray lines) defined as 2.25 times the low amplitude. [Figure from [139]].

8.5.6 Experimental Validation of the Theory Proposed by Using the Direct Computation

The following application [35] gives an experimental validation of the theory proposed, which is devoted to the nonlinear static analysis of the post-buckling of a thin cylindrical shell for which the geometry is uncertain (the experimental data are from [138] and the dynamical aspects can be found in [35]).

The geometry of the cylindrical shell is characterized by the mean radius $0.125\ m$, thickness $2.7 \times 10^{-4}\ m$, and height of $0.125\ m$. The shell is composed of nickel that is assumed to be linear elastic in the range of stresses considered. The measured Young modulus is $1.8 \times 10^{11}\ N/m^2$ and the measured Poisson coefficient is 0.3. The bottom of the cylindrical shell is clamped and the upper ring is rigid with three DOFs in translation. The process employed for the manufacturing of this thin shell cylinder yields an almost uniformity in the geometry but the behavior of the cylinder stays very sensitive to geometric imperfections for the post-buckling behavior. Concerning the experimental data, a constant traction load of $8\,500\ N$ is applied in the longitudinal direction in order to delay the onset of the post-buckling. An external load, defined as a static shear point load with maximum magnitude of $9\,750\ N$ is applied at the top of the shell along the transversal direction (this force F, in Newton, is expressed as a function of a load factor s such that $s = F/10\,000$). The observation is the displacement in the direction of the external load and at the point at which the external load is applied. Two different mechanical behaviors can be observed: an elastic domain until the critical external load of $7\,450\ N$ ($s = 0.7450$) and an approximately linear elastic behavior in the postbuckling domain.

Fig. 8.25 Horizontal axis: load factor $s = F/10\,000$ in which F is the force in Newton. Vertical axis: displacement at observation point (m). Experimental data (o markers). Observed displacement computed with the nonlinear mean computational model (thick dashed line). Confidence region with a probability level 0.95 computed with the nonlinear SROM for $\delta = 0.45$ (gray region). [Figure from [35]].

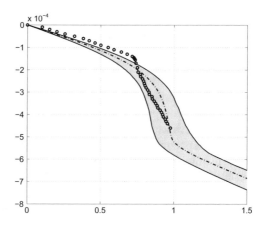

The finite element model has $4\,230\,003$ DOFs corresponding to a regular mesh composed of $712\,500$ solid finite elements with 8 nodes and with 8 Gauss integration points. The nonlinear ROM is constructed by using an ROB made up of $N = 27$ POD modes (proper orthogonal decomposition) for which the convergence is reached. The nonlinear SROM is constructed with the optimal value $\delta_{K_B}^{\mathrm{opt}} = 0.45$ of the hyperparameter for the pdf of the random matrix $[\mathbf{K}_B]$, which is identified with the experiments by solving a statistical inverse problem based on the Maximum likelihood method. For an applied external load $F = s \times 10\,000\,N$ with $s \in [0\,, 1.5]$, and for the observed displacement, Figure 8.25 displays the result given by the nonlinear mean computational model and displays the confidence region with a probability level 0.95 computed with the nonlinear SROM. This figure also shows the experimental measurements. Despite the slight underestimation provided by the nonlinear SROM in the linear range, the results are consistent with the experimental nonlinear response, what validates the nonlinear SROM that could still be improved by introducing uncertain boundary conditions as explained in Section 8.2.9.

8.6 A Nonparametric Probabilistic Approach for Quantifying Uncertainties in Low- and High-Dimensional Nonlinear Models

In Section 8.5, we have presented an efficient extension of the nonparametric probabilistic approach of uncertainties for a particular type of nonlinear operator encountered in the boundary value problems for elastodynamics with geometric nonlinearities. However, the strong properties that have been used for constructing the nonlinear SROM cannot generally be developed for general nonlinear boundary value problems.

We present a nonparametric probabilistic approach of the modeling errors, which can be used for any nonlinear high-dimensional computational model (HDM) for which a nonlinear ROM can be constructed from the nonlinear HDM. The details of theory can be found in [209]. The method presented concerns the model uncertainties induced by the modeling errors in computational sciences and engineering (such as in computational structural dynamics, fluid–structure interaction, vibroacoustics, etc.) for which a parametric nonlinear HDM is used for addressing optimization problems (such as a robust design optimization problem), which are solved by introducing a parametric nonlinear ROM that is constructed by using an adapted ROB derived from the parametric nonlinear HDM. The methodology proposed consists in substituting the deterministic ROB with a stochastic ROB (SROB) for which the probability distribution is constructed on a subset of a compact Stiefel manifold, which is introduced for representing some constraints that must be verified. The constructed nonlinear SROM depends on a small number of hyperparameters for which the identification is performed by solving a statistical inverse problem. An application is presented in nonlinear computational structural dynamics for validating this nonparametric probabilistic approach for nonlinear problems.

8.6.1 Problem to Be Solved and Approach Proposed

With more details than in Figure 1.1 of Section 1.4, Figure 8.26 summarizes the construction of a μ-parametric nonlinear ROM deduced from a μ-parametric nonlinear HDM, and shows the sources of uncertainties and of variabilities. As it can be seen in this figure, in addition to the usual presence of the variabilities in the real mechanical system (due to the manufacturing process and due to small differences in the configurations that differ from the designed mechanical system) and in presence of usual model-parameter uncertainties, there are two contributions in the modeling errors: (1) the HDM modeling errors that are introduced during the construction of the nonlinear high-dimensional computational model that does not exactly represent the real system and (2) the ROM modeling errors that are induced by the use of the nonlinear ROM instead of the nonlinear HDM.

□ *Usefulness of introducing a nonlinear ROM.* The time-dependent numerical simulations based on a μ-parametric nonlinear HDM remain so cost-prohibitive that it cannot be used as often as needed. For this reason, μ-parametric nonlinear ROM has emerged as a promising numerical tool for parametric applications such as design optimization and simulation-based decision making. Among numerous works devoted to the construction of a μ-parametric nonlinear ROM from a μ-parametric nonlinear HDM, we refer the reader to [4, 38, 39, 40, 60, 78, 147, 224].

□ *Construction of a μ-parametric nonlinear ROM.* The projection of the μ-parametric nonlinear HDM of dimension m onto a subspace of dimension $N \ll m$ is done by using an $(m \times N)$ reduced-order basis $[V]$ independent of μ, which yields the μ-parametric nonlinear ROM (note that we have used the notation m instead of m_{DOF} contrarily to the notation used in this chapter, because we need to simplify the reading of the mathematical symbols that we use hereinafter). The knowledge

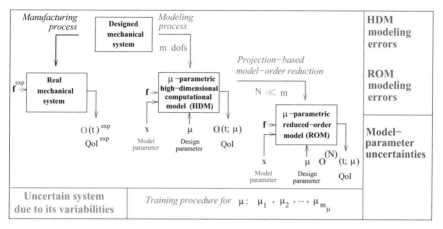

Fig. 8.26 Construction of a μ-parametric nonlinear ROM deduced from a μ-parametric nonlinear HDM and, sources of uncertainties and of variabilities.

about the HDM response is obtained during a training procedure: the design parameter vector μ is sampled at a few points using an effective sampling strategy [5, 79, 160]. A set of problems related to the HDM are solved to obtain a set of parametric solution snapshots that are compressed using, for example, the singular value decomposition to construct the ROB $[V]$ independent of μ.

□ *Computational feasibility and modeling errors.* Despite its low dimension, the resulting μ-parametric nonlinear ROM does not necessarily guarantee computational feasibility because the construction of the nonlinear ROM does not scale only with its size N, but also with that of the underlying μ-parametric nonlinear HDM, $m >> N$. A remedy consists in equipping the nonlinear ROM with a procedure for approximating the resulting reduced operators whose computational complexity scales only with the small size N of the ROM hyper-reduction method) [78, 79, 173, 91, 45]. It follows that a μ-parametric nonlinear ROM inherits the modeling errors induced by the construction of the ROM and also of the modeling errors introduced during the construction of the μ-parametric nonlinear HDM.

8.6.2 Nonparametric Probabilistic Approach for Taking into Account the Modeling Errors in a Nonlinear ROM

□ *Main objectives.* Independently of the type of nonlinearities, the objective is to take into account the modeling errors that are responsible for the distance between the predictions given by the μ-parametric nonlinear ROM and the available data. As previously explained, there are two sources of modeling errors,

- the modeling errors associated with the construction of the nonlinear ROM from the nonlinear HDM due to the training procedure on $\boldsymbol{\mu}$ for constructing an ROB that is independent of $\boldsymbol{\mu}$, and due to the choice of the reduction order N.
- the modeling errors introduced in the construction of the nonlinear HDM (with respect to the real dynamical system and its variabilities). If experimental data are available, then these uncertainties can be quantified.

□ *Methodology of the nonparametric probabilistic approach proposed.* The methodology consists in substituting the deterministic ROB represented by the $(m \times N)$ matrix $[V]$ with a SROB represented by an $(m \times N)$ random matrix $[\mathbf{W}]$. The probability distribution of the SROB $[\mathbf{W}]$ is constructed on a subset of a compact Stiefel manifold in order to preserve some important properties of the ROB. In addition, the construction is performed with a small number of hyperparameters in order that they can be identified by solving a statistical inverse problem.

□ *Example of a $\boldsymbol{\mu}$-parametric nonlinear HDM.* Let $\mathcal{C}_{\boldsymbol{\mu}}$ be the admissible set of the parameter $\boldsymbol{\mu}$, which is a subset of $\mathbb{R}^{m_{\boldsymbol{\mu}}}$. For $\boldsymbol{\mu}$ in $\mathcal{C}_{\boldsymbol{\mu}}$, the $\boldsymbol{\mu}$-parametric nonlinear HDM on $\mathbb{R}^m$ is written as

$$[M]\ddot{\mathbf{y}}(t) + \mathbf{g}(\mathbf{y}(t), \dot{\mathbf{y}}(t); \boldsymbol{\mu}) = \mathbf{f}(t; \boldsymbol{\mu}) \quad , \quad t > t_0 , \tag{8.117}$$

$$\mathbf{y}(t_0) = \mathbf{y}_0 \quad , \quad \dot{\mathbf{y}}(t_0) = \mathbf{y}_1 , \tag{8.118}$$

with $m_{\mathrm{CD}} < m$ constraint equations,

$$[B]^T \mathbf{y}(t) = \mathbf{0}_{m_{\mathrm{CD}}} \quad \text{with} \quad [B]^T[B] = [I_{m_{\mathrm{CD}}}] . \tag{8.119}$$

At time t, the quantity of interest (QoI) is a vector in $\mathbb{R}^{m_o}$, which is written as

$$\mathbf{o}(t; \boldsymbol{\mu}) = \mathbf{h}(\mathbf{y}(t; \boldsymbol{\mu}), \dot{\mathbf{y}}(t; \boldsymbol{\mu}), \mathbf{f}(t; \boldsymbol{\mu}), t; \boldsymbol{\mu}) . \tag{8.120}$$

□ *Construction of the associated $\boldsymbol{\mu}$-parametric nonlinear ROM.* The ROB $[V]$ in $\mathbb{M}_{m,N}$ (with $N \ll m$) is independent of $\boldsymbol{\mu}$ and satisfies

$$[V]^T[M][V] = [I_N] \quad , \quad [B]^T[V] = [0_{m_{\mathrm{CD}}, N}] . \tag{8.121}$$

For $\boldsymbol{\mu}$ in $\mathcal{C}_{\boldsymbol{\mu}}$, the $\boldsymbol{\mu}$-parametric nonlinear ROM on $\mathbb{R}^N$ is written as

$$\mathbf{y}^N(t) = [V]\mathbf{q}(t) , \tag{8.122}$$

$$\ddot{\mathbf{q}}(t) + [V]^T \mathbf{g}([V]\mathbf{q}(t), [V]\dot{\mathbf{q}}(t); \boldsymbol{\mu}) = [V]^T \mathbf{f}(t; \boldsymbol{\mu}) \quad , \quad t > t_0 , \tag{8.123}$$

$$\mathbf{q}(t_0) = [V]^T[M]\mathbf{y}_0 \quad , \quad \dot{\mathbf{q}}(t_0) = [V]^T[M]\mathbf{y}_1 , \tag{8.124}$$

$$\mathbf{o}^N(t; \boldsymbol{\mu}) = \mathbf{h}(\mathbf{y}^N(t; \boldsymbol{\mu}), \dot{\mathbf{y}}^N(t; \boldsymbol{\mu}), \mathbf{f}(t; \boldsymbol{\mu}), t; \boldsymbol{\mu}) . \tag{8.125}$$

□ *Construction of the associated μ-parametric nonlinear SROM.* The SROB is the $\mathbb{M}_{m,N}$-valued random matrix $[\mathbf{W}]$, independent of μ, which must satisfy (subset of a compact Stiefel manifold)

$$[\mathbf{W}]^T [M] [\mathbf{W}] = [I_N] \quad , \quad [B]^T [\mathbf{W}] = [0_{m_{\mathrm{CD}},N}] \quad , \quad \text{a.s.} \tag{8.126}$$

For N fixed and for $\mu \in \mathcal{C}_\mu$, the μ-parametric nonlinear SROM on $\mathbb{R}^N$ is written as

$$\mathbf{Y}(t) = [\mathbf{W}] \mathbf{Q}(t), \tag{8.127}$$

$$\ddot{\mathbf{Q}}(t) + [\mathbf{W}]^T \mathbf{g}([\mathbf{W}] \mathbf{Q}(t), [\mathbf{W}] \dot{\mathbf{Q}}(t); \mu) = [\mathbf{W}]^T \mathbf{f}(t; \mu) \quad , \quad t > t_0, \tag{8.128}$$

$$\mathbf{Q}(t_0) = [\mathbf{W}]^T [M] \mathbf{y}_0 \quad , \quad \dot{\mathbf{Q}}(t_0) = [\mathbf{W}]^T [M] \mathbf{y}_1, \tag{8.129}$$

$$\mathbf{O}(t; \mu) = \mathbf{h}(\mathbf{Y}(t; \mu), \dot{\mathbf{Y}}(t; \mu), \mathbf{f}(t; \mu), t; \mu). \tag{8.130}$$

□ *Identification of the hyperparameters of the probability distribution of $[\mathbf{W}]$.* The probability distribution of $[\mathbf{W}]$ depends on a hyperparameter α in $\mathcal{C}_\alpha \subset \mathbb{R}^{m_\alpha}$, which is constructed for that m_α has a small value in order that the identification of α be feasible by solving a statistical inverse problem. The identification of hyperparameter α is performed by solving the optimization problem,

$$\alpha^{\mathrm{opt}} = \min_{\alpha \in \mathcal{C}_\alpha} J(\alpha), \tag{8.131}$$

in which $J(\alpha)$ measures a distance between the stochastic QoI $\mathbf{O}$ constructed with the SROM and a corresponding target $\mathbf{o}^{\mathrm{target}}$. The target is chosen as $\mathbf{o}^{\mathrm{target}} = \mathbf{o}$ for taking into account the modeling errors induced by the use of the nonlinear ROM instead of the nonlinear HDM. The target must be chosen as $\mathbf{o}^{\mathrm{target}} = \mathbf{o}^{\mathrm{exp}}$ by using experimental data for taking into account the two types of the modeling errors, the one induced by the use of the nonlinear ROM instead of the nonlinear HDM and the other one introduced during the construction of the nonlinear HDM.

8.6.3 Construction of the Stochastic Model of the SROB on a Subset of a Compact Stiefel Manifold

□ *Definition of the compact Stiefel manifold $\mathbb{S}_{m,N}$ and of its subspace $\mathcal{S}_{m,N}$.* The compact Stiefel manifold $\mathbb{S}_{m,N}$ is defined by

$$\mathbb{S}_{m,N} = \{ [W] \in \mathbb{M}_{m,N} , [W]^T [M] [W] = [I_N] \} \subset \mathbb{M}_{m,N}. \tag{8.132}$$

The subspace $\mathcal{S}_{m,N} \subset \mathbb{S}_{m,N}$, which is associated with the constraint $[B]^T [W] = [0_{m_{\mathrm{CD}},N}]$ with $[B]^T [B] = [I_{m_{\mathrm{CD}}}]$, is defined by

$$\mathcal{S}_{m,N} = \{ [W] \in \mathbb{S}_{m,N} , [B]^T [W] = [0_{m_{\mathrm{CD}},N}] \}. \tag{8.133}$$

□ *Parameterization of* $\mathcal{S}_{m,N}$ *adapted to the high-dimension.* For $[V]$ given in $\mathcal{S}_{m,N}$, a nonclassical parameterization of $\mathcal{S}_{m,N}$ has been constructed as the mapping $\mathcal{R}_{s,V}$ from $\mathbb{M}_{m,N}$ into $\mathcal{S}_{m,N}$,

$$[U] \mapsto [W] = \mathcal{R}_{s,V}([U]) , \tag{8.134}$$

which depends on a parameter s that controls the distance of $[W]$ to $[V]$, which satisfies the desired property

$$[V] = \mathcal{R}_{s,V}([0_{m,N}]) \in \mathcal{S}_{m,N} , \tag{8.135}$$

and which avoids the usual construction of a big matrix in $\mathbb{M}_{m,m-N}$. This construction is based on the use of the polar decomposition of the mapping that maps the tangent vector space $T_V\mathbb{S}_{m,N}$ of $\mathbb{S}_{m,N}$ at the given point $[V]$, into $\mathbb{S}_{m,N}$.

8.6.4 Construction of a Stochastic Reduced-Order Basis (SROB)

□ *Stochastic construction that must be done.* Since the ROB $[V]$ is given in $\mathcal{S}_{m,N} \subset \mathbb{S}_{m,N}$, we have $[V]^T[M][V] = [I_N]$ and $[B]^T[V] = [0_{m_{\text{CD}},N}]$. Consequently, the associated SROB is the random matrix $[\mathbf{W}]$, defined on a probability space $(\Theta, \mathcal{T}, \mathcal{P})$, with values in $\mathcal{S}_{m,N} \subset \mathbb{S}_{m,N}$, that is to say, must be such that,

$$[\mathbf{W}]^T[M][\mathbf{W}] = [I_N] \quad , \quad [B]^T[\mathbf{W}] = [0_{m_{\text{CD}},N}] \qquad \text{a.s.} \tag{8.136}$$

We then have to construct the probability distribution $P_{[\mathbf{W}]}$ on $\mathbb{M}_{m,N}$ of the random matrix $[\mathbf{W}]$ for which its support is the manifold $\mathcal{S}_{m,N}$,

$$\text{supp}\, P_{[\mathbf{W}]} = \mathcal{S}_{m,N} \subset \mathbb{S}_{m,N} \subset \mathbb{M}_{m,N} , \tag{8.137}$$

and which must be such that, if the statistical fluctuations of $[\mathbf{W}]$ goes to zero, then $[\mathbf{W}]$ goes to $[V]$ in probability distribution.

□ *Stochastic representation of random matrix* $[\mathbf{W}]$. The details of the construction are given in [209]. The random matrix $[\mathbf{W}]$ is a second-order non-Gaussian and not centered random matrix with values in the manifold $\mathcal{S}_{m,N} \subset \mathbb{S}_{m,N} \subset \mathbb{M}_{m,N}$, which is written as

$$[\mathbf{W}] = R_{s,V}([\mathbf{Z}]) = ([V] + s\,[\mathbf{Z}])\,[H_s(\mathbf{Z})] , \tag{8.138}$$

$$[H_s(\mathbf{Z})] = ([I_N] + s^2\,[\mathbf{Z}]^T[M]\,[\mathbf{Z}])^{-1/2} , \tag{8.139}$$

$$[\mathbf{Z}] = [\mathbf{A}] - [V]\,[\mathbf{D}] , \tag{8.140}$$

$$[\mathbf{D}] = ([V]^T[M]\,[\mathbf{A}] + [\mathbf{A}]^T[M]\,[V])/2 , \tag{8.141}$$

$$[\mathbf{A}] = [\mathbf{U}] - [B]\,\{[B]^T\,[\mathbf{U}]\} , \tag{8.142}$$

$$[\mathbf{U}] = [\mathbf{G}(\beta)] [\sigma], \tag{8.143}$$

with the following properties:

. s, β, and $[\sigma]$ are the hyperparameters of the probability distribution $P_{[\mathbf{W}]}$, which are described after.
. $[\mathbf{G}(\beta)]$ is a non-Gaussian centered $\mathbb{M}_{m,N}$-valued random matrix, defined on probability space $(\Theta, \mathcal{T}, \mathcal{P})$, for which the generator of independent realizations are detailed in [209].
. $[\mathbf{A}]$ is an $\mathbb{M}_{m,N}$-valued random matrix such that $[B]^T[\mathbf{A}] = [0_{m_{\mathrm{CD}},N}]$ a.s.
. $[\mathbf{D}]$ is an $\mathbb{M}_N^S$-valued random matrix.
. $[\mathbf{Z}]$ is an random matrix with values in the tangent vector space $T_V \mathbb{S}_{m,N}$ of $\mathbb{S}_{m,N}$ at the given point $[V]$.
. $[H_s(\mathbf{Z})]$ is a $\mathbb{M}_N^+$-valued random matrix.
. $[\mathbf{W}]$ is a second-order non-Gaussian and not centered random matrix with values in the manifold $\mathcal{S}_{m,N}$.

□ *Hyperparameters of the stochastic model.* For $[V]$ given in $\mathcal{S}_{m,N}$, the $m_\alpha = 2 + N(N+1)/2$ hyperparameters of the stochastic model of random matrix $[\mathbf{W}]$ with values in $\mathcal{S}_{m,N}$ are the following:

. $s \geq 0$ that allows for controlling the level of the statistical fluctuations of $[\mathbf{W}]$ around $[V]$ in $\mathcal{S}_{m,N}$ (if $s = 0$, then $[\mathbf{W}] = [V]$ a.s).
. $\beta > 0$ that allows for controlling the correlation of the random components of each column of $[\mathbf{W}]$.
. $[\sigma]$ that is an $(N \times N)$ upper triangular matrix with positive diagonal entries, which allows for controlling the correlation between the columns of $[\mathbf{W}]$.

The hyperparameter is thus $\alpha = (s, \beta, \{[\sigma]_{kk'}, 1 \leq k \leq k' \leq N\})$ with length $m_\alpha = 2 + N(N+1)/2$, which belongs to the admissible set $\mathcal{C}_\alpha$. It should be noted that the number of hyperparameters can be reduced in taking a sparse version of matrix $[\sigma]$ with the particular case of $[\sigma] = [I_N]$ (no correlation between the columns of $[\mathbf{W}]$).

8.6.5 Numerical Validation in Nonlinear Structural Dynamics

An application in nonlinear structural dynamics is presented, which gives an element of validation of the theory proposed. Other applications are presented in [209], which show the capability of the theory proposed.

□ *Description of the mechanical system.* The mechanical system is a 3D slender damped linear elastic bounded medium with two nonlinear elastic barriers that induce impact nonlinearities in the dynamical system. The geometry, the boundary conditions, the nonlinear elastic barriers, the applied forces are defined in Figure 8.27 with $L_1 = 1.2\ m$, $L_2 = 0.12\ m$, and $L_3 = 0.24\ m$. The elastic medium is

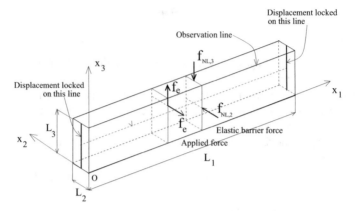

Fig. 8.27 Definition of the dynamical system with nonlinear elastic barriers. [Figure from [209]].

made of a homogeneous and isotropic elastic material for which the Young modulus is $10^{10}\ N/m^2$, the Poisson coefficient is 0.15, and the mass density is $1500\ Kg/m^3$. A damping term is added and is described by a global damping rate of $\xi_d = 0.01$ for each elastic mode of the structure without the elastic barriers, and is introduced at the ROM level. The time-dependent applied forces has an energy located in a narrow frequency band $B_e = [320\,,620]\ Hz$ within a broad frequency band of analysis $B_a = [0\,,1500]\ Hz$. The observation is the Fourier transform $\hat{o}_j(\omega)$ of the time-dependent x_2-acceleration at the observation node Obs_{51} that belongs to the observation line for which the x_1-coordinate is $1.00\ m$.

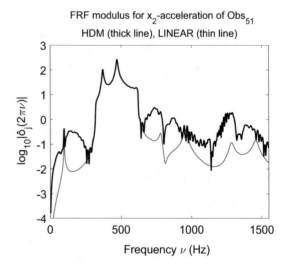

Fig. 8.28 Graph $\nu \mapsto \log_{10}(|\hat{o}_j(2\pi\nu)|)$ computed with the nonlinear HDM (thick line) and with the linear HDM (blue thin line) for the observation. [Figure from [209]].

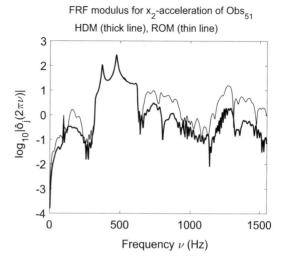

Fig. 8.29 Graph $\nu \mapsto \log_{10}(|\widehat{o}_j(2\pi\nu)|)$ computed with the nonlinear HDM (thick line) and the one computed with the nonlinear ROM (thin line). [Figure from [209]].

□ *Nonlinear HDM, numerical solver, quantification of the effects of the nonlinearities.* The finite element mesh of the nonlinear HDM is made up of $60 \times 6 \times 12 = 4\,320$ three-dimensional 8-nodes solid elements. There are $5\,551$ nodes and $m = 16\,653$ DOFs. The number of zero Dirichlet conditions is $m_{\text{CD}} = 78$ (the displacements are zero for 2×13 nodes). The implicit Newmark time-integration scheme is used with a fixed point method at each sampling time and with a local adaptive time step (for the shocks). Figure 8.28 displays the graph $\nu \mapsto \log_{10}(|\widehat{o}_j(2\pi\nu)|)$ computed with the nonlinear HDM and the one computed with the linear HDM (that is to say, in removing the nonlinear elastic barriers in the HDM). This figure shows the effects of the nonlinear elastic barriers on the response, in particular, it can be seen an important transfer of the energy in the frequency band that is outside the main frequency band $[320\,,620]\ Hz$ of the excitation.

□ *Nonlinear ROM, quantification of the errors due to the use of the nonlinear ROM instead of the nonlinear HDM.* The nonlinear ROM is constructed with the ROB constituted of $N = 20$ elastic modes (11 in $[0\,,1550]\ Hz$ and 9 in $[1550\,,3100]\ Hz$). The numerical solver is the same one that the one used for the nonlinear HDM. Figure 8.29 displays the graph $\nu \mapsto \log_{10}(|\widehat{o}_j(2\pi\nu)|)$ computed with the nonlinear HDM and the one computed with the nonlinear ROM. This figure shows that the differences between the nonlinear HDM and the nonlinear ROM are very small in the frequency band $[320\,,620]\ Hz$ of the excitation, but are significant outside this frequency band. Such differences could be reduced in increasing dimension N of the nonlinear ROM, but as we have explained above, the reduced-order dimension N is chosen in order that significant differences exist between the nonlinear ROM and the nonlinear HDM. The large uncertainties outside B_e (due to the energy transfer outside B_e) are associated with second-order contributions and consequently, the prediction for such a situation is a challenging problem in UQ.

Fig. 8.30 Graph $\nu \mapsto$ $\log_{10}(|\hat{o}_j(2\pi\nu)|)$ computed with the nonlinear HFM (thick line), computed with the nonlinear ROM (thin line), and confidence region with a probability level 0.98 (region in yellow color or in gray, with red upper envelope and red lower envelope) computed with the nonlinear SROM. [Figure from [209]].

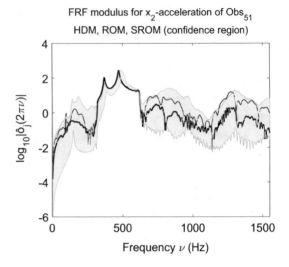

□ *SROM, stochastic solver, and result.* The length of the hyperparameter $\alpha = (s, \beta, \{[\sigma]_{kk'}, 1 \leq k \leq k' \leq N\})$ is $m_\alpha = 212$. The SROM defined by Equations (8.127) to (8.130) with $[\mathbf{W}]$ given by (8.138) to (8.143) is solved by using the Monte Carlo stochastic solver with $1\,000$ realizations. For identifying hyperparameter α, the optimization problem defined by Equation (8.131) is solved by using the interior-point algorithm with constraints. Figure 8.30 summarizes the results obtained with the nonlinear SROM that allows for taking into account the error induced by the use of the nonlinear ROM instead of the nonlinear HFM. This figure displays the graph $\nu \mapsto \log_{10}(|\hat{o}_j(2\pi\nu)|)$ computed with the nonlinear HFM (the target for the mean), the one computed with the nonlinear ROM, and the confidence region (with a probability level of 0.98) constructed with the nonlinear SROM. It can be seen that the results obtained are very good and that the nonlinear SROM allows for generating a confidence region, which is not centered around the responses computed with the nonlinear ROM, but which is approximatively well centered around the responses computed with the nonlinear HFM, which is a relatively difficult problem for taking into account contributions of second-order. Such a result demonstrates the capability of the method proposed.

Chapter 9
Robust Analysis with Respect to the Uncertainties for Analysis, Updating, Optimization, and Design

Abstract *The objective of this chapter is to present some applications in different areas in order to show the importance to perform robust computation with respect to the uncertainties that exist in the computational models. The developments given in this chapter use all the tools and methods presented in the previous chapters and in particular, the random matrix theory (Chapter 5), the stochastic solvers (Chapters 4 and 6), the statistical inverse methods (Chapter 7), and the parametric probabilistic approach of model-parameter uncertainties, the nonparametric probabilistic approach of both the model-parameter uncertainties and the model uncertainties induced by the modeling errors, and the generalized probabilistic approaches of uncertainties, that have been presented in Chapter 8.*

9.1 Role of Statistical and Physical Reduction Techniques

In this section, we briefly summarize the role played by the statistical reduction and by the physical reduction for constructing stochastic reduced-order models that must be used in uncertainty quantification and in particular to obtain feasible computation in stochastic optimization such as for the robust updating and for the robust design.

© Springer International Publishing AG 2017   217
C. Soize, *Uncertainty Quantification*, Interdisciplinary Applied Mathematics 47,
DOI 10.1007/978-3-319-54339-0_9

9.1.1 Why Taking into Account Uncertainties in the Computational Models Induces an Extra Numerical Cost

☐ With respect to the usual dimensions in classical physics (time or frequency, 3 spatial coordinates for space), the uncertainty quantification introduces an extra "statistical dimension" and consequently, roughly speaking, "adds a loop" in the computation. Such an additional loop considerably increases the CPU time for the identification of the stochastic computational models, for the robust updating, for the robust optimization, and for the robust design with respect to the uncertainties.

☐ On another hand, the numerical cost induced by the "statistical dimension" can also be very high if the required stochastic model that is introduced has a high stochastic dimension.

☐ The stochastic model of uncertainties involves hyperparameters that allow for controlling the stochastic model and for which their admissible sets must be explored in order to perform their statistical identification and to carry out robust analyses with respect to uncertainties.

9.1.2 What Type of Reduction Must Be Performed

☐ The classical reduction must be done (if possible) with respect to the usual physics dimension in order to decrease the unitary numerical cost for each deterministic computation inside the statistical loop. In Chapter 8, we have presented such methodologies for constructing ROM and SROM by using a ROB or a SROB.

☐ A statistical reduction must systematically be done in order to reduce the statistical dimension, which allows for reducing the numerical cost of the "statistical loop" for solving the statistical inverse problems in the identification process and also for increasing the rate of convergence of the stochastic solver with respect to the stochastic dimension (concerning the stochastic solvers that differ from the Monte Carlo solver for which the rate of the stochastic convergence is independent of the stochastic dimension). The statistical reduction methods have been presented in Chapter 5.

☐ A minimal hyperparameterization must be favored when constructing the stochastic model of uncertainties in order to reduce the dimension of the set of hyperparameters that have to be explored for a robust analysis or for their statistical inverse identification (for instance, by constructing an algebraic prior stochastic model with a minimum of hyperparameters).

9.1.3 Reduction Techniques Must Be Taken Into Account

Even if the use of the parallel computing and the increasing power of the computers allow for solving huge calculations (present and next future), in the framework of uncertainty quantification, the reduction techniques for robust analysis, robust updating, robust design, and robust optimization are still a challenging problem in many cases and require additional research.

9.2 Applications in Robust Analysis

In this section, the following applications are presented:

1. Robust analysis of the thermomechanics of a complex multilayer composite plate submitted to a thermal loading.
2. Robust analysis of the vibrations of an automobile in the low-frequency range.
3. Robust analysis of the vibrations of a spatial structure.
4. Robust analysis for the computational nonlinear dynamics of a reactor coolant system.
5. Robust analysis of the dynamic stability of a pipe conveying fluid.

We also refer the reader to other applications concerning robust analysis based on the use of the nonparametric and of the generalized probabilistic approaches of uncertainties:

– for a robust analysis of rigid multibody system dynamics with uncertain rigid bodies, [18].
– for a robust analysis of the flutter speed of a wing with uncertain coupling between substructures, [141].
– for a robust analysis of an uncertain computational model to predict well integrity for geologic CO_2 sequestration, [68].
– for a robust post-buckling nonlinear static and dynamical analyses of uncertain very thin cylindrical shells, see Section 8.5.6 and [35].
– for a mistuning analysis and uncertainty quantification of an industrial bladed disk with geometrical nonlinearity, [36].
– for robust analysis and uncertainty quantification of wave propagation in soil, urban cities, and biological tissues, [41, 42, 43, 53, 121, 127].
– for robust analysis in computational structural dynamics with applications in engineering mechanics and space engineering, [16, 17, 20, 48, 124, 161, 168, 169].

9.2.1 Robust Analysis of the Thermomechanics of a Complex Multilayer Composite Plate Submitted to a Thermal Loading

In this section, we present a stochastic modeling of uncertainties for performing a robust thermomechanical analysis of a complex multilayer composite plate made up of a cardboard-plaster-cardboard (CPC) and submitted to a thermal loading [174, 175].

□ *Objective.* A robust simulation model is required for justifying the thermal resistance of CPC multilayer plates submitted to a thermal loading, taking into account all the physical phenomena that cannot be taken into account in the mean (or nominal) computational model. A fundamental key is the development of a stochastic nonlinear thermomechanical model, which takes into account the model uncertainties induced by the modeling errors that are unavoidable given the complexity of the thermomechanical behavior of such a multilayer composite.

□ *CPC multilayer thermomechanical mean computational model.* A schematic representation of the CPC plate is shown in Figure 9.1. The CPC multilayer is composed of three physical layers (indexed by the subscript j). For the mean model, in the linear elastic part of the constitutive equation of each layer, cardboard 1 ($j = 1$) is assumed to be orthotropic, the plaster ($j = 2$) is assumed to be isotropic, and cardboard 2 ($j = 3$) is assumed to be orthotropic. The statistical fluctuations of the mechanical properties in these three layers are assumed to be anisotropic. A homogenization of the thermomechanical behavior through the thickness is constructed and a cut-off damage model is used for the cardboard while a brittle damage model is used for the plasterboard consequently, the thermomechanical model is nonlinear.

□ *CPC multilayer thermomechanical stochastic computational model.* The nonparametric probabilistic approach of uncertainties is used and consists in substituting the mean stiffness matrix $[\underline{A}^j]$, which belongs to $\mathbb{M}_5^+(\mathbb{R})$, of each physical layer j by a random matrix $[\mathbf{A}^j]$ with values in $\mathbb{M}_5^+(\mathbb{R})$, which belongs to ensemble $\mathrm{SE}_\varepsilon^+$ of random matrices (see Section 5.4.8.2). The homogenization is performed throughout the thickness of the CPC, which yields the random matrix $[\mathbf{A}^g]$, with values in $\mathbb{M}_5^+(\mathbb{R})$, of the random homogenized constitutive equation of the composite plate.

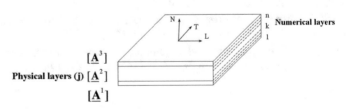

Fig. 9.1 Schematic representation of CPC plates. [Figure from [174]].

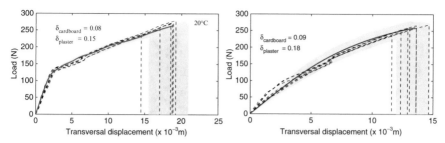

Fig. 9.2 Results for $20°C$ (left figure) and for $120°C$ (right figure). Mechanical applied load (vertical axis in N) as a function of the transversal displacement (horizontal axis in mm). Experiments (black dashed thin lines), numerical simulations with the nonlinear mean computational model (blue thick solid line), confidence region with the probability level 0.98 computed with the nonlinear stochastic model (region in yellow color or in gray). [Figure from [174]].

For a given mean vector $\underline{\mathbf{T}}$ of the temperature field inside the composite, the random vector $\mathbf{Y}$ of the displacements is solution of the following nonlinear static equation,

$$[\mathbf{K}(\mathbf{Y}, \underline{\mathbf{T}})]\,\mathbf{Y} = \mathbf{f}, \tag{9.1}$$

where the stiffness matrix is constructed by the assembly of the random elementary matrices as a function of $[\mathbf{A}^g]$. The damage is updated during the internal iterations of the nonlinear stochastic solver that is based on the use of the Monte Carlo method (see Chapter 6). The dispersion parameters (the hyperparameters) have been identified with the experimental measurements and are the following. For $20°C$, $\delta = 0.15$ in the plaster layer and $\delta = 0.08$ in the two cardboard layers. For $120°C$, $\delta = 0.18$ in the plaster layer and $\delta = 0.09$ in the two cardboard layers.

□ *Results for $20°C$ and for $210°C$ with experimental comparisons.* Figure 9.2 displays the mechanical applied load as a function of the transversal displacement at $20°C$ (left figure) and for $120°C$ (right figure) computed with the nonlinear mean computational model and with the nonlinear stochastic computational model. The comparison with the experiments shows that the prevision given with the stochastic model is good.

9.2.2 Robust Analysis of the Vibrations of an Automobile in the Low-Frequency Range

This section deals with the robust analysis of the linear vibrations of an automobile in the low-frequency range for which the details can be found in [9].

□ *Objective.* The structure of an automobile is generally composed of several structural levels that induce the presence of numerous local elastic modes intertwined with global modes in the low-frequency range. The approach used is based on the construction of a stochastic reduced-order computational model in low-frequency

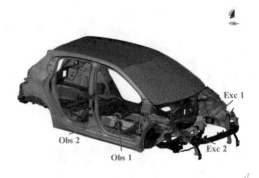

Fig. 9.3 Finite element model
of the automotive vehicle.
[Figure from [9]].

dynamics in presence of these numerous local elastic modes. Since all these local modes do not give a significant contribution for the responses in the stiffness
part of the structure (for the low-frequency range), a two-scale methodology has
been proposed for constructing an adapted ROB that allows for filtering the local
modes. Such an adapted ROM will be referenced as the G-ROB. The construction
of the stochastic ROM by using the G-ROB will be referenced as the G-SROM (see
[202]). A generalization of the construction of such a G-ROB has been proposed
for constructing a multi-scale ROB adapted to the low- and the medium-frequency
ranges [76, 77].

□ *Mean computational model.* The mean computational model of the structure is
shown in Figure 9.3. It is constituted of a finite element model made up of $m_{\text{DOF}} =$
$1\,462\,698$ DOFs. The structure is composed of several structural levels. The two

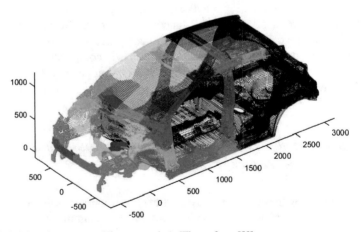

Fig. 9.4 Subdomains generated (one per color). [Figure from [9]].

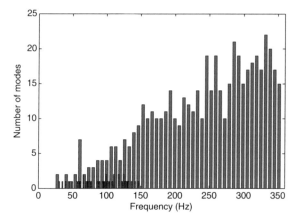

Fig. 9.5 Distributions of the eigenvalues of the global displacements eigenvectors (red ray or dark gray) and of the eigenvalues of the local displacements eigenvectors (blue ray or gray). [Figure from [9]].

observation points Obs 1 and Obs 2, and the two excitation points Exc 1 and Exc 2 are shown in Figure 9.3. Figure 9.4 displays the 90 subdomains generated by the Fast Marching Method for performing a spatial filtering of the local elastic modes. Figure 9.5 displays the distribution of the eigenvalues corresponding to the global displacements eigenvectors used for constructing the G-ROB and the distribution of the eigenvalues corresponding to the local displacements eigenvectors that are filtered. The low-frequency band of analysis is defined as $\mathcal{B} = [0, 120]$ Hz. At convergence in the frequencies in the band $\mathcal{B}$, the ROB (that is constructed with all the global and the local elastic modes) is made up of 160 elastic modes (whose 128 belong to $\mathcal{B}$). The G-ROB is constructed by using the first 36 global displacements eigenvectors intertwined with the first 124 local displacements eigenvectors that are filtered. The constructed G-ROB approximatively spans the same subspace than the one spans by the ROB.

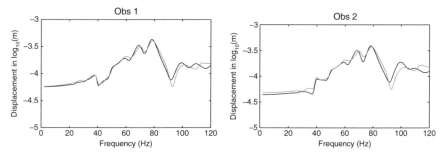

Fig. 9.6 For the two observations Obs 1 (left figure) and Obs 2 (right figure), graph of $\log_{10}$ of the modulus of the displacement (in m) as the function of the frequency (in Hz) computed with the ROM (black line) and with the G-ROM (red line or dark gray). [Figure from [9]].

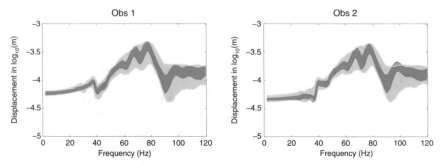

Fig. 9.7 For the two observations Obs 1 (left figure) and Obs 2 (right figure), confidence region with the probability level 0.95 for $\log_{10}$ of the modulus of the displacement (in m) as the function of the frequency (in Hz) computed with the SROM (dark gray region) and computed with the G-SROM (light gray region). [Figure from [9]].

□ *Robust analysis.* For the two observations localized in the stiff part of the structure and for the LF band $\mathcal{B}$, Figure 9.6 displays the $\log_{10}$ of the modulus of the displacement as a function of the frequency, which is computed with the ROM and with the G-ROM. Figure 9.7 displays the same quantities computed with the SROM and with the G-SROM. The SROM is constructed with the ROM by using the nonparametric probabilistic approach presented in Section 8.2. The reduced mass matrix and the reduced stiffness matrix are modeled by random matrices that belong to ensemble $\mathrm{SE}_\varepsilon^+$ of random matrices (see Section 5.4.8.2). The dispersion parameters δ_M and δ_K (the hyperparameters) are those that have been identified in [72]. The G-SROM is constructed with the G-ROB by using again the nonparametric probabilistic approach for which the reduced mass matrix and the reduced stiffness matrix are modeled by random matrices that belong to ensemble $\mathrm{SE}_\varepsilon^+$ and for which the dispersion parameters δ_M^{opt} and δ_K^{opt} have been identified by using the maximum likelihood (see Section 7.3.3). With the G-SROM, there is an additional modeling error (with respect to the SROM) induced by the use of the G-ROB instead of the ROB. Consequently, the level of uncertainties is larger for the G-ROM than for the ROM, and therefore, the confidence regions predicted by using the G-SROM is larger than the confidence regions predicted by the SROM in band $\mathcal{B}$.

9.2.3 Robust Analysis of the Vibrations of a Spatial Structure

This section deals with the robust analysis of the vibrations of a spatial structure for which the details can be found in [30].

□ *Objective.* The objective is to perform a robust analysis in computational linear dynamics of a spatial structure with respect to

1. the model-parameter uncertainties,
2. both the model-parameter uncertainties and the modeling errors,

in order to identify the frequency band of analysis in which there is a lost of robustness with respect to the modeling errors.

□ *Mean computational model.* The mean computational model is made up of a finite element model of the satellite that has 120 000 DOFs and of the satellite-launcher that has 168 800 DOFs (see Figure 9.8). The excitation is a time-dependent force applied to the satellite base in the transversal direction to the launcher axis for which its Fourier transform is flat in the frequency band $[5, 54]$ Hz. The observation is the displacement at the end section of the beam connected to the solar panel. The ROB that is used for constructing the ROM and the SROM is constituted of elastic modes.

□ *Parametric-SROMs and nonparametric-SROMs for the satellite.* The parametric probabilistic approach presented in Section 8.1 is used for the model-parameter uncertainties. The nonparametric probabilistic approach presented in Section 8.2 is used for both the model-parameter uncertainties and the modeling errors. The steps of the uncertainty analysis of the satellite are defined as follows:

1. Development of a parametric-SROMs for the model-parameter uncertainties in the computational model of the satellite using the parametric probabilistic approach, for which the prior probability model is completely defined. There are 1 319 uncertain parameters that are related to section dimensions, to plate and membrane thicknesses, to concentrated masses, to nonstructural masses, to mass densities, to structural dampings, to spring-element stiffnesses, to Young's moduli, to Poisson's ratios, to shear moduli related to structural elements of the satellite. These uncertain parameters are modeled by 1 319 independent real-valued

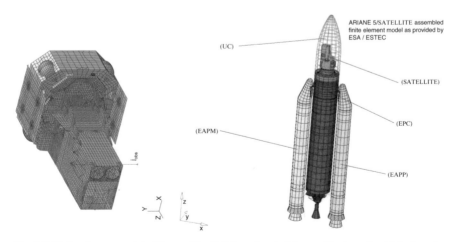

Fig. 9.8 Finite element model of the satellite (left figure) and of the satellite-launcher (right figure). [Figure from [30]].

random variables. The pdf of each random variable is adapted to the available information that are constituted of the knowledge of the admissible subset in which the random variable takes its values (support of the pdf), of the mean value that is chosen as the nominal value of the uncertain parameter, and of the coefficient of variation that controls the level of uncertainties. All the given coefficient of variations belong to the interval $[0.03, 0.4]$.

2. Development of a nonparametric-SROM$_S$ for both the model-parameter uncertainties and the modeling errors in the computational model of the satellite. The reduced mass matrix, the reduced damping matrix, and the reduced stiffness matrix are modeled by independent random matrices that belong to ensemble SE_ε^+ of random matrices (see Section 5.4.8.2), for which the dispersion parameters δ_M, δ_D, and δ_K (the hyperparameters of the stochastic model) are unknown and must be identified.

3. Hyperparameters δ_M and δ_K are identified by minimizing the distance between the pdf of the random lowest eigenfrequency that is constructed with the parametric-SROM$_S$ and the one that is constructed with the nonparametric-SROM$_S$. This identification gives $\delta_M = 0.14$ and $\delta_K = 0.13$. It should be noted that such a criterion (choice of the first eigenfrequency of the satellite) is realistic because this first eigenfrequency is not really sensitive to the modeling errors.

4. Hyperparameter δ_D is identified by minimizing the distance of the two random reduced damping matrices that are constructed with the parametric-SROM$_S$ and with the nonparametric-SROM$_S$. The identification gives $\delta_D = 0.42$.

□ *Parametric-SROML$_{SL}$ and nonparametric-SROML$_{SL}$ for the satellite coupled with the launcher.* It is assumed that the uncertainties in the launcher computational model are not significant with respect to the uncertainties in the satellite computational model. Consequently, the uncertainties are taken into account only in the satellite. For constructing the stochastic reduced-order computational model of the satellite coupled with the launcher, the Craig-Bampton dynamic substructuring method is used. The steps of the analysis are as follows:

1. Construction of the parametric-SROML$_{SL}$ for the satellite coupled with the launcher by using the parametric probabilistic approach of uncertainties. The probability model of the uncertain parameters in the satellite is the one used for constructing the parametric-SROML$_{SL}$ (1 319 independent real-valued random variables).

2. Construction of the nonparametric-SROML$_{SL}$ for the satellite coupled with the launcher by using the nonparametric probabilistic approach of uncertainties described in Section 8.2.8 based on the use of the Craig-Bampton substructuring technique. The dispersion parameters of the random reduced matrices for the satellite substructure are those estimated for the nonparametric-SROM$_S$ ($\delta_M = 0.14$, $\delta_D = 0.42$, and $\delta_K = 0.13$). The launcher substructure is deterministic (no uncertainties).

□ *Analysis of robustness for the satellite coupled with the launcher by using the nonparametric-SROML$_{SL}$.* Figure 9.9 displays the confidence region, with a prob-

Fig. 9.9 The horizontal axis is the frequency f in Hz and the vertical axis is the frequency response function (FRF) $f \mapsto 20 \log_{10}(\|\mathbf{U}(2\pi f)\|)$. Deterministic FRF computed with the mean computational model (thick line). Random FRF computed with the nonparametric-SROML$_{SL}$: mean value (thin line), confidence region with a probability level of 0.96 (dark grey zone). [Figure from [30]].

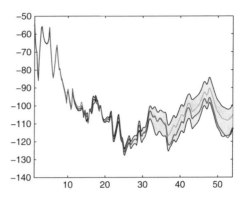

ability level of 0.96, of the norm $\|\mathbf{U}(\omega)\|$ of the random displacement $\mathbf{U}(\omega)$ with values in $\mathbb{C}^3$ at the observation point as a function of the frequency ω. Figure 9.9 shows that the larger the frequency, the larger the confidence region, because the sensitivity to the uncertainties increases with frequency. For frequencies lower than 30 Hz, the computational model of the coupled system is robust to both the model-parameter uncertainties and to the model uncertainties in the satellite. The response is sensitive to uncertainties for frequencies greater than 30 Hz. The role played by modeling errors is analyzed hereinafter.

□ *Identification of the role played by the modeling errors in comparing the nonparametric probabilistic approach with the parametric one.* For the satellite coupled with the launcher and over the frequency band $[30 , 54]$ Hz, Figure 9.10 displays the random response computed with the nonparametric-SROML$_{SL}$ that takes into account both the model-parameter uncertainties and the modeling errors (left figure) and the random response computed with the parametric-SROML$_{SL}$ that takes into account only the model-parameter uncertainties (right figure). This two figures allow for

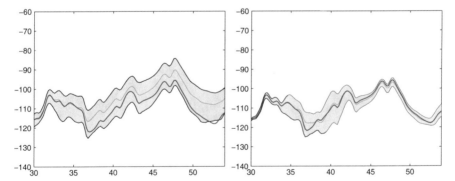

Fig. 9.10 The horizontal axis is frequency f in Hz and the vertical axis is the frequency response function (FRF) $f \mapsto 20 \log_{10}(\|\mathbf{U}(2\pi f)\|)$. Deterministic FRF computed with the mean computational model (thick line). Random FRF: mean value (thin line) and confidence region with a probability level of 0.96 (dark grey zone), computed with the nonparametric-SROML$_{SL}$ (left figure) and computed with the parametric-SROML$_{SL}$ (right figure). [Figure from [30]].

concluding that the computational model is less robust with respect to the modeling errors than with respect to the model-parameter uncertainties in the frequency band $[30\,,54]\,Hz$.

9.2.4 Robust Analysis for the Computational Nonlinear Dynamics of a Reactor Coolant System

In this section, we present the robust analysis detailed in [62] concerning the computational nonlinear dynamics of a multisupported reactor coolant system subjected to seismic loads.

□ *Objective*. The objective is to perform a robust analysis in transient nonlinear dynamics of a multisupported structure constituted of a reactor coolant system (PWR) inside a building subjected to seismic loads, in order to quantify the design margins with respect to both the model-parameter uncertainties and the model uncertainties induced by the modeling errors.

□ *Real system*. The nonlinear dynamical system (see Figure 9.11) is made up of a linear damped elastic structure that represents a reactor cooling system. Each loop is constituted of a reactor, a reactor coolant pump, and a steam generator. The nonlinearities are due to restoring forces induced by elastic stops modeling the supports of the reactor cooling system. For given gaps, these elastic stops limit the vibration amplitudes of the steam generator system. The displacement field of this structure is constrained by several time-dependent Dirichlet conditions corresponding to seismic loads operating on the anchors of the reactor cooling system and elastic stops.

□ *Mean computational model*. The curvilinear finite element model is displayed in Figure 9.12 and has 5 022 DOFs and 828 Lagrange's DOFs. There are 36 time-dependent Dirichlet conditions for which the seismic accelerograms are applied to

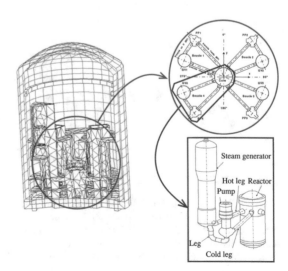

Fig. 9.11 Reactor coolant system in the building (left part of the figure), four loops (right up part of the figure), one loop (right down part of the figure). [Figure from [62]].

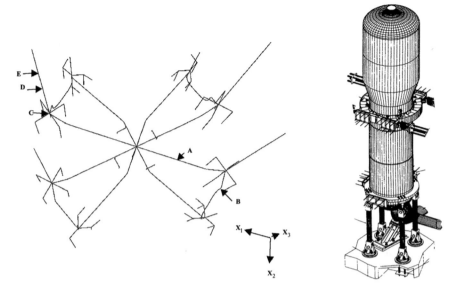

Fig. 9.12 Curvilinear finite element model for the four loops reactor coolant system (left figure) and for a steam generator (right figure). [Figure from [62]].

the 36 supports that consist of anchors located under the reactor coolant pumps, the steam generators, and the cold legs. The nonlinearities are due to elastic stops inducing nonlinear restoring forces for 28 DOFs. The lowest eigenfrequency of the underlying linear dynamical system (without the stops) is 1.4 Hz and the eigenfrequency of rank 200 is 164 Hz.

□ *Nonlinear SROM.* The nonlinear ROM is constructed by using as the ROB the first $N = 200$ elastic modes of the underlying linear system. The nonlinear SROM is constructed from the nonlinear ROM by using the generalized probabilistic approach of uncertainties presented in Section 8.4. For the linear part of the computational model, the modeling errors are taken into account with the nonparametric probabilistic approach for which the reduced mass matrix, the reduced damping matrix, and the reduced stiffness matrix are modeled by random matrices that belong to ensemble SE_ε^+ (see Section 5.4.8.2), for which the dispersion parameters δ_M, δ_D, and δ_K (the hyperparameters) are such that $\delta_M = \delta_D = \delta_K = 0.2$. The model-parameter uncertainties concern the 28 stiffnesses of the 28 elastic stops, for which the parametric probabilistic approach is used. The pdf's of these random variables are detailed in [62]. The dispersion parameter (hyperparameter) of each random variable is 0.2. The stochastic solver is the Monte Carlo numerical method (see Section 6.4) for which 700 realizations have been used.

□ *Robust analysis.* The stochastic observation is the random Shock Response Spectrum (SRS) (see [52, 62]) represented by the random variable $\text{dB}(f) = \log_{10}(S(2\pi f))$ in which $S(2\pi f)$ is the SRS at frequency f, for the normalized ac-

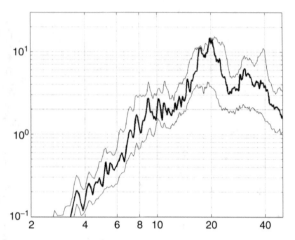

Fig. 9.13 The horizontal axis is the frequency f in Hz and the vertical axis is $dB(f)$. Prediction with the nonlinear mean computational model (thick line). Upper envelope (upper thin line) and lower envelope (lower thin line) of the confidence region predicted with the nonlinear SROM. [Figure from [62]].

celeration at the A-marked point (middle point of a hot leg in x_3 direction), which is shown in Figure 9.12 (left). Figure 9.13 displays the confidence region of $dB(f)$ for a probability level of 0.98, which permits the analysis of the design margins with respect to both the model-parameter uncertainties and the modeling errors. For instance, it can be seen that the mean model is very robust for the prediction of the maximum of the response in the frequency band $[17, 21]$ Hz and is less robust in the other part of the frequency band.

9.2.5 Robust Analysis of the Dynamic Stability of a Pipe Conveying Fluid

This section deals with the robust analysis of the dynamic stability of a pipe conveying fluid, which is detailed in [170].

□ *Objective*. The objective is to perform a robust analysis of the linear dynamic stability of a pipe conveying fluid in the classical Paidoussis formulation [158], with respect to model uncertainties induced by the modeling errors in the computational model. The model uncertainties induced by the modeling errors are mainly due to

Fig. 9.14 Fluid-structure coupled system. The arrow represents the direction of the internal fluid flow with a constant speed U).[Figure from [170]].

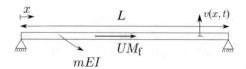

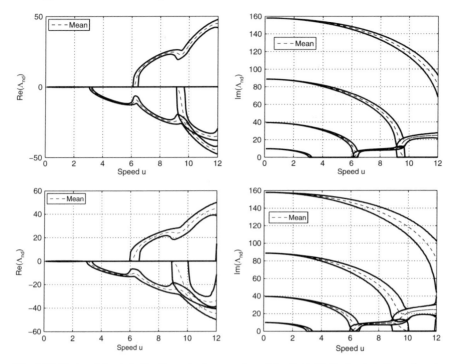

Fig. 9.15 Horizontal axis is the dimensionless speed u of the fluid. Vertical axis is the real parts and the imaginary parts of one half of the dimensionless eigenvalues, Λ_{nd}. Prediction with the mean computational model (dashed line). Confidence region with a probability level of 0.95 for the real parts (left figures) and for the imaginary parts (right figures), for $\delta = 0.05$ (the two up figures) and for $\delta = 0.1$ (the two down figures). [Figure from [170]].

the noninertial coupled fluid forces (related to the damping and to the stiffness) and are modeled by using the nonparametric probabilistic approach of uncertainties presented in Section 8.2. The resulting random eigenvalue problem is used for performing a robust analysis of the flutter and of the divergence unstable modes of the fluid-structure coupled system for different values of the dimensionless speed of the fluid as a function of the level of uncertainties.

□ *Mean computational model.* The fluid-structure coupled system is shown in Figure 9.14. The model [159] consists in an Euler-Bernoulli beam, without damping, simply supported and coupled with an internal flow for which the plug flow model is used. The length of the beam is L, its Young modulus is E, and its area moment of inertia is I. The fluid mass per unit length is M_f. The internal fluid flow has a constant speed U. The dimensionless speed u of the internal fluid flow is introduced such that $u = UL\sqrt{M_f/(EI)}$. The beam is discretized in 40 beam finite elements.

□ *Deterministic divergence and flutter instabilities predicted by using the mean computational model.* The first divergence mode occurs at $u = \pi$ and the second one at $u = 2\pi$. The first flutter instability occurs around $u = 6.29$ and the third divergence mode occurs around $u = 3\pi$.

□ *SROM and robust analysis of the divergence and of the flutter instabilities.* The prior probability model of the reduced matrix of the generalized fluid forces is constructed similarly to the construction of ensemble SE^{rect} of random real matrices presented in Section 5.4.8.4, but for complex matrices as explained in [132, 170]. There is only one random real matrix germ $[\mathbf{G}]$ belonging to ensemble SG_ε^+ (see Section 5.4.7.2 for which the dispersion parameter δ (the hyperparameter) allows for controlling the level of model uncertainties of the generalized fluid forces. The independent realizations of the random dimensionless eigenvalues Λ_{nd} of the SROM for the fluid-structure coupled system (with no structural damping) are computed by using the Monte Carlo method (See Section 6.4). The robust analysis of the divergence instability and of the flutter instabilities is performed by constructing, as a function of the dimensionless speed u of the fluid, the confidence regions with a probability level of 0.95 for the real part $\mathrm{Re}(\Lambda_{nd})$ and for the imaginary part $\mathrm{Im}(\Lambda_{nd})$ of each random eigenvalue Λ_{nd}. Figure 9.15 displays these confidence regions for 5% and for 10% of uncertainties ($\delta = 0.05$ and $\delta = 0.1$) in the computational model.

9.3 Application in Robust Updating

This section is devoted to the robust updating of the deterministic parameters of a mean damping model, in the low frequencies and in the medium frequencies, and in presence of uncertainties in the mean computational model of a multilayered sandwich panel. The details of this work can be found in [32, 195]. For other examples of robust updating of uncertain computational models using experimental modal analysis, we also refer the reader to [21, 194].

□ *Objective.* The analysis that is performed is the updating of the nominal parameters of the damping in the mean computational model of the multilayered composite sandwich panel that has been presented in Section 8.2.7, for which the available data are made up of the experimental FRFs that are measured for 8 panels manufactured following the same design and by using with the same manufacturing process. The robust analysis consists in performing such an updating by using the nonparametric probabilistic approach presented in Section 8.2 in order to take into account both the model-parameter uncertainties and the model uncertainties that exist in the mean computational model.

□ *Why the mean damping model must be updated.* For the normal acceleration $\gamma_{j_{obs}}(\omega)$ to the panel at a given point and in the frequency domain, the associated observation is defined, in dB, as $w_{j_{obs}}(\omega) = 20\log_{10}(|\gamma_{j_{obs}}(\omega)|)$. For such an observation, Figure 9.16 shows the FRFs $f \mapsto w_{j_{obs}}^{exp,\ell}(2\pi f)$ that are measured for the 8 composite panels ($\ell = 1, \ldots, 8$) and the FRF $f \mapsto \underline{w}_{j_{obs}}(2\pi f)$ that is calculated with the mean computational model. In the medium-frequency band $[1500, 4500]$ Hz, there are significant differences between the frequency response calculated by using the mean computational model and the frequency responses that are measured.

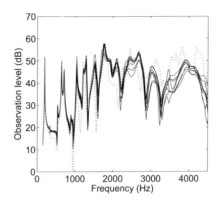

Fig. 9.16 Horizontal axis is the frequency in Hz. Graph $f \mapsto \underline{w}_{j_{obs}}(2\pi f)$ of the FRF calculated with the mean computational model in dB (dotted line). Graphs $f \mapsto w_{j_{obs}}^{exp,\ell}(2\pi f)$ for $\ell = 1, \dots, 8$ of the measured FRFs in dB (8 solid lines). [Figure from [195]].

The differences are mainly induced by the nominal damping model that is not sufficiently good and the presence of modeling errors in the mean computational model.

□ *Robust identification of the mean damping model in presence of model uncertainties.* The elastic modes are used for constructing the ROB. The following parameterized algebraic model of the reduced damping matrix $[\underline{D}(\omega)] = [D(\mathbf{x}, \omega)]$ is introduced:

$$[D(\mathbf{x}, \omega)]_{\alpha\beta} = 2\,\xi(\mathbf{x}, \omega_\alpha)\,\mu_\alpha\,\omega_\alpha\,\delta_{\alpha\beta}\,,$$

in which μ_α, ω_α, and $\xi(\mathbf{x}, \omega_\alpha)$ are the generalized mass, the eigenfrequency (in rad/s), and the damping rate of the elastic modes α. The parameterized algebraic model of the damping rate is written as

$$\xi(\mathbf{x}, \omega) = \xi_0 + (\xi_1 - \xi_0)\frac{\omega^a}{\omega^a 10^b}\,,$$

in which $\mathbf{x} = (\xi_0, \xi_1, a, b)$, with $\xi_0 \leq \xi_1$, is the vector of the values of the mean model parameters that has to be updated. In the mean computational model, the modeling errors are taken into account with the nonparametric probabilistic approach for which the reduced mass matrix, the reduced damping matrix, and the reduced stiffness matrix are modeled by random matrices that belong to ensemble SE_ε^+ (see Section 5.4.8.2), for which the dispersion parameters δ_M, δ_D (that is taken independent of ω), and δ_K (the hyperparameters). The vector $\boldsymbol{\delta}_G = (\delta_M, \delta_D, \delta_K)$ of the hyperparameters has to be identified. Consequently, we have to identify the vector-valued parameter $\mathbf{s}$ defined by

$$\mathbf{s} = (\mathbf{x}, \boldsymbol{\delta}_G) \in \mathcal{C}_{ad} \subset \mathbb{R}^7\,,$$

in which the admissible set $\mathcal{C}_{ad}$ is such that $\xi_0 \in [0.0095\,, 0.0105]$, $\xi_1 \in [0.05\,, 0.15]$, $a \in [5\,, 20]$, $b \in [30\,, 50]$, δ_M, δ_D, and δ_K belong to $[0.05\,, 0.5]$.

□ *Identification of parameter s by using a statistical inverse method.* For fixed angular frequency ω, the components of the random observation vector $\mathbf{W}(\mathbf{s}, \omega) = (W_1(\mathbf{s}, \omega), \ldots, W_{24}(\mathbf{s}, \omega))$ are the moduli in dB of the normal accelerations measured in 24 points. The frequency band $\mathcal{B} = 2\pi \times [150, 4500]$ rad/s is sampled in $\omega_1, \ldots, \omega_m$ frequency sample points with $m = 584$. The corresponding experimental data are the vectors $\mathbf{w}^{\exp,1}(\omega_k), \ldots, \mathbf{w}^{\exp,8}(\omega_k)$ with $k = 1, \ldots, m$. The identification of $\mathbf{s}$ is carried out by solving a statistical inverse problem of the type described in Section 7.3.2 but for which a non-differentiable objective function and the Monte Carlo method with $\nu_s = 400$ realizations (see Section 6.4) are used. The corresponding optimization problem is thus written as

$$\mathbf{s}^{\mathrm{opt}} = \arg \min_{\mathbf{s} \in \mathcal{C}_{\mathrm{ad}}} \mathcal{J}^{\mathrm{ND}}(\mathbf{s}),$$

in which the non-differentiable objective function $\mathcal{J}^{\mathrm{ND}}(\mathbf{s})$ is constructed in order to minimize, over frequency band $\mathcal{B}$, the area defined by the yellow region in Figure 9.17. In this figure, for a given component j of the observation vector, and for $\mathbf{s}$ fixed in $\mathcal{C}_{\mathrm{ad}}$, the functions $\omega \mapsto w_j^-(\mathbf{s}, \omega)$ and $\omega \mapsto w_j^+(\mathbf{s}, \omega)$ represent the lower and the upper envelopes of the confidence region (for a probability level P_c) of the random observation $\omega \mapsto W_j(\mathbf{s}, \omega)$ such that, for all $k = 1, \ldots, m$,

$$\mathrm{Proba}\{w_j^-(\mathbf{s}, \omega_k) < W_j(\mathbf{s}, \omega_k) \leq w_j^+(\mathbf{s}, \omega_k)\} = P_c,$$

and, for the experimental measurements, the function $\omega \mapsto w_j^{-\exp}(\omega)$ and $\omega \mapsto w_j^{+\exp}(\omega)$ are defined, for all $k = 1, \ldots, m$, by

$$w_j^{-\exp}(\omega_k) = \min_\ell w_j^{\exp,\ell}(\omega_k) \quad , \quad w_j^{+\exp}(\omega_k) = \max_\ell w_j^{\exp,\ell}(\omega_k).$$

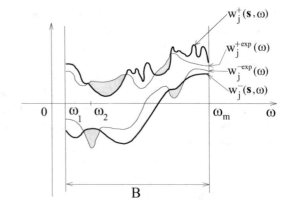

Fig. 9.17 Horizontal axis is the angular frequency ω in rad/s. The non-differentiable objective function $\mathcal{J}^{\mathrm{ND}}(\mathbf{s})$ is constructed in order to minimize the area defined by the region in yellow color (or in gray). [Figure from [195]].

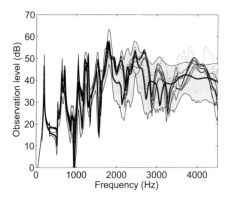

Fig. 9.18 Horizontal axis is the frequency f in Hz. Graph $f \mapsto \underline{w}_{j_{obs}}(2\pi f)$ (in dB) of the FRF calculated with the mean computational model (dotted line). Graphs $f \mapsto w_{j_{obs}}^{exp,\ell}(2\pi f)$ for $\ell = 1,\ldots,8$ (in dB) of the measured FRFs (8 solid lines). Confidence region with $P_c = 0.95$ of the random frequency response $f \mapsto W_{j_{obs}}(\mathbf{s}^{opt}, 2\pi f)$ (in dB) computed with the SROM for the identified parameter $\mathbf{s}^{opt} = (\underline{\mathbf{x}}^{opt}, \boldsymbol{\delta}_G^{opt})$ (region in yellow color or in gray). [Figure from [195]].

The optimization problem is solved by using a Genetic Algorithm with constraints defined by $\mathcal{C}_{ad}$ and yields $\mathbf{s}^{opt} = (\underline{\mathbf{x}}^{opt}, \boldsymbol{\delta}_G^{opt})$ with $\underline{\mathbf{x}}^{opt} = (0.01, 0.081, 10.9, 47)$ and $\boldsymbol{\delta}_G^{opt} = (0.23, 0.07, 0.24)$.

□ *Random frequency response* $\omega \mapsto W_{j_{obs}}(\mathbf{s}^{opt}, \omega)$ *computed with the SROM for the identified parameter* $\mathbf{s}^{opt} = (\underline{\mathbf{x}}^{opt}, \boldsymbol{\delta}_G^{opt})$. For the observation j_{obs}, Figure 9.18 displays the graph $f \mapsto \underline{w}_{j_{obs}}(2\pi f)$ of the FRF calculated with the mean computational model, the graphs $f \mapsto w_{j_{obs}}^{exp,\ell}(2\pi f)$ for $\ell = 1, \ldots, 8$ of the measured FRFs, and the confidence region with $P_c = 0.95$ of the random FRF $f \mapsto W_{j_{obs}}(\mathbf{s}^{opt}, 2\pi f)$ computed with the SROM for the identified parameter $\mathbf{s}^{opt} = (\underline{\mathbf{x}}^{opt}, \boldsymbol{\delta}_G^{opt})$. It can be seen that the robust updating of the parameters of the mean damping model in presence of the model uncertainties in the mean computational model significatively improves the quality of the predictions with respect to the measurements.

9.4 Applications in Robust Optimization and in Robust Design

Robust optimization or robust design mean that the optimization problem must be solved by taking into account uncertainties in the computational model, that is to say, by using a stochastic computational model. The two following applications are presented for illustrating the robust optimization and the robust design.

1. Robust design in computational dynamics of turbomachines.
2. Robust design in computational vibroacoustics.

For a general approach devoted to the robust design optimization in computational mechanics, we refer the reader to [31], and for an application devoted to a robust crash design analysis using a high-dimensional nonlinear uncertain computational mechanical model, we refer the reader to [67].

9.4.1 Robust Design in Computational Dynamics of Turbomachines

□ *Problem definition.* It is known that the phenomenon related to mistuning in the dynamics of turbomachines can induce strong vibrations for the forced response of the bladed disk and can produce spatial localization in the dynamic response of the blades as it was demonstrated by Whitehead in 1966 [75, 218]. In the framework of unsteady aeroelasticity, a small variation in the blades in terms of geometry (due to the manufacturing tolerances), of boundary conditions, and of mechanical properties of the material induces a rupture of the structural cyclic symmetry, called mistuning, which can imply a strong dynamic amplification of the responses. A method for reducing this type of dynamic amplification consists in intentionally detune the mistuned bladed disk by slightly modifying a few blade shapes of the bladed disk.

□ *Objectives.* Two aspects are presented in this section:

— the robust analysis of the mistuning that is induced by the manufacturing tolerances in a bladed disk [28, 29].
— the robust design that is based on a detuning technique by modifying some blade shapes in presence of mistuning with rotation and unsteady flow [132].

9.4.1.1 Robust Analysis of the Mistuning Induced by the Manufacturing Tolerances in a Bladed Disk [28, 29]

□ *Mean computational model.* The bladed disk is a wide chord supersonic fan geometry called SGC1. The fan has 22 blades that are numbered as $j = 1, \ldots, 22$. The bladed disk rotates around its revolution axis with a constant velocity. The mean computational model is shown in Figure 9.19 and has $m_{\text{DOF}} = 504\,174$ DOFs. We are interested in analyzing the forced response over the frequency band $\mathcal{B} = [515\,, 535]\ Hz$, corresponding to a given engine order excitation, for which the mistuning induces a large dynamic amplification factor for a few blades.

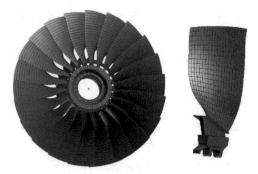

Fig. 9.19 Finite element model of the bladed disk and of a blade. [Figure from [29]].

□ *Construction of the ROM.* The reduced-order model is constructed by using the Benfield-Hruda dynamic substructuring technique [22] for which 10×22 Craig-Bampton modes [54] are used with 440 additional elastic modes for the loaded disk. Therefore the total number of elastic modes used for constructing the ROB is $N = 660$.

□ *Construction of the SROM for the mistuned bladed disk.* For the bladed disk that is considered, the blades uncertainties are due to the blade manufacturing tolerances, which induce mistuning. The stochastic model of uncertainties is constructed by using the nonparametric probabilistic approach of uncertainties presented in Section 8.2. The amount of uncertainty is mainly distributed on the stiffness because the value of the mass dispersion parameter is less than $1\,000$ times the stiffness dispersion parameter. For each blade $j \in \{1, \ldots, 22\}$, its random reduced stiffness matrix is taken in ensemble SG_ε^+ of random matrices, defined in Section 5.4.7.2. The statistics of the blades are assumed to be mutually independent, and each blade is assumed to have the same level of uncertainties. Consequently, the hyperparameter δ_K^j related to the level of uncertainties of blade j is chosen as 0.05 for all $j \in \{1, \ldots, 22\}$. We then introduce the notation, $\delta_K = \delta_K^1 = \ldots = \delta_K^{22} = 0.05$. For the stochastic solver, the Monte Carlo numerical simulation method is used with $\nu_s = 1\,500$ realizations.

□ *Definition of the frequency-dependent dynamic amplification factor (DAF).* For ω fixed in $2\pi \times \mathcal{B}$, the random DAFs are defined as follows. For j fixed in $\{1, \ldots, 22\}$, let $\underline{e}_j(\omega)$ be the elastic energy of blade j computed with the ROM of the tuned bladed disk. Due to the cyclic symmetry of the tuned bladed disk, we have $\underline{e}_1(\omega) = \ldots = \underline{e}_{22}(\omega)$ denoted as $\underline{e}(\omega)$. Similarly, the random elastic energy of blade j is denoted by $E_j(\omega)$ and is computed with the SROM for the mistuned bladed disk.

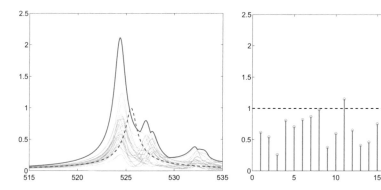

Fig. 9.20 Left figure: the horizontal axis is the frequency f in Hz; graph $f \mapsto \underline{b}(2\pi f)$ for the tuned bladed disk (blue dashed line); for $j = 1, \ldots, 22$, graphs $f \mapsto B_j(2\pi f, \theta)$ for the mistuned bladed disk (thin lines and thick line for $j = 17$). Right figure: the horizontal axis is the index j of blades; graph $j \mapsto \max_{f \in 2\pi \times \mathcal{B}} \underline{b}(2\pi f)$ (horizontal dashed line); graph $j \mapsto \max_{f \in 2\pi \times \mathcal{B}} B_j(2\pi f, \theta)$ (vertical lines).

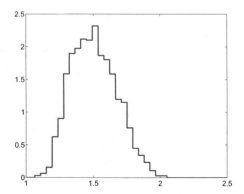

Fig. 9.21 Graph $b \mapsto$ $p_{B_\infty}(b)$ of the pdf of random variable B_∞ computed with the SROM of the mistuned bladed disk.

The random DAF of blade j, denoted by $B_j(\omega)$, and the one for the bladed disk, denoted by $B(\omega)$, are defined by

$$B_j(\omega) = \sqrt{\frac{E_j(\omega)}{\max_{\omega \in B} \underline{e}(\omega)}} \quad , \quad B(\omega) = \max_{j=1,\dots,22} B_j(\omega).$$

For the tuned bladed disk, the DAF $\underline{b}_j(\omega)$ of blade j is obtained by taking $E_j(\omega) = \underline{e}_j(\omega)$ in the equation above. Due to the cyclic symmetry of the tuned bladed disk, we have $\underline{b}_1(\omega) = \dots = \underline{b}_{22}(\omega)$ denoted as $\underline{b}(\omega)$.

□ *Illustration of the localization of vibrations for the mistuned bladed disk.* Let us consider a given realization θ of the random response constructed with the SROM, and let $B_j(\omega, \theta))$ and $B(\omega, \theta))$ be the corresponding realization of the DAFs. Figure 9.20 allows for analyzing the values of the frequency-dependent DAFs for each one of the 22 blades for the mistuned bladed disk and for the tuned one. The left figure displays the graph $f \mapsto \underline{b}(2\pi f)$ of the DAF for the tuned bladed disk computed with the ROM (same graph for each blade) and, for $j = 1, \dots, 22$, the graphs $f \mapsto B_j(2\pi f, \theta)$ of the given realization computed with the SROM (different graph for each blade). The right figure displays the graph $j \mapsto \max_{f \in 2\pi \times B} \underline{b}(2\pi f)$, which is the same for each blade, and consequently, which is a horizontal line, and the graph $j \mapsto \max_{f \in 2\pi \times B} B_j(2\pi f, \theta)$. The right figure clearly shows that the vibrations are spatially localized in blade number 17 for which the DAF is greater than 2.

□ *Probability density function of the random DAF for the mistuned system.* For the mistuned bladed disk, the random dynamic magnification factor B_∞, over frequency band B, is defined by

$$B_\infty = \max_{\omega \in B} B(\omega).$$

Figure 9.21 displays the graph $b \mapsto p_{B_\infty}(b)$ of the pdf of the random variable B_∞ that is computed using the SROM of the mistuned bladed disk. This figure shows that there exist configurations of the random geometry of the bladed disk that yield amplification factors higher than 2.

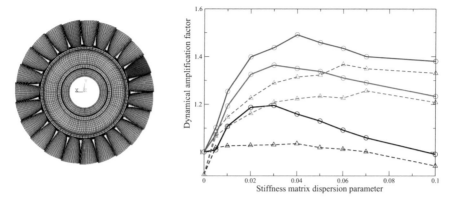

Fig. 9.22 Left figure: finite element model of the bladed disk for which the shape of two blades has been modified for obtaining an intentional detuning. Right figure: the horizontal axis is the dispersion parameter δ_K that controls the level of mistuning; the vertical axis is the value a of the DAF such that $\mathrm{Proba}\{\mathrm{DAF} \leq a\} = P_c$; the solid lines correspond to the mistuned system without intentional detuning; the dashed lines correspond to the mistuned system with intentional detuning; the lower, the middle, and the upper lines correspond to the level of probability $P_c = 0.50, 0.95$, and 0.99. [Figure from [132]]

9.4.1.2 Robust Design Based on a Detuning Technique by Modifying Some Blade Shapes in Presence of Mistuning with Rotation and Unsteady Flow [132]

We consider the bladed disk shown in Figure 9.22 (left) that is in rotation, with upstream flow at Mach 0.8 and an unsteady flow with a Mach field from 0.1 to 1.5.

□ *Methodology.* The computational model is shown in Figure 9.22 (left) in which the intentional detuning is realized by introducing a modification of the shape for two blades. The detuned ROM is constructed by using a projection basis that is adapted to the different sector types [131]. The mistuning is taken into account by using the nonparametric probabilistic approach of uncertainties (see Section 8.2). The construction of the SROM is similar to the one presented in Section 9.4.1.1. The random reduced stiffness matrix is constructed for each sector by using the ensemble $\mathrm{SG}_\varepsilon^+$ of random matrices (see Section 5.4.7.2) for which the level of mistuning is controlled with the hyperparameter δ_K that is assumed to be the same of each sector (as in Section 9.4.1.1). The frequency-dependent random generalized unsteady aeroelastic matrix is constructed by using the ensemble $\mathrm{SE}^{\mathrm{rect}}$ adapted to the case of complex matrices (see Section 5.4.8.4), for which the level of uncertainties is controlled by the hyperparameter δ_A.

□ *Robust design based on the analysis of the mistuning rate induced by the level of uncertainties on the value of the DAF for an intentionally detuned bladed disk.* Figure 9.22 (right) presents an analysis of the role played by the level of mistuning on the dynamic amplification factor (DAF) of the bladed disk that is intentionally detuned. The calculation has been performed with the SROM. This figure shows that the intentional detuning is very effective for low mistuning levels, but for high mistuning levels, the DAF for the intentionally detuned bladed disk tends towards the one without intentional detuning.

9.4.2 Robust Design in Computational Vibroacoustics.

In this section, we present a robust design performed with an uncertain computational vibroacoustic model for which the details are given in [33].

□ *Objective*. The problem concerns the robust design of a structural element belonging to a structure coupled to an acoustic cavity in which there is an acoustic source that excites the vibroacoustic system. The design is performed with a computational vibroacoustic model for which the stiffness uncertainties are taken into account with the nonparametric probabilistic approach.

□ *Mean computational vibroacoustic model and its ROM*. The structure is made up of steel thin plates and of frames (beams) coupled to an acoustic cavity by one of the structural plates. The acoustic cavity is filled with air and is excited by a deterministic acoustic source whose spectral density is flat in the frequency band of analysis $\mathcal{B} = 2\pi \times [1190 , 1260] \, Hz$. The mean computational vibroacoustic model is shown in Figure 9.23. It is made up of a finite element model with $m_{DOF} = 10\,927$ DOFs for the structural displacement field and with $m_f = 2\,793$ DOFs for the acoustic pressure field. The mesh is compatible on the coupling interface. The ROM is constructed as explained in Section 8.3.2. At convergence, the ROB is made up of 90 elastic modes for the structure and of 41 acoustic modes for the acoustic cavity.

□ *SROM of the vibroacoustic system*. It is assumed that the model uncertainties induced by the modeling errors concern only the stiffness of the structure. The nonparametric probabilistic approach is used (see Section 8.3.3) for which only the

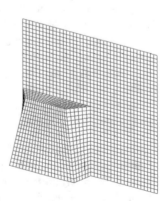

Fig. 9.23 Finite element model of the vibroacoustic system. [Figure from [33]]

reduced stiffness matrix is modeled by a random matrix that belongs to the ensemble SE_ε^+ defined in Section 5.4.8.2. The dispersion parameter δ_{K_s} (the hyperparameter) allows for controlling the level of uncertainties, for which the value has been chosen to 0.25.

□ *Robust design optimization of the vibroacoustic system with uncertain stiffness.* The design parameter is the thickness r of the structural plate that is coupled to the acoustic cavity. The admissible set is defined by $C_r = \{r \in [0.005\,,\,0.007]\,m\}$. The random acoustic observation $W(r, \omega)$ is the random spectral acoustic energy that is defined, for the angular frequency ω in $\mathcal{B}$ by

$$W(r, \omega) = \frac{|\Omega_a|}{\rho_a\,c_a^2}\,\overline{P(r, \omega)^2} \quad \text{with} \quad \overline{P(r, \omega)^2} = \frac{1}{m_f}\sum_{j=1}^{m_f}|P_j(r, \omega)|^2\,,$$

in which $\mathbf{P}(r, \omega) = (P_1(r, \omega), \ldots, P_{m_f}(r, \omega))$ is the random vector of the finite element discretization of the random acoustic pressure fields, where $|\Omega_a|$ is the volume of the internal acoustic cavity, where ρ_a is the mass density of air, and where c_a is the speed of sound in air at equilibrium. The robust design optimization problem consists in minimizing the maximum of the 99 % quantile of the random spectral acoustic energy density with respect to r, over the frequency band. For r fixed in C_r and for ω fixed in $\mathcal{B}$, let $w^+(r, \omega)$ be the 99 % quantile of random $W(r, \omega)$, such that

$$\text{Proba}\{W(r, \omega) \leq w^+(r, \omega))\} = 0.99\,,$$

The objective function is defined by

$$\mathcal{J}(r) = \frac{\max_{\omega \in \mathcal{B}}\,w^+(r, \omega)}{\max_{\omega \in \mathcal{B}}\,\underline{w}(r_0, \omega)}\,,$$

in which $\underline{w}(r_0, \omega)$ is the corresponding observation of the mean computational vibroacoustic model for the nominal value $r_0 = 0.005\,m$ (before optimization) of design parameter r. For given δ_{K_s}, the optimization problem consists in finding r^{RD} in C_r such that

$$\mathcal{J}(r^{\text{RD}}) \leq \mathcal{J}(r) \quad \text{for all} \quad r \quad \text{in} \quad C_r\,.$$

The stochastic solver is the Monte Carlo numerical simulation method (see Section 6.4) for which $n_s = 500$ independent realizations are used.

□ *Optimum value of the design parameter obtained with the design optimization and with the robust design optimization.* Figure 9.24 (left) displays the graph $f \mapsto 10\,\log_{10}(\underline{w}(r_0, 2\pi f))$ computed with the mean computational model for the nominal value r_0 and the confidence region for a probability level 0.98 of $f \mapsto 10\,\log_{10}(W(r_0, 2\pi f))$. Figure 9.24 (right) displays the graph of function $r \mapsto 10\,\log_{10}(\max_{\omega \in \mathcal{B}}\,\underline{w}(r, \omega))$, which shows that the optimum r^{D} of design parameter r, computed by solving the design optimization problem, is $r^{\text{D}} = 5.9 \times 10^{-3}\,m$.

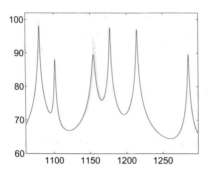 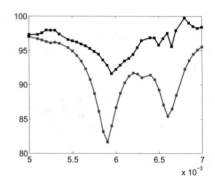

Fig. 9.24 Left figure: the horizontal axis is the frequency in Hz; graph $f \mapsto 10 \log_{10}(\underline{w}(r_0, 2\pi f))$ computed with the mean computational model for the nominal value r_0 (red solid line); confidence region for a probability level 0.98 of $f \mapsto 10 \log_{10}(W(r_0, 2\pi f))$ (region in yellow color or in gray). Right figure: the horizontal axis is the design parameter r; graph $r \mapsto 10 \log_{10}(\max_{\omega \in \mathcal{B}} \underline{w}(r, \omega))$ (lower blue line); graph $r \mapsto 10 \log_{10}(\max_{\omega \in \mathcal{B}} w^+(r, \omega))$ (upper black line). [Figure from [33]]

This figure also displays the graph $r \mapsto 10 \log_{10}(\max_{\omega \in \mathcal{B}} w^+(r, \omega))$, which shows that the optimum r^{RD} of design parameter r, computed by solving the robust design optimization problem, is $r^{\mathrm{RD}} = 5.95 \times 10^{-3}\ m$.

□ *Confidence region of the acoustic observation computed with the design optimization and with the robust design optimization.* Figure 9.25 displays the confidence region of the acoustic observation that corresponds to the optimal design point r^{D} of design parameter r computed by solving the design optimization problem (left figure) and to the optimal design point r^{RD} of design parameter r computed by solving the robust design optimization (right figure). It can be seen that the resonance peaking of the spectral acoustic energy density $\underline{w}(r^{\mathrm{D}}, \omega)$ has been considerably reduced. For the optimal design parameter r^{D}, the structure has an elastic mode coupled with an acoustic mode of the cavity yielding an elastoacoustic mode. At this resonance, the transfer of energy from the internal acoustic cavity to the structure is maximum and this energy pumping phenomenon is then very sensitive to the design parameter r. With respect to uncertainties, the robustness of the structural-acoustic system for $r = r^{\mathrm{RD}}$ decreases in comparison to the robustness of the nominal structural-acoustic system ($r = r_0$). Considering the acoustic gains normalized with respect to the acoustic level corresponding to the upper envelope of the confidence region of the nominal structural-acoustic system ($r = r_0$), we obtain an acoustic gain of 4.5 dB for the robust design optimization and an acoustic gain of 15.7 dB for the design optimization. Clearly, a real structural-acoustic system manufactured from the optimal robust design yields the most optimal performance.

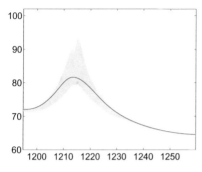 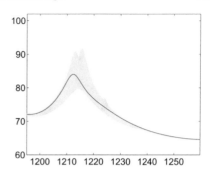

Fig. 9.25 The horizontal axis is the frequency in *Hz*. Design optimization (left figure): graph $f \mapsto 10 \log_{10}(\underline{w}(r^{\mathrm{D}}, 2\pi f))$ (black solid line); confidence region for a probability level 0.98 of $f \mapsto 10 \log_{10}(W(r^{\mathrm{D}}, 2\pi f))$ (region in yellow color or in gray). Robust design optimization (right figure): graph $f \mapsto 10 \log_{10}(\underline{w}(r^{\mathrm{RD}}, 2\pi f))$ (black solid line); confidence region for a probability level 0.98 of $f \mapsto 10 \log_{10}(W(r^{\mathrm{RD}}, 2\pi f))$ (region in yellow region of in gray). [Figure from [33]]

Chapter 10
Random Fields and Uncertainty Quantification in Solid Mechanics of Continuum Media

Abstract *The statistical inverse problem for the experimental identification of a non-Gaussian matrix-valued random field that is the model parameter of a boundary value problem, using some partial and limited experimental data related to a model observation, is a very difficult and challenging problem. A complete advanced methodology is presented and is based on the use of all the developments presented in the previous chapters and in particular, the random matrix theory (Chapter 5), the stochastic solvers (Chapters 4 and 6), the statistical inverse methods (Chapter 7). However, we will start this chapter by presenting new mathematical results concerning the random fields and their polynomial chaos representations, which constitute the extension in infinite dimension of the tools presented in Sections 5.5 to 5.7 for the finite dimension, and which are necessary for solving the statistical inverse problems related to the non-Gaussian random fields.*

245

C. Soize, *Uncertainty Quantification*, Interdisciplinary Applied Mathematics 47,
DOI 10.1007/978-3-319-54339-0_10

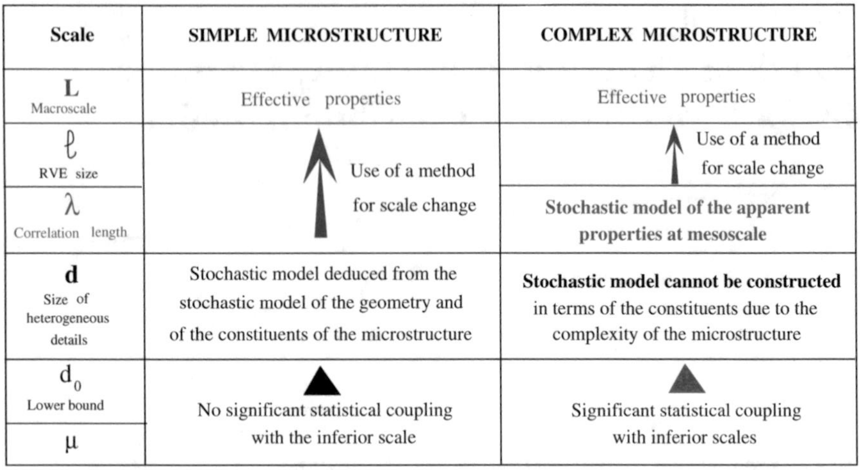

Scale	SIMPLE MICROSTRUCTURE	COMPLEX MICROSTRUCTURE
L Macroscale	Effective properties	Effective properties
ℓ RVE size	▲ Use of a method	▲ Use of a method for scale change
λ Correlation length	for scale change	Stochastic model of the apparent properties at mesoscale
d Size of heterogeneous details	Stochastic model deduced from the stochastic model of the geometry and of the constituents of the microstructure	**Stochastic model cannot be constructed** in terms of the constituents due to the complexity of the microstructure
d_0 Lower bound μ	▲ No significant statistical coupling with the inferior scale	▲ Significant statistical coupling with inferior scales

Fig. 10.1 Definition of the scales for a simple microstructure and for a complex microstructure, and types of stochastic modeling.

What Is a Microstructure for a Material and Which Kind of Stochastic Modeling Can Be Done

☐ *Simple microstructure.* A simple microstructure of a material is a microstructure that can be described in terms of its constituents such as a composite material made up of a polymer matrix with long carbon fibers.

☐ *Complex microstructure.* A complex microstructure of a material is a microstructure that cannot be described in terms of its constituents such as a live tissue. For instance, the hierarchy of the structure of a tendon can be described by using six scales from 1.5×10^{-9} m to 10^{-3} m that correspond to the tropocollagen, to the microfibril, to the subfibril, to the fibril, to the fascicle, and to the tendon.

Figure 10.1 shows which kind of stochastic modeling can be done in the framework of multiscale modeling of materials. In particular, it can be seen that, for a heterogeneous complex microstructure, a stochastic model of the apparent properties of the microstructure can be constructed at the mesoscale that corresponds to the scale of the spatial correlation length of the microstructure. This case constitutes the framework of the applications that will be given in this chapter for illustrating the general methodologies that will be presented.

10.1 Random Fields and Their Polynomial Chaos Representations

A non-Gaussian second-order random field is completely defined by its system of marginal probability distributions, which is an uncountable family of probability distributions on sets of finite dimension, and not only by its mean function and its covariance function as for a Gaussian random field. The experimental identification of such a non-Gaussian random field then requires the introduction of an adapted representation in order to be in capability to solve the statistical inverse problem. For any non-Gaussian second-order random field, an important type of representation is based on the use of the polynomial chaos expansion [27], for which the development and the use in computational sciences and engineering was initiated in [84, 85]. An efficient construction is proposed, which consists in combining a Karhunen-Loève expansion (that allows using a statistical reduced model) with a polynomial chaos expansion of the statistical reduced model. This type of construction has then been re-analyzed and used for solving boundary value problems using the spectral approach (see, for instance, [167, 59, 69, 87, 88, 145, 123, 152]). The polynomial chaos expansion has also been extended for an arbitrary probability measure [222, 122, 126, 188, 217, 74] and for sparse representation [24]. New algorithms have been proposed for obtaining a robust computation of realizations of high degrees polynomial chaos [199, 162]. This type of representation has also been extended for the case of the polynomial chaos expansion with random coefficients [196], for the construction of a basis adaptation in homogeneous chaos spaces [213], for an arbitrary multimodal multidimensional probability distribution [206], and for data-driven probability concentration and sampling on manifold [210].

10.1.1 Definition of a Random Field

Let Ω be any part of $\mathbb{R}^d$ with $d \geq 1$ (possibly with $\Omega = \mathbb{R}^d$). It is assumed that Ω in an uncountable set. A *random field* $\{\mathbf{V}(\mathbf{x}), \mathbf{x} \in \Omega\}$, defined on a probability space $(\Theta, \mathcal{T}, \mathcal{P})$, indexed by Ω, with values in $\mathbb{R}^\mu$, is defined by a mapping

$$\mathbf{x} \mapsto \mathbf{V}(\mathbf{x}) = (V_1(\mathbf{x}), \dots, V_\mu(\mathbf{x})), \qquad (10.1)$$

from Ω into the set of all the random variables from Θ into $\mathbb{R}^\mu$. For all $\mathbf{x}$ fixed in Ω, $\mathbf{V}(\mathbf{x})$ is a random variable $\theta \mapsto \mathbf{V}(\mathbf{x}, \theta)$ from Θ into $\mathbb{R}^\mu$. For all θ fixed in Θ, $\mathbf{V}(\mathbf{x}, \theta)$ is a *realization* (or a *sample*) of $\mathbb{R}^\mu$-valued random variable $\mathbf{V}(\mathbf{x})$, and $\mathbf{x} \mapsto \mathbf{V}(\mathbf{x}, \theta)$ is a *trajectory* (or a *realization path*) of the random field $\mathbf{V}$.

10.1.2 System of Marginal Probability Distributions

Let $\mathbf{x}^1, \dots, \mathbf{x}^\nu$ be any finite subset of Ω and let $\mathbb{V} = (\mathbf{V}(\mathbf{x}^1), \dots, \mathbf{V}(\mathbf{x}^\nu))$ be the random variable with values in $\mathbb{R}^n$ with $n = \nu \times \mu$. The probability distribution $P_{\mathbb{V}}(d\mathbf{v})$ on $\mathbb{R}^n$ with $\mathbf{v} = (\mathbf{v}^1, \dots, \mathbf{v}^\nu)$ of random variable $\mathbb{V}$ is defined as the *marginal* probability distribution of random field $\mathbf{V}$ related to the finite subset $\mathbf{x}^1, \dots, \mathbf{x}^\nu$. The *system of marginal* probability distributions of random field $\mathbf{V}$ is made up of the uncountable family of all the finite probability distributions $P_{\mathbb{V}}(d\mathbf{v})$ obtained for all the possible finite subset of Ω.

▷ *Particular case of a Gaussian random field.* Random field **V** is Gaussian if every probability distribution $P_\mathbf{V}(d\mathbf{v})$ of the system of marginal probability distributions is Gaussian, that is to say, is defined by a Gaussian probability density function.

10.1.3 Summary of the Karhunen-Loeve Expansion of the Random Field

For readability, we summarize the main results for the Karhunen-Loeve expansion of random field **V** by adapting the notations that we have introduced in Section 5.7.

☐ *Second-order moments of random field V.* Random field $\{\mathbf{V}(\mathbf{x}), \mathbf{x} \in \Omega\}$ is a second-order random field if $E\{\|\mathbf{V}(\mathbf{x})\|^2\} < +\infty$ for all $\mathbf{x}$ in Ω. The second-order quantities of this random field are defined as follows:

– *Mean function:* $\mathbf{x} \mapsto \mathbf{m}_\mathbf{V}(\mathbf{x}) = E\{\mathbf{V}(\mathbf{x})\}$ from Ω into $\mathbb{R}^\mu$.
– *Autocorrelation function:* $(\mathbf{x}, \mathbf{x}') \mapsto [R_\mathbf{V}(\mathbf{x}, \mathbf{x}')] = E\{\mathbf{V}(\mathbf{x})\,\mathbf{V}(\mathbf{x}')^T\}$ from $\Omega \times \Omega$ into $\mathbb{M}_\mu(\mathbb{R})$.
– *Covariance function:* $(\mathbf{x}, \mathbf{x}') \mapsto [C_\mathbf{V}(\mathbf{x}, \mathbf{x}')] = E\{(\mathbf{V}(\mathbf{x}) - \mathbf{m}_\mathbf{V}(\mathbf{x})) \times (\mathbf{V}(\mathbf{x}') - \mathbf{m}_\mathbf{V}(\mathbf{x}'))^T\}$ from $\Omega \times \Omega$ into $\mathbb{M}_\mu(\mathbb{R})$, which is assumed to be such that

$$\int_\Omega \int_\Omega \|\,[C_\mathbf{V}(\mathbf{x}, \mathbf{x}')]\,\|_F^2 \; d\mathbf{x}' \, d\mathbf{x}' < +\infty. \qquad (10.2)$$

☐ *Eigenvalue problem and Hilbert basis.* Under the hypothesis defined by Equation (10.2), the eigenvalue problem,

$$\mathbf{C}_\mathbf{V}\,\phi = \lambda\,\phi \quad : \quad \int_\Omega [C_\mathbf{V}(\mathbf{x}, \mathbf{x}')]\,\phi(\mathbf{x}')\,d\mathbf{x}' = \lambda\,\phi(\mathbf{x}), \qquad (10.3)$$

admits a decreasing sequence of positive eigenvalues $\lambda_1 \geq \lambda_2 \geq \ldots \to 0$ such that $\sum_{i=1}^{+\infty} \lambda_i^2 < +\infty$ and the family $\{\phi^i, i \in \mathbb{N}^*\}$ of the eigenvectors constitutes a Hilbert basis of $L^2(\Omega, \mathbb{R}^\mu)$. Consequently, we have

$$< \phi^i, \phi^{i'} >_{L^2} = \int_\Omega < \phi^i(\mathbf{x}), \phi^{i'}(\mathbf{x}) > d\mathbf{x} = \delta_{ii'}, \qquad (10.4)$$

$$\|\phi^i\|_{L^2}^2 = \int_\Omega \|\phi^i(\mathbf{x})\|^2 \, d\mathbf{x} = 1. \qquad (10.5)$$

☐ *KL expansion.* The KL expansion of $\{\mathbf{V}(\mathbf{x}), \mathbf{x} \in \Omega\}$ is written as

$$\mathbf{V}(\mathbf{x}) = \mathbf{m_V}(\mathbf{x}) + \sum_{i=1}^{\infty} \sqrt{\lambda_i}\, \eta_i\, \boldsymbol{\phi}^i(\mathbf{x}), \tag{10.6}$$

in which the random variables $\{\eta^i, i \in \mathbb{N}^*\}$ are given by

$$\eta_i = \frac{1}{\sqrt{\lambda_i}} < \mathbf{V} - \mathbf{m_V}, \boldsymbol{\phi}^i >_{L^2}$$

$$= \frac{1}{\sqrt{\lambda_i}} \int_{\Omega} < \mathbf{V}(\mathbf{x}) - \mathbf{m_V}(\mathbf{x}), \boldsymbol{\phi}^i(\mathbf{x}) > d\mathbf{x}, \tag{10.7}$$

are centered, and are uncorrelated

$$E\{\eta_i\} = 0 \quad , \quad E\{\eta_i\, \eta_{i'}\} = \delta_{ii'}. \tag{10.8}$$

It should be noted that the random variables $\{\eta^i, i \in \mathbb{N}^*\}$ are statistically dependent and are, generally, not Gaussian. Nevertheless, if the random field $\mathbf{V}$ was Gaussian, then the random variables $\{\eta^i, i \in \mathbb{N}^*\}$ would be Gaussian (see Theorem 2.1 in Section 2.5.2), and as there are centered and uncorrelated, they would statistically be independent.

□ *Finite approximation (statistical reduction).* The reduced statistical model of $\{\mathbf{V}(\mathbf{x}), \mathbf{x} \in \Omega\}$ is then written as $\{\mathbf{V}^{(m)}(\mathbf{x}), \mathbf{x} \in \Omega\}$ such that

$$\mathbf{V}^{(m)}(\mathbf{x}) = \mathbf{m_V}(\mathbf{x}) + \sum_{i=1}^{m} \sqrt{\lambda_i}\, \eta_i\, \boldsymbol{\phi}^i(\mathbf{x}). \tag{10.9}$$

The $\mathbb{R}^m$-valued random variable $\boldsymbol{\eta} = (\eta_1, \ldots, \eta_m)$ is such that

$$E\{\boldsymbol{\eta}\} = 0 \quad , \quad E\{\boldsymbol{\eta}\, \boldsymbol{\eta}^T\} = [I_m]. \tag{10.10}$$

The L^2-convergence of the reduced statistical model is controlled by

$$E\{\|\mathbf{V} - \mathbf{V}^{(m)}\|_{L^2}^2\} = \sum_{i=m+1}^{\infty} \lambda_i = E\{\|\mathbf{V} - \mathbf{m_V}\|_{L^2}^2\} - \sum_{i=1}^{m} \lambda_i. \tag{10.11}$$

10.1.4 Polynomial Chaos Expansion of a Random Field

□ *Methodology.* The method [85] consists in performing a finite PCE (polynomial chaos expansion) $\boldsymbol{\eta}^{\text{chaos}}(N_d, N_g)$ (see Section 5.5) of the $\mathbb{R}^m$-valued random variable $\boldsymbol{\eta}$, yielding the finite approximation

$$\mathbf{V}^{(m, N_d, N_g)}(\mathbf{x}) = \mathbf{m_V}(\mathbf{x}) + \sum_{i=1}^{m} \sqrt{\lambda_i}\, \eta_i^{\text{chaos}}(N_d, N_g)\, \boldsymbol{\phi}^i(\mathbf{x}), \tag{10.12}$$

$$E\{\boldsymbol{\eta}^{\text{chaos}}(N_d, N_g)\} = 0 \; , \;\; E\{\boldsymbol{\eta}^{\text{chaos}}(N_d, N_g)\, \boldsymbol{\eta}^{\text{chaos}}(N_d, N_g)^T\} = [I_m] \,, \quad (10.13)$$

and then, analyzing the convergence of $\{\mathbf{V}^{(m, N_d, N_g)}(\mathbf{x}), \mathbf{x} \in \Omega\}_{(m, N_d, N_g)}$ in $L^2(\Omega, L^2(\Theta, \mathbb{R}^\mu))$.

□ *Summary of the finite PCE introduced in Section 5.5.* Let $\boldsymbol{\Xi}$ be a given $\mathbb{R}^{N_g}$-valued random variable, defined on $(\Theta, \mathcal{T}, \mathcal{P})$, with a given pdf $p_{\boldsymbol{\Xi}}(\boldsymbol{\xi})$ with respect to $d\boldsymbol{\xi}$, with support $\text{Supp}\, p_{\boldsymbol{\Xi}} = s \subset \mathbb{R}^{N_g}$ (possibly with $s = \mathbb{R}^{N_g}$), and such that

$$\int_{\mathbb{R}^{N_g}} \|\boldsymbol{\xi}\|^{\widetilde{m}}\, p_{\boldsymbol{\Xi}}(\boldsymbol{\xi})\, d\boldsymbol{\xi} < +\infty \quad , \quad \forall \widetilde{m} \in \mathbb{N}. \tag{10.14}$$

Let $\boldsymbol{\alpha} = (\alpha_1, \ldots, \alpha_{N_g}) \in \mathcal{A} = \mathbb{N}^{N_g}$ be a multi-index. For $N_d \geq 1$, let $\mathcal{A}_{N_d} \subset \mathcal{A}$ be the finite subset of multi-indices,

$$\mathcal{A}_{N_d} = \{\boldsymbol{\alpha} = (\alpha_1, \ldots, \alpha_{N_g}) \in \mathbb{N}^{N_g} \mid 0 \leq \alpha_1 + \ldots + \alpha_{N_g} \leq N_d\}. \tag{10.15}$$

The $1 + N$ elements of $\mathcal{A}_{N_d}$ are denoted by $\boldsymbol{\alpha}^{(0)}, \ldots \boldsymbol{\alpha}^{(N)}$ with $\boldsymbol{\alpha}^{(0)} = (0, \ldots, 0)$, and where the integer $N = \mathsf{h}(N_d, N_g)$ is defined as a function of N_d and N_g by

$$N = \mathsf{h}(N_d, N_g) := \frac{(N_g + N_d)!}{N_g!\, N_d!} - 1. \tag{10.16}$$

For m fixed, the PCE of $\boldsymbol{\eta}$ is then written as

$$\boldsymbol{\eta} = \lim_{N_g \to m, N_d \to +\infty} \boldsymbol{\eta}^{\text{chaos}}(N_d, N_g) \quad \text{in} \quad L^2(\Theta, \mathbb{R}^m)\,, \tag{10.17}$$

$$\boldsymbol{\eta}^{\text{chaos}}(N_d, N_g) = \sum_{j=1}^{N} \mathbf{y}^j\, \Psi_{\boldsymbol{\alpha}^{(j)}}(\boldsymbol{\Xi}) \quad , \quad \mathbf{y}^j = E\{\boldsymbol{\eta}\, \Psi_{\boldsymbol{\alpha}^{(j)}}(\boldsymbol{\Xi})\} \in \mathbb{R}^m\,. \tag{10.18}$$

Taking into account the second equation in Equation (10.13), it can be deduced that the coefficients of the PCE must satisfy the following constraint equation,

$$\mathbf{y}^0 = \mathbf{0} \quad , \quad \sum_{j=1}^{N} \mathbf{y}^j (\mathbf{y}^j)^T = [I_m]\,. \tag{10.19}$$

The L^2-error of the finite PCE can be estimated by using the following equation:

$$E\{\|\boldsymbol{\eta} - \boldsymbol{\eta}^{\text{chaos}}(N_d, N_g)\|^2\} = E\{\|\boldsymbol{\eta}\|^2\} - \sum_{j=0}^{N} \|\mathbf{y}^j\|^2\,. \tag{10.20}$$

□ *Finite polynomial chaos expansion of the random field.* The finite PCE denoted by $\{\mathbf{V}^{(m, N_d, N_g)}(\mathbf{x}), \mathbf{x} \in \Omega\}$ of random field $\{\mathbf{V}(\mathbf{x}), \mathbf{x} \in \Omega\}$ can be summarized by the following equations:

$$\mathbf{V}^{(m,N_d,N_g)}(\mathbf{x}) = \mathbf{m_V}(\mathbf{x}) + \sum_{i=1}^{m} \sqrt{\lambda_i}\, \eta_i^{\text{chaos}}(N_d,N_g)\, \boldsymbol{\phi}^i(\mathbf{x})\,, \tag{10.21}$$

$$\boldsymbol{\eta}^{\text{chaos}}(N_d,N_g) = \sum_{j=1}^{N} \mathbf{y}^j\, \Psi_{\boldsymbol{\alpha}(j)}(\boldsymbol{\Xi})\quad,\quad \mathbf{y}^j \in \mathbb{R}^m\,, \tag{10.22}$$

$$\mathbf{y}^0 = \mathbf{0}\quad,\quad \sum_{j=1}^{N} \mathbf{y}^j(\mathbf{y}^j)^T = [I_m]\,, \tag{10.23}$$

which can be rewritten, by introducing $\mathbf{v}^j(\mathbf{x}) = \sum_{i=1}^{m} \sqrt{\lambda_i}\, y_i^j\, \boldsymbol{\phi}^i(\mathbf{x})$, as

$$\mathbf{V}^{(m,N_d,N_g)}(\mathbf{x}) = \mathbf{m_V}(\mathbf{x}) + \sum_{j=1}^{N} \mathbf{v}^j(\mathbf{x})\, \Psi_{\boldsymbol{\alpha}(j)}(\boldsymbol{\Xi})\,, \tag{10.24}$$

$$\mathbf{v}^j(\mathbf{x}) = E\{(\mathbf{V}(\mathbf{x}) - \mathbf{m_V}(\mathbf{x}))\, \Psi_{\boldsymbol{\alpha}(j)}(\boldsymbol{\Xi})\} \in \mathbb{R}^\mu\,, \tag{10.25}$$

$$\mathbf{v}^0 = \mathbf{0}\quad,\quad <\mathbf{v}^j, \mathbf{v}^{j'}>_{L^2} = \sum_{i=1}^{m} \lambda_i\, y_i^j\, y_i^{j'}\,. \tag{10.26}$$

10.2 Setting the Statistical Inverse Problem to Be Solved for the High Stochastic Dimensions

The statistical inverse problem related to the identification of a non-Gaussian random field, which is the matrix-valued coefficient of a partial differential equation of a boundary value problem, is a very challenging problem, due to the non-Gaussian character (which cannot be summarized in a simple identification of a mean function and of a covariance function), and also due to the fact that the stochastic problem is generally in high dimension.

The use of the polynomial chaos expansion for constructing a parameterized representation of a non-Gaussian random field that models the model parameter of a boundary value problem in order to identify it using a statistical inverse method has been initialized in [63, 64], used in [92, 93], and revisited in [57]. In [56], the construction of the probability model of the random coefficients of the polynomial chaos expansion is proposed by using the asymptotic sampling Gaussian distribution constructed with the Fisher information matrix, and used for model validation [87]. This work has been developed for statistical inverse problems that are rather in low stochastic dimension, and new ingredients have been introduced in [200, 162, 203, 163] for statistical inverse problems in high stochastic dimension. In using the reduced chaos decomposition with random coefficients of random fields [196], a Bayesian approach for identifying the posterior probability model of the random coefficients of the polynomial chaos expansion of the model parameter of the BVP has been proposed in [10] for the low stochastic dimension and in [201] for the high stochastic dimension. Uncertainty quantification in computational stochastic multiscale analysis of nonlinear elastic materials has been carried out by using PCE of random fields [50, 51]. The experimental identification of a non-Gaussian positive matrix-valued random field in high stochastic dimension, using partial and limited experimental data for a model observation related to the random solution of a stochastic BVP, is a difficult problem that requires both adapted representations and methodologies [197, 201, 203, 153].

This chapter is mainly focussed on the stochastic elliptic operators that are frequently encountered in computational mechanics. We present the methodologies, the stochastic modeling, and the statistical inverse problems.

10.2.1 Stochastic Elliptic Operator and Boundary Value Problem

□ *Notation.* Let d be equal to 2 or 3. Let $n \geq 1$ and $1 \leq N_u \leq n$ be integers. Let Ω be a bounded open domain of $\mathbb{R}^d$, with generic point $\mathbf{x} = (x_1, \ldots, x_d)$, with boundary $\partial\Omega$, and let $\overline{\Omega} = \Omega \cup \partial\Omega$ be its boundary for which $\mathbf{n}(\mathbf{x}) = (n_1(\mathbf{x}), \ldots, n_d(\mathbf{x}))$ is the outward unit normal to $\partial\Omega$.

□ *Definition of the random field as the coefficient of an elliptic operator.* Let $[\mathbf{K}] = \{[\mathbf{K}(\mathbf{x})], \mathbf{x} \in \Omega\}$ be a non-Gaussian random field, defined on $(\Theta, \mathcal{T}, \mathcal{P})$, indexed by Ω, and with values in $\mathbb{M}_n^+(\mathbb{R})$. Since $[\mathbf{K}]$ is with values in $\mathbb{M}_n^+(\mathbb{R})$, then random field $[\mathbf{K}]$ cannot be a Gaussian random field. Random field $[\mathbf{K}]$ is used for modeling the coefficients of a given stochastic elliptic operator $\mathbf{u} \mapsto \mathcal{D}_{\mathbf{x}}(\mathbf{u})$ that applies to the field $\mathbf{u}(\mathbf{x}) = (u_1(\mathbf{x}), \ldots, u_{N_u}(\mathbf{x}))$ defined on Ω and with values in $\mathbb{R}^{N_u}$. The boundary value problem that is formulated in $\mathbf{u}$ involves the stochastic elliptic operator $\mathcal{D}_{\mathbf{x}}$ with the boundary conditions given on $\partial\Omega = \Gamma_0 \cup \Gamma \cup \Gamma_1$ in which, a Dirichlet condition is given on Γ_0, a zero Neumann condition is given on Γ, and a nonzero Neumann condition is given on Γ_1.

□ *Example of a stochastic elliptic operator: diffusion operator.* For a three-dimensional anisotropic diffusion problem, the stochastic elliptic differential operator $\mathcal{D}_{\mathbf{x}}$ relative to the density u of the diffusing medium is written as

$$\{\mathcal{D}_{\mathbf{x}}(u)\}(\mathbf{x}) = -\mathrm{div}_{\mathbf{x}}([\mathbf{K}(\mathbf{x})]\,\nabla_{\mathbf{x}}u(\mathbf{x})) \quad, \quad \mathbf{x} \in \Omega, \tag{10.27}$$

in which $d = n = 3$, where $N_u = 1$, where $\mathrm{div}_{\mathbf{x}} = \{\nabla_{\mathbf{x}} \cdot\}$ is the divergence operator, and where $\{[\mathbf{K}(\mathbf{x})], \mathbf{x} \in \Omega\}$ is the $\mathbb{M}_n^+(\mathbb{R})$-valued random field of the medium.

□ *Example of stochastic elliptic operator: elasticity operator.* For the wave propagation inside a three-dimensional random heterogeneous anisotropic linear elastic medium, we have $d = 3$, $n = 6$, $N_u = 3$, and the stochastic elliptic differential operator $\mathcal{D}_{\mathbf{x}}$ relative to the displacement field $\mathbf{u}$ is written as

$$\{\mathcal{D}_{\mathbf{x}}(\mathbf{u})\}(\mathbf{x}) = -[D_{\mathbf{x}}]^T[\mathbf{K}(\mathbf{x})][D_{\mathbf{x}}]\,\mathbf{u}(\mathbf{x})\quad,\quad \mathbf{x} \in \Omega\,, \tag{10.28}$$

in which $\{[\mathbf{K}(\mathbf{x})], \mathbf{x} \in \Omega\}$ is the $\mathbb{M}_n^+(\mathbb{R})$-valued elasticity random field of the medium deduced from the fourth-order tensor-valued elasticity field $\{\mathbf{C}_{ijkh}(\mathbf{x}), \mathbf{x} \in \Omega\}$ by the following equation

$$[\mathbf{K}] = \begin{bmatrix} \mathbf{C}_{1111} & \mathbf{C}_{1122} & \mathbf{C}_{1133} & \sqrt{2}\mathbf{C}_{1112} & \sqrt{2}\mathbf{C}_{1113} & \sqrt{2}\mathbf{C}_{1123} \\ \mathbf{C}_{2211} & \mathbf{C}_{2222} & \mathbf{C}_{2233} & \sqrt{2}\mathbf{C}_{2212} & \sqrt{2}\mathbf{C}_{2213} & \sqrt{2}\mathbf{C}_{2223} \\ \mathbf{C}_{3311} & \mathbf{C}_{3322} & \mathbf{C}_{3333} & \sqrt{2}\mathbf{C}_{3312} & \sqrt{2}\mathbf{C}_{3313} & \sqrt{2}\mathbf{C}_{3323} \\ \sqrt{2}\mathbf{C}_{1211} & \sqrt{2}\mathbf{C}_{1222} & \sqrt{2}\mathbf{C}_{1233} & 2\mathbf{C}_{1212} & 2\,\mathbf{C}_{1213} & 2\mathbf{C}_{1223} \\ \sqrt{2}\mathbf{C}_{1311} & \sqrt{2}\mathbf{C}_{1322} & \sqrt{2}\mathbf{C}_{1333} & 2\mathbf{C}_{1312} & 2\,\mathbf{C}_{1313} & 2\mathbf{C}_{1323} \\ \sqrt{2}\mathbf{C}_{2311} & \sqrt{2}\mathbf{C}_{2322} & \sqrt{2}\mathbf{C}_{2333} & 2\mathbf{C}_{2312} & 2\,\mathbf{C}_{2313} & 2\mathbf{C}_{2323} \end{bmatrix} \tag{10.29}$$

and where $[D_{\mathbf{x}}]$ is the differential operator,

$$[D_{\mathbf{x}}] = [M^{(1)}]\frac{\partial}{\partial x_1} + [M^{(2)}]\frac{\partial}{\partial x_2} + [M^{(3)}]\frac{\partial}{\partial x_3}\,, \tag{10.30}$$

in which $[M^{(1)}]$, $[M^{(2)}]$, and $[M^{(3)}]$ are the $(n \times N_u)$ real matrices defined by

$$[M^{(1)}] = \begin{bmatrix} 1 & 0 & 0 \\ 0 & 0 & 0 \\ 0 & 0 & 0 \\ 0 & \frac{1}{\sqrt{2}} & 0 \\ 0 & 0 & \frac{1}{\sqrt{2}} \\ 0 & 0 & 0 \end{bmatrix}, [M^{(2)}] = \begin{bmatrix} 0 & 0 & 0 \\ 0 & 1 & 0 \\ 0 & 0 & 0 \\ \frac{1}{\sqrt{2}} & 0 & 0 \\ 0 & 0 & 0 \\ 0 & 0 & \frac{1}{\sqrt{2}} \end{bmatrix}, [M^{(3)}] = \begin{bmatrix} 0 & 0 & 0 \\ 0 & 0 & 0 \\ 0 & 0 & 1 \\ 0 & 0 & 0 \\ \frac{1}{\sqrt{2}} & 0 & 0 \\ 0 & \frac{1}{\sqrt{2}} & 0 \end{bmatrix}.$$

☐ *Example of a time-independent stochastic boundary value problem in linear elasticity.* The stochastic boundary value problem for a linear elastostatic of a 3D random heterogeneous anisotropic linear elastic medium is written as

$$-[D_{\mathbf{x}}]^T[\mathbf{K}(\mathbf{x})][D_{\mathbf{x}}]\,\mathbf{u}(\mathbf{x}) = \mathbf{0}\quad \text{in}\quad \Omega\,, \tag{10.31}$$

$$\mathbf{u} = \mathbf{0}\quad \text{on}\quad \Gamma_0\,, \tag{10.32}$$

$$[\mathcal{M}_{\mathbf{n}}(\mathbf{x})]^T\,[\mathbf{K}(\mathbf{x})][D_{\mathbf{x}}]\,\mathbf{u}(\mathbf{x}) = \mathbf{0}\ \text{on}\ \Gamma\,, \tag{10.33}$$

$$[\mathcal{M}_{\mathbf{n}}(\mathbf{x})]^T\,[\mathbf{K}(\mathbf{x})][D_{\mathbf{x}}]\,\mathbf{u}(\mathbf{x}) = \mathbf{f}_{\Gamma_1}\ \text{on}\ \Gamma_1\,, \tag{10.34}$$

in which $[\mathcal{M}_{\mathbf{n}}(\mathbf{x})] = [M^{(1)}]\,n_1(\mathbf{x}) + [M^{(2)}]\,n_2(\mathbf{x}) + [M^{(3)}]\,n_3(\mathbf{x})$, and where $\mathbf{f}_{\Gamma_1}$ is a given surface force field applied to Γ_1.

10.2.2 Stochastic Computational Model and Available Data Set

Let $\mathcal{I} = \{\mathbf{x}^1, \ldots, \mathbf{x}^{N_p}\} \subset \Omega$ be the finite subset of Ω made up of all the integrating points that are introduced by the numerical integration formulas used by the finite elements of the mesh of Ω. The random observation of the computational model is the random vector $\mathbf{W}$, defined on a probability space $(\Theta, \mathcal{T}, \mathcal{P})$, with values in $\mathbb{R}^{N_W}$, for which there exist available data that are partial and limited. The components of $\mathbf{W}$ are constituted of the N_W DOFs of the nodal values of the random field $\{\mathbf{U}(\mathbf{x}), \mathbf{x} \in \Omega\}$ at all the nodes of the mesh located on Γ. Consequently, random vector $\mathbf{W}$ is the unique deterministic nonlinear transformation of the finite family of the N_p dependent random matrices $[\mathbf{K}(\mathbf{x}^1)], \ldots, [\mathbf{K}(\mathbf{x}^{N_p})]$ such that

$$\mathbf{W} = \mathbf{h}([\mathbf{K}(\mathbf{x}^1)], \ldots, [\mathbf{K}(\mathbf{x}^{N_p})]), \tag{10.35}$$

in which $([K^1], \ldots, [K^{N_p}]) \mapsto \mathbf{h}([K^1], \ldots, [K^{N_p}])$ is a given deterministic transformation from $\mathbb{M}_n^+(\mathbb{R}) \times \ldots \times \mathbb{M}_n^+(\mathbb{R})$ into $\mathbb{R}^{N_W}$. This transformation $\mathbf{h}$ (which is generally nonlinear) is constructed by solving the computational model that corresponds to the spatial discretization by the finite element method of the boundary value problem. The available data, which correspond to the random observation $\mathbf{W}$ of the stochastic computational model, are made up of the $\nu_{\exp}$ vectors $\mathbf{w}^1_{\exp}, \ldots, \mathbf{w}^{\nu_{\exp}}_{\exp}$ in $\mathbb{R}^{N_W}$, and which are assumed to be $\nu_{\exp}$ independent realizations of a random vector $\mathbf{W}_{\exp}$ that is defined on a probability space $(\Theta^{\exp}, \mathcal{T}^{\exp}, \mathcal{P}^{\exp})$.

10.2.3 Statistical Inverse Problem to Be Solved

The statistical inverse problem that is, *a priori*, in high stochastic dimension, consists in identifying the non-Gaussian matrix-valued random field $\{[\mathbf{K}(\mathbf{x})], \mathbf{x} \in \Omega\}$ by using the partial and limited experimental data $\mathbf{w}^1_{\exp}, \ldots, \mathbf{w}^{\nu_{\exp}}_{\exp}$ that are related to the random observation $\mathbf{W}$ of the stochastic computational model. More exactly, we have to identify the statistically dependent random matrices $[\mathbf{K}(\mathbf{x}^1)], \ldots, [\mathbf{K}(\mathbf{x}^{N_p})]$ such that $\mathbf{w}^1_{\exp}, \ldots, \mathbf{w}^{\nu_{\exp}}_{\exp}$ are $\nu_{\exp}$ independent realizations of the random vector $\mathbf{W}$ defined by Equation (10.35).

10.3 Parametric Model-Based Representation for the Model Parameters and for the Model Observations

A parametric model-based representation of random observation $\mathbf{W}$ is built to make possible the resolution of the statistical inverse problem that allows for identifying random field $[\mathbf{K}]$ by using the available data set. Such a construction is a fundamental question in order to have the capacity to solve the statistical inverse problem in high dimension.

Introduction a Class of Lower-Bounded Random Fields for $[\mathbf{K}]$ and Normalization

☐ *Introduction of a normalization function.* In order to normalize the random field $[\mathbf{K}]$, a deterministic function $\mathbf{x} \mapsto [\underline{K}(\mathbf{x})]$ from Ω into $\mathbb{M}_n^+(\mathbb{R})$ is introduced such that for all $\mathbf{x}$ in Ω and for all $\mathbf{z}$ in $\mathbb{R}^n$,

$$\underline{k}_0 \|\mathbf{z}\|^2 \; \leq \; <[\underline{K}(\mathbf{x})]\,\mathbf{z}\,,\mathbf{z}> \; \leq \; \underline{k}_1 \|\mathbf{z}\|^2 \,, \tag{10.36}$$

in which $0 < \underline{k}_0 < \underline{k}_1 < +\infty$ are independent of $\mathbf{x}$, and which corresponds to the uniform deterministic ellipticity for the operator whose coefficient is $[\underline{K}]$. For all $\mathbf{x}$ fixed in Ω, the Cholesky factorization of $[\underline{K}(\mathbf{x})]$ in $\mathbb{M}_n^+(\mathbb{R})$ yields

$$[\underline{K}(\mathbf{x})] = [\underline{L}(\mathbf{x})]^T \, [\underline{L}(\mathbf{x})] \,. \tag{10.37}$$

☐ *Definition of a normalized class of random fields with a lower bound.* We introduce a normalized class of non-Gaussian positive-definite matrix-valued random fields $[\mathbf{K}]$, which admits a positive-definite matrix-valued lower bound, such that

$$[\mathbf{K}(\mathbf{x})] = \frac{1}{1+\varepsilon}[\underline{L}(\mathbf{x})]^T \{\varepsilon[I_n] + [\mathbf{K}_0(\mathbf{x})]\} [\underline{L}(\mathbf{x})] \quad , \quad \forall \mathbf{x} \in \Omega \,, \tag{10.38}$$

in which $\varepsilon > 0$ is any fixed positive real number, and where $[\mathbf{K}_0] = \{[\mathbf{K}_0(\mathbf{x})], \mathbf{x} \in \Omega\}$ is a random field indexed by Ω, with values in $\mathbb{M}_n^+(\mathbb{R})$. The representation defined by Equation (10.38) can be inverted and yields

$$[\mathbf{K}_0(\mathbf{x})] = (1+\varepsilon)[\underline{L}(\mathbf{x})]^{-T} [\mathbf{K}(\mathbf{x})] [\underline{L}(\mathbf{x})]^{-1} - \varepsilon [I_n] \quad , \quad \forall \mathbf{x} \in \Omega \,. \tag{10.39}$$

There is a lower bound such that, for all $\mathbf{x}$ in Ω,

$$0 < [K_\varepsilon(\mathbf{x})] \leq [\mathbf{K}(\mathbf{x})] \quad \text{a.s.}\,, \quad [K_\varepsilon(\mathbf{x})] = \frac{\varepsilon}{1+\varepsilon}[\underline{K}(\mathbf{x})] \in \mathbb{M}_n^+(\mathbb{R})\,, \tag{10.40}$$

and Equation (10.38) can be rewritten as

$$[\mathbf{K}(\mathbf{x})] = [K_\varepsilon(\mathbf{x})] + \frac{1}{1+\varepsilon}[\underline{L}(\mathbf{x})]^T [\mathbf{K}_0(\mathbf{x})] [\underline{L}(\mathbf{x})] \quad , \quad \forall \mathbf{x} \in \Omega \,. \tag{10.41}$$

If for all $\mathbf{x}$ in Ω, the deterministic matrix $[\underline{K}(\mathbf{x})]$ was chosen as $[\underline{K}(\mathbf{x})] = E\{[\mathbf{K}(\mathbf{x})]\}$, then Equations (10.37) and (10.38) would give $E\{[\mathbf{K}_0(\mathbf{x})]\} = [I_n]$, and consequently, the random field $[\mathbf{K}_0]$ would be normalized. In practice, matrix $[\underline{K}(\mathbf{x})]$ will not be chosen as the mean value but will be close to the mean value and therefore, the random field $[\mathbf{K}_0]$ will be normalized. Finally, from Equations (10.36) and (10.40), it can easily be proved (see [153]) that we have the following invertibility property,

$$\forall p \geq 1 \,, \; \forall \mathbf{x} \in \Omega \,, \; E\{\|[\mathbf{K}(\mathbf{x})]^{-1}\|_F^p\} < +\infty \,. \tag{10.42}$$

▷ The class of random fields $[\mathbf{K}] = \{[\mathbf{K}(\mathbf{x})], \mathbf{x} \in \Omega\}$ defined by Equation (10.38), in which $\varepsilon > 0$ is any fixed positive real number and where

$[\mathbf{K}_0] = \{[\mathbf{K}_0(\mathbf{x})], \mathbf{x} \in \Omega\}$ is any random field indexed by Ω, with values in $\mathbb{M}_n^+(\mathbb{R})$, yields a uniform stochastic elliptic operator $\mathcal{D}_\mathbf{x}$ that allows for studying the existence and uniqueness of a second-order random solution of a stochastic boundary value problem involving $\mathcal{D}_\mathbf{x}$.

Construction of a Nonlinear Transformation $\mathcal{G}$

We have now to complete the construction of the representation in order to release the constraint related to the positiveness that induces some difficulties with respect to the polynomial chaos expansion in order to be in capability to represent any non-Gaussian random field. For that, we propose two types of nonlinear transformation $\mathcal{G}$, an *exponential-type representation* and a *square-type representation*, such that

$$\forall \mathbf{x} \in \Omega \quad , \quad [\mathbf{K}(\mathbf{x})] = \mathcal{G}([\mathbf{G}(\mathbf{x})]) , \tag{10.43}$$

in which $\mathcal{G}$ is independent of $\mathbf{x}$. These two types of representation do not give the same mathematical properties that are slightly different and for which the computational aspects will be different. The better class is the one based on the square-type representation because, for this class, the second-order property of the random field $[\mathbf{G}]$ exists while it does not exist for the exponential-type representation. The exponential-type representation is easier to implement than the square-type representation. This is a good illustration of the modeling effort that must be done in order to construct an efficient stochastic modeling for the statistical inverse problem devoted to its identification.

□ *Exponential-type representation of random field* $[\mathbf{K}_0]$. For all second-order random field $[\mathbf{G}] = \{[\mathbf{G}(\mathbf{x})], \mathbf{x} \in \Omega\}$ with values in $\mathbb{M}_n^S(\mathbb{R})$, which is not assumed to be Gaussian, the random field $[\mathbf{K}_0]$ defined by

$$[\mathbf{K}_0(\mathbf{x})] = \exp_M([\mathbf{G}(\mathbf{x})]) \quad , \quad \forall \mathbf{x} \in \Omega , \tag{10.44}$$

is an $\mathbb{M}_n^+(\mathbb{R})$-valued random field, where $\exp_M$ is the matrix exponential from $\mathbb{M}_n^S(\mathbb{R})$ into $\mathbb{M}_n^+(\mathbb{R})$. If $[\mathbf{K}_0]$ is any $\mathbb{M}_n^+(\mathbb{R})$-valued random field, then there exists a unique $\mathbb{M}_n^S(\mathbb{R})$-valued random field $[\mathbf{G}]$, which, in general, is not a second-order random field, such that

$$[\mathbf{G}(\mathbf{x})] = \log_M([\mathbf{K}_0(\mathbf{x})]) \quad , \quad \forall \mathbf{x} \in \Omega . \tag{10.45}$$

in which $\log_M$ is the principal matrix logarithm, the inverse of $\exp_M$.

□ *Square-type representation of random field* $[\mathbf{K}_0]$. For all second-order random field $[\mathbf{G}] = \{[\mathbf{G}(\mathbf{x})], \mathbf{x} \in \Omega\}$ with values in $\mathbb{M}_n^S(\mathbb{R})$, which is not assumed to be Gaussian, the random field $[\mathbf{K}_0]$ defined by

$$[\mathbf{K}_0(\mathbf{x})] = \mathbb{L}([\mathbf{G}(\mathbf{x})]) = \mathcal{L}([\mathbf{G}(\mathbf{x})])^T \mathcal{L}([\mathbf{G}(\mathbf{x})]) \quad , \quad \forall \mathbf{x} \in \Omega , \tag{10.46}$$

is an $\mathbb{M}_n^+(\mathbb{R})$-valued random field, where $\mathcal{L}$ is a mapping from $\mathbb{M}_n^S(\mathbb{R})$ into $\mathbb{M}_n^U(\mathbb{R})$ (upper triangular matrices with positive diagonal entries), and where

$\mathbb{L}$ is a mapping from $\mathbb{M}_n^S(\mathbb{R})$ into $\mathbb{M}_n^+(\mathbb{R})$ that admits a unique inverse function $\mathbb{L}^{-1}$ from $\mathbb{M}_n^+(\mathbb{R})$ into $\mathbb{M}_n^S(\mathbb{R})$. If $[\mathbf{K}_0]$ is any random field with values in $\mathbb{M}_n^+(\mathbb{R})$, then $\mathbb{L}$ can be constructed for that there exists a unique second-order $\mathbb{M}_n^S(\mathbb{R})$-valued random field $[\mathbf{G}]$, such that

$$[\mathbf{G}(\mathbf{x})] = \mathbb{L}^{-1}([\mathbf{K}_0(\mathbf{x})]) \quad , \quad \forall \mathbf{x} \in \Omega, \tag{10.47}$$

in which $\mathbb{L}^{-1}$ is the mapping from $\mathbb{M}_n^+(\mathbb{R})$ into $\mathbb{M}_n^S(\mathbb{R})$, which is the unique inverse function of $\mathbb{L}$. The general theory is detailed in [153] and a classical example (derived from [191]) is presented in Section 10.6.

□ *Construction of the transformation $\mathcal{G}$ and its inverse $\mathcal{G}^{-1}$ for the exponential-type representation.* From Equations (10.38) and (10.44), it can be deduced that, for all $\mathbf{x}$ in Ω,

$$[\mathbf{K}(\mathbf{x})] = \mathcal{G}([\mathbf{G}(\mathbf{x})])$$
$$= \frac{1}{1+\varepsilon}[\underline{L}(\mathbf{x})]^T \left\{ \varepsilon[I_n] + \exp_{\mathbb{M}}([\mathbf{G}(\mathbf{x})]) \right\} [\underline{L}(\mathbf{x})]. \tag{10.48}$$

From Equations (10.39) and (10.45), it can be deduced that, for all $\mathbf{x}$ in Ω,

$$[\mathbf{G}(\mathbf{x})] = \mathcal{G}^{-1}([\mathbf{K}(\mathbf{x})])$$
$$= \log_{\mathbb{M}}\left\{ (1+\varepsilon)[\underline{L}(\mathbf{x})]^{-T}[\mathbf{K}(\mathbf{x})][\underline{L}(\mathbf{x})]^{-1} - \varepsilon[I_n] \right\}. \tag{10.49}$$

□ *Construction of the transformation $\mathcal{G}$ and its inverse $\mathcal{G}^{-1}$ for the square-type representation.* From Equations (10.38) and (10.46), it can be deduced that, for all $\mathbf{x}$ in Ω,

$$[\mathbf{K}(\mathbf{x})] = \mathcal{G}([\mathbf{G}(\mathbf{x})])$$
$$= \frac{1}{1+\varepsilon}[\underline{L}(\mathbf{x})]^T \left\{ \varepsilon[I_n] + [\mathcal{L}([\mathbf{G}(\mathbf{x})])]^T [\mathcal{L}([\mathbf{G}(\mathbf{x})])] \right\} [\underline{L}(\mathbf{x})]. \tag{10.50}$$

From Equations (10.39) and (10.47), it can be deduced that, for all $\mathbf{x}$ in Ω,

$$[\mathbf{G}(\mathbf{x})] = \mathcal{G}^{-1}([\mathbf{K}(\mathbf{x})])$$
$$= \mathcal{L}^{-1}\left\{ (1+\varepsilon)[\underline{L}(\mathbf{x})]^{-T}[\mathbf{K}(\mathbf{x})][\underline{L}(\mathbf{x})]^{-1} - \varepsilon[I_n] \right\}. \tag{10.51}$$

Truncated Reduced Representation of Second-Order Random Field $[\mathbf{G}]$ and Its Polynomial Chaos Expansion

□ *Construction of a finite approximation of the non-Gaussian random field $[\mathbf{G}]$.* The finite approximation $[\mathbf{G}^{(m,N_d,N_g)}]$ of the non-Gaussian second-order random field $[\mathbf{G}]$ is introduced by combining the KL expansion (see Section 10.1.3) with the finite PCE (see Section 10.1.4) such that,

$$[\mathbf{G}^{(m,N_d,N_g)}(\mathbf{x})] = [G_0(\mathbf{x})] + \sum_{i=1}^{m} \sqrt{\lambda_i} \, [G_i(\mathbf{x})] \, \eta_i^{\text{chaos}}(N_d, N_g) \,, \qquad (10.52)$$

$$\eta_i^{\text{chaos}}(N_d, N_g) = \sum_{j=1}^{N} y_i^j \, \Psi_j(\boldsymbol{\Xi}) \,, \qquad (10.53)$$

in which $\{\Psi_j\}_{j=1}^{N}$ are the multivariate polynomial Gaussian chaos composed of the normalized multivariate Hermite polynomials such that $E\{\Psi_j(\boldsymbol{\Xi}) \, \Psi_{j'}(\boldsymbol{\Xi})\} = \delta_{jj'}$. The integer $N = \hbar(N_d, N_g)$ is such that (see Equation (10.16)),

$$N = (N_d + N_g)! / (N_d! \, N_g!) - 1 \,, \qquad (10.54)$$

where N_d is the maximum degree of the normalized multivariate Hermite polynomials. The coefficients y_i^j are such that

$$\sum_{j=1}^{N} y_i^j \, y_{i'}^j = \delta_{ii'} \,. \qquad (10.55)$$

□ *Rewriting the constraints equation for the PCE coefficients.* The relation defined by Equation (10.55) between the coefficients can be rewritten as

$$[z]^T \, [z] = [I_m] \,, \qquad (10.56)$$

in which $[z] \in \mathbb{M}_{N,m}(\mathbb{R})$ is such that

$$[z]_{ji} = y_i^j \quad, \quad 1 \le i \le m \quad, \quad 1 \le j \le N \,. \qquad (10.57)$$

Introducing the random vectors $\boldsymbol{\Psi}(\boldsymbol{\Xi}) = (\Psi_1(\boldsymbol{\Xi}), \ldots, \Psi_N(\boldsymbol{\Xi}))$ and $\boldsymbol{\eta}^{\text{chaos}}(N_d, N_g) = (\eta_1^{\text{chaos}}(N_d, N_g), \ldots, \eta_m^{\text{chaos}}(N_d, N_g))$ yield

$$\boldsymbol{\eta}^{\text{chaos}}(N_d, N_g) = [z]^T \, \boldsymbol{\Psi}(\boldsymbol{\Xi}) \,, \qquad (10.58)$$

in which $[z]$ belongs to the compact Stiefel manifold,

$$\mathbb{V}_m(\mathbb{R}^N) = \{[z] \in \mathbb{M}_{N,m}(\mathbb{R}) \,; \; [z]^T \, [z] = [I_m]\} \,. \qquad (10.59)$$

Parameterization of the Compact Stiefel Manifold $\mathbb{V}_m(\mathbb{R}^N)$

For solving the statistical inverse problem related to the identification of the coefficients of the PCE of the non-Gaussian random field $[\mathbf{K}]$, we need to introduce a parameterization of $\mathbb{V}_m(\mathbb{R}^N)$ and to propose an algorithm for numerically exploring this manifold.

□ *Objective of the parameterization.* For all $[z_0]$ fixed in $\mathbb{V}_m(\mathbb{R}^N)$, let $T_{[z_0]}$ be the tangent space to $\mathbb{V}_m(\mathbb{R}^N)$ at $[z_0]$. Let $[v] \mapsto [z] = \mathcal{R}_{[z_0]}([v])$ be the mapping from $T_{[z_0]}$ into $\mathbb{V}_m(\mathbb{R}^N)$ such that $\mathcal{R}_{[z_0]}([0]) = [z_0]$. If $[v]$ belongs to a subset of

$T_{[z_0]}$, which is centered in $[v] = [0]$, and which has a sufficiently small diameter, then $[z] = \mathcal{R}_{[z_0]}([v])$ belongs to a subset of $\mathbb{V}_m(\mathbb{R}^N)$, which is approximatively centered in $[z] = [z_0]$.

□ *Algorithm with a small complexity* [208]. For $N > m$ and for $[z_0]$ in $\mathbb{V}_m(\mathbb{R}^N)$, the mapping $\mathcal{R}_{[z_0]}$ is constructed as follows:

$$[z] = \mathcal{R}_{[z_0]}([v])$$
$$:= \mathrm{qr}([z_0] + \sigma\,[v]) \quad , \quad [v] \in T_{[z_0]}, \tag{10.60}$$

in which the mapping qr corresponding to the QR economy-size decomposition of the matrix $[z_0] + \sigma\,[v]$, for which only the first m columns of matrix $[q]$ such that $[z_0] + \sigma\,[v] = [q]\,[r]$ are computed and such that $[z]^T\,[z] = [I_m]$. Parameter σ allows the diameter of the subset of $\mathbb{M}_{N,m}(\mathbb{R})$, which is centered in $[0]$, to be controlled.

Parameterized Representation of the Non-Gaussian Random Field [K]

The parameterized representation of the non-Gaussian positive-definite matrix-valued random field $\{[\mathbf{K}(\mathbf{x})], \mathbf{x} \in \Omega\}$ is denoted by $\{[\mathbf{K}^{(m,N_d,N_g)}(\mathbf{x})], \mathbf{x} \in \Omega\}$ and is rewritten, for all $\mathbf{x}$ in Ω, as

$$[\mathbf{K}^{(m,N_d,N_g)}(\mathbf{x})] = \mathcal{K}^{(m,N_d,N_g)}(\mathbf{x}, \mathbf{\Xi}, [z]), \tag{10.61}$$

in which $(\mathbf{x}, \boldsymbol{\xi}, [z]) \mapsto \mathcal{K}^{(m,N_d,N_g)}(\mathbf{x}, \boldsymbol{\xi}, [z])$ is a deterministic mapping defined on $\Omega \times \mathbb{R}^{N_g} \times \mathbb{V}_m(\mathbb{R}^N)$ with values in $\mathbb{M}_n^+(\mathbb{R})$ such that

$$\mathcal{K}^{(m,N_d,N_g)}(\mathbf{x}, \boldsymbol{\xi}, [z]) = \mathcal{G}(\,[G_0(\mathbf{x})] + \sum_{i=1}^{m} \sqrt{\lambda_i}\,[G_i(\mathbf{x})]\,\{[z]^T\,\mathbf{\Psi}(\boldsymbol{\xi})\}_i\,). \tag{10.62}$$

Parametric Model-Based Representation of the Random Observation W

The parametric model-based representation of the random observation $\mathbf{W}$ with values in $\mathbb{R}^{N_W}$, corresponding to the representation $\{[\mathbf{K}^{(m,N_d,N_g)}(\mathbf{x})], \mathbf{x} \in \Omega\}$ of random field $\{[\mathbf{K}(\mathbf{x})], \mathbf{x} \in \Omega\}$, is

$$\mathbf{W}^{(m,N_d,N_g)} = \mathcal{B}^{(m,N_d,N_g)}(\mathbf{\Xi}, [z]), \tag{10.63}$$

in which $(\boldsymbol{\xi}, [z]) \mapsto \mathcal{B}^{(m,N_d,N_g)}(\boldsymbol{\xi}, [z])$ is a deterministic mapping defined on $\mathbb{R}^{N_g} \times \mathbb{V}_m(\mathbb{R}^N)$ with values in $\mathbb{R}^{N_W}$ such that

$$\mathcal{B}^{(m,N_d,N_g)}(\boldsymbol{\xi}, [z]) = \mathbf{h}(\mathcal{K}^{(m,N_d,N_g)}(\mathbf{x}^1, \boldsymbol{\xi}, [z]), \dots, \mathcal{K}^{(m,N_d,N_g)}(\mathbf{x}^{N_p}, \boldsymbol{\xi}, [z])\,). \tag{10.64}$$

For N_p fixed, the sequence $\{\mathbf{W}^{(m,N_d,N_g)}\}_{m,N_d,N_g}$ of $\mathbb{R}^{N_W}$-valued random variables converge to $\mathbf{W}$ in $L^2(\Theta, \mathbb{R}^{N_W})$.

10.4 Methodology for Solving the Statistical Inverse Problem for the High Stochastic Dimensions

The general methodology presented in Chapter 7 is used with a model-based representation for the model parameters and for the model observations. In this section, this methodology is detailed for the random fields and allows for identifying the non-Gaussian matrix-valued random field $\{[\mathbf{K}(\mathbf{x})], \mathbf{x} \in \Omega\}$ by using the partial and limited experimental data $\mathbf{w}_{\exp}^1, \ldots, \mathbf{w}_{\exp}^{\nu_{\exp}}$ relative to the random observation vector $\mathbf{W}$ of the stochastic computational model (see also [200, 201, 153, 208]).

For the methodology that is proposed for identifying a non-Gaussian random field in high stochastic dimension, an important step is the construction of a parameterized representation for which the number of hyperparameters is generally very large due to the high stochastic dimension. In the framework of hypotheses for which only partial and limited data are available, such an identification is difficult if there is no information concerning the region of the admissible set (in high dimension), in which the optimal values of these hyperparameters must be searched. The optimization process, related to the statistical inverse problem, requires to localize the region in which the algorithms must search for an optimal value. The method consists in previously identifying the "center" of such a region, which corresponds to the value of the hyperparameters of the parameterized representation using a set of realizations generated with an *algebraic prior stochastic model* (APSM) that is specifically constructed on the basis of the available information associated with all the mathematical properties of the non-Gaussian random field that has to be identified. This APSM allows for enriching the information in order to overcome the lack of experimental data (since only partial experimental data are assumed to be available). This is particularly crucial for the identification of the non-Gaussian matrix-valued random field encountered, for instance, in three-dimensional linear elasticity. Some works have been performed in order to obtain some properties related to the symmetry, to the positiveness, to the boundedness, and to the invertibility, but also in order to introduce the capability of the prior stochastic model to exhibit simultaneously, anisotropic statistical fluctuations and some statistical fluctuations in a symmetry class such as isotropic, cubic, transversely isotropic, orthotropic, etc., and to develop the corresponding generators of realizations. Such a very important aspect is taken into account in the methodology that is proposed after.

Step 1: Introduction of a Family $\{[\mathbf{K}^{\text{APSM}}(\mathbf{x}; \mathbf{s})], \mathbf{x} \in \Omega\}$ of Algebraic Prior Stochastic Models (APSM) or the Non-Gaussian Random Field $[\mathbf{K}]$

The notion of algebraic prior stochastic model (APSM) has been presented in Section 5.6. In this framework, a family $\{[\mathbf{K}^{\text{APSM}}(\mathbf{x}; \mathbf{s})], \mathbf{x} \in \Omega\}$ of APSM is introduced for representing the non-Gaussian second-order random field $[\mathbf{K}]$, defined on $(\Theta, \mathcal{T}, \mathcal{P})$, indexed by Ω, with values in $\mathbb{M}_n^+(\mathbb{R})$, depending on an unknown hyperparameter $\mathbf{s} \in \mathcal{C}_{\text{ad}} \subset \mathbb{R}^{N_s}$, for which the dimension, N_s, is small, while the stochastic dimension of $[\mathbf{K}^{\text{APSM}}]$ is high. Such a construction is detailed in Sections 10.5 and 10.6.

Example for the hyperparameter. Hyperparameter **s** can be made up of both the mean function, a matrix-valued lower bound, some spatial-correlation lengths, some parameters controlling the statistical fluctuations, and the shape of the tensor-valued correlation function.

For **s** fixed in $\mathcal{C}_{\text{ad}}$, the system of marginal probability distributions of random field $[\mathbf{K}^{\text{APSM}}]$ and the corresponding generator of independent realizations are assumed to have been constructed and consequently, are assumed to be known (see Sections 10.5 to 10.7).

Step 2: Identification of an Optimal Algebraic Prior Stochastic Model (OAPSM) for Non-Gaussian Random Field [K]

The identification of an optimal value $\mathbf{s}^{\text{opt}} \in \mathcal{C}_{\text{ad}}$ of hyperparameter **s** is carried out by using the maximum likelihood method (see Section 7.3.3) with the available data set $\mathbf{w}_{\text{exp}}^1, \ldots, \mathbf{w}_{\text{exp}}^{\nu_{\text{exp}}}$ relative to the random observation **W** of the stochastic computational model,

$$\mathbf{W} = \mathbf{h}([\mathbf{K}^{\text{APSM}}(\mathbf{x}^1; \mathbf{s})], \ldots, [\mathbf{K}^{\text{APSM}}(\mathbf{x}^{N_p}; \mathbf{s})]) . \tag{10.65}$$

We then have to solve the following optimization problem:

$$\mathbf{s}^{\text{opt}} = \arg \max_{\mathbf{s} \in \mathcal{C}_{\text{ad}}} \sum_{\ell=1}^{\nu_{\text{exp}}} \log p_{\mathbf{W}}(\mathbf{w}_{\text{exp}}^{\ell}; \mathbf{s}) , \tag{10.66}$$

in which the value $p_{\mathbf{W}}(\mathbf{w}_{\text{exp}}^{\ell}; \mathbf{s})$ of the pdf $p_{\mathbf{W}}$ of random vector **W** at point $\mathbf{w}_{\text{exp}}^{\ell}$ is estimated as explained in Section 7.2. Note that the least-square method, with a control of the statistical fluctuations, could also be used (see Section 7.3.2) instead of the maximum likelihood method.

Using the generator of realizations of the optimal APSM presented in Section 10.6, ν_{KL} independent realizations are computed,

$$\{[K^{(\ell)}], \mathbf{x} \in \Omega\} = \{[\mathbf{K}^{\text{OAPSM}}(\mathbf{x}; \theta_{\ell})], \mathbf{x} \in \Omega\} , \ \theta_{\ell} \in \Theta , \ \ell = 1, \ldots, \nu_{\text{KL}} , \tag{10.67}$$

at points $\mathbf{x}^1, \ldots, \mathbf{x}^{N_p}$ (or at any other points).

Step 3: Choice of an Adapted Representation for Non-Gaussian Random Field [K] and OAPSM for Non-Gaussian Random Field [G]

For a fixed representation of random field $[\mathbf{K}]$ (exponential type or square type), the corresponding optimal algebraic prior model $\{[\mathbf{G}^{\text{OAPSM}}(\mathbf{x})], \mathbf{x} \in \Omega\}$ of $[\mathbf{G}]$ is written as

$$[\mathbf{G}^{\text{OAPSM}}(\mathbf{x})] = \mathcal{G}^{-1}([\mathbf{K}^{\text{OAPSM}}(\mathbf{x})]) \quad , \quad \forall \mathbf{x} \in \Omega . \tag{10.68}$$

It is assumed that random field $[\mathbf{G}^{\text{OAPSM}}]$ is a second-order random field (this hypothesis is automatically verified for the square-type representation, but not for the exponential-type representation, see Section 10.3). From the ν_{KL} realizations $[K^{(1)}], \ldots, [K^{(\nu_{\text{KL}})}]$ of random field $[\mathbf{K}^{\text{OAPSM}}]$, it can be deduced the ν_{KL} independent realizations $[G^{(1)}], \ldots, [G^{(\nu_{\text{KL}})}]$ of random field $[\mathbf{G}^{\text{OAPSM}}]$ such that

$$[G^{(\ell)}(\mathbf{x})] = \mathcal{G}^{-1}([K^{(\ell)}(\mathbf{x})]) \quad , \quad \forall \mathbf{x} \in \Omega \quad , \quad \ell = 1, \ldots, \nu_{\mathrm{KL}} . \tag{10.69}$$

Step 4: Construction of a Truncated Reduced Representation of Second-Order Random Field $[\mathbf{G}^{\mathrm{OAPSM}}]$

The empirical estimates of the mean function $[G_0(\mathbf{x})]$ and of the tensor-valued cross-covariance function $C_{\mathbf{G}^{\mathrm{OAPSM}}}(\mathbf{x}, \mathbf{x}')$ of random field $[\mathbf{G}^{\mathrm{OAPSM}}]$ are calculated by using the ν_{KL} independent realizations $[G^{(1)}], \ldots, [G^{(\nu_{\mathrm{KL}})}]$ of $[\mathbf{G}^{\mathrm{OAPSM}}]$. Then the truncated reduced representation of random field $[\mathbf{G}^{\mathrm{OAPSM}}]$ is constructed as explained in Section 10.1.3,

$$[\mathbf{G}^{\mathrm{OAPSM}\,(m)}(\mathbf{x})] = [G_0(\mathbf{x})] + \sum_{i=1}^{m} \sqrt{\lambda_i} \, [G_i(\mathbf{x})] \, \eta_i^{\mathrm{OAPSM}} \quad , \quad \forall \mathbf{x} \in \Omega , \tag{10.70}$$

in which m is chosen in order to obtain a given accuracy. This truncated reduced representation allows for computing ν_{KL} independent realizations $\boldsymbol{\eta}^{(1)}, \ldots, \boldsymbol{\eta}^{(\nu_{\mathrm{KL}})}$ of random vector $\boldsymbol{\eta}^{\mathrm{OAPSM}} = (\eta_1^{\mathrm{OAPSM}}, \ldots, \eta_m^{\mathrm{OAPSM}})$ that are written, for all $i = 1, \ldots, m$ and $\ell = 1, \ldots, \nu_{\mathrm{KL}}$, as

$$\eta_i^{(\ell)} = \frac{1}{\sqrt{\lambda_i}} \int_{\Omega} \ll [G^{(\ell)}(\mathbf{x})] - [G_0(\mathbf{x})] , [G_i(\mathbf{x}] \gg d\mathbf{x} . \tag{10.71}$$

Step 5: Construction of a Truncated PCE of $\boldsymbol{\eta}^{\mathrm{OAPSM}}$ and Representation of Random Field $[\mathbf{K}^{\mathrm{OAPSM}}]$

□ *Truncated PCE.* By using the independent realizations $\boldsymbol{\eta}^{(1)}, \ldots, \boldsymbol{\eta}^{(\nu_{\mathrm{KL}})}$ of the random vector $\boldsymbol{\eta}^{\mathrm{OAPSM}}$, the truncated PCE of random vector $\boldsymbol{\eta}^{\mathrm{OAPSM}}$, denoted by $\boldsymbol{\eta}^{\mathrm{chaos}}(N_d, N_g) = (\eta_1^{\mathrm{chaos}}(N_d, N_g), \ldots, \eta_m^{\mathrm{chaos}}(N_d, N_g))$, is constructed by using Equation (10.58),

$$\boldsymbol{\eta}^{\mathrm{OAPSM}} \simeq \boldsymbol{\eta}^{\mathrm{chaos}}(N_d, N_g) \quad , \quad \boldsymbol{\eta}^{\mathrm{chaos}}(N_d, N_g) = [z]^T \, \boldsymbol{\Psi}(\boldsymbol{\Xi}) , \tag{10.72}$$

in which the symbol "$\simeq$" means that the mean-square convergence is reached for N_d and N_g (with $N_g \leq m$) sufficiently large and where the matrix $[z]$ belongs to compact Stiefel manifold $\mathbb{V}_m(\mathbb{R}^N)$ (that is to say, $[z] \in \mathbb{M}_{N,m}(\mathbb{R})$ and $[z]^T [z] = [I_m]$). The integer $N = \mathrm{h}(N_d, N_g)$ is defined (see Equation (10.16)) by

$$N = (N_d + N_g)! / (N_d! \, N_g!) - 1 . \tag{10.73}$$

□ *Identification of an optimal value $[z_0]$ of $[z]$ for a fixed value of N_d and N_g.* For a fixed value of N_d and N_g such that $N_d \geq 1$ and $1 \leq N_g \leq m$, the identification of $[z]$ is performed by using the maximum likelihood method (see Section 7.3.3),

$$[z_0(N_d, N_g)] = \arg \max_{[z] \in \mathbb{V}_m(\mathbb{R}^N)} \mathcal{L}([z]) , \tag{10.74}$$

$$\mathcal{L}([z]) = \sum_{\ell=1}^{\nu_{\mathrm{KL}}} \log p_{\boldsymbol{\eta}^{\mathrm{chaos}}(N_d,N_g)}(\boldsymbol{\eta}^{(\ell)} \, ; [z]) \,. \tag{10.75}$$

For $[z]$ fixed in $\mathbb{V}_m(\mathbb{R}^N)$, the pdf $p_{\boldsymbol{\eta}^{\mathrm{chaos}}(N_d,N_g)}$ is estimated by the multidimensional kernel density estimation method (see Section 7.2) by using ν_{chaos} independent realizations $\boldsymbol{\eta}^{\mathrm{chaos}\,(1)}, \dots, \boldsymbol{\eta}^{\mathrm{chaos}\,(\nu_{\mathrm{chaos}})}$ of $\boldsymbol{\eta}^{\mathrm{chaos}}(N_d, N_g)$, which are such that $\boldsymbol{\eta}^{\mathrm{chaos}\,(\ell)} = [z]^T \boldsymbol{\Psi}(\boldsymbol{\xi}^{(\ell)})$ in which $\boldsymbol{\xi}^{(1)}, \dots, \boldsymbol{\xi}^{(\nu_{\mathrm{chaos}})}$ are ν_{chaos} independent realizations of $\boldsymbol{\Xi}$.

☐ *Identification of truncation parameters* N_d *and* N_g. For $[z_0(N_d, N_g)]$ in $\mathbb{V}_m(\mathbb{R}^N)$, the quantification of the convergence of $\boldsymbol{\eta}^{\mathrm{chaos}}(N_d, N_g)$ towards $\boldsymbol{\eta}^{\mathrm{OAPSM}}$ with respect to N_d and N_g is carried out by using an L^1-log error function, which allows for measuring the errors of the small values of the pdf (the tails of the pdf) [200],

$$\mathrm{err}(N_d, N_g) = \frac{1}{m} \sum_{i=1}^{m}$$

$$\int_{\mathrm{BI}_i} |\log_{10} p_{\eta_i^{\mathrm{OAPSM}}}(e) - \log_{10} p_{\eta_i^{\mathrm{chaos}}(N_d,N_g)}(e; [z_0(N_d, N_g)])| \, de \,, \tag{10.76}$$

in which BI_i is a bounded interval of the real line, which is defined as the support of the one-dimensional kernel density estimator of random variable η_i^{OAPSM}, and which is then adapted to independent realizations $\boldsymbol{\eta}^{(1)}, \dots, \boldsymbol{\eta}^{(\nu_{\mathrm{KL}})}$ of $\boldsymbol{\eta}^{\mathrm{OAPSM}}$. The optimal values N_d^{opt} and N_g^{opt} of the truncation parameters N_d and N_g are computed by solving the following optimization problem [162],

$$(N_d^{\mathrm{opt}}, N_g^{\mathrm{opt}}) = \arg \min_{(N_d,N_g) \in \mathcal{C}_\varepsilon} (N_d + N_g)! \, / (N_d! \, N_g!) \,, \tag{10.77}$$

in which the admissible set $\mathcal{C}_\varepsilon$ is defined by

$$\mathcal{C}_\varepsilon = \{(N_d, N_g) \in \mathcal{C}_{N_d,N_g} \mid \mathrm{err}(N_d, N_g) \leq \varepsilon\} \,, \tag{10.78}$$

where the set $\mathcal{C}_{N_d,N_g}$ is written as

$$\mathcal{C}_{N_d,N_g} = \{(N_d, N_g) \in \mathbb{N}^2 \mid N_g \leq m, \, (N_d + N_g)! \, / (N_d! \, N_g!) - 1 \geq m\} \,.$$

The corresponding optimal value N^{opt} of N is thus given by

$$N^{\mathrm{opt}} = (N_d^{\mathrm{opt}} + N_g^{\mathrm{opt}})! \, / (N_d^{\mathrm{opt}}! \, N_g^{\mathrm{opt}}!) - 1 \,. \tag{10.79}$$

Remark. More the values of N_d and N_g are high, more big is the matrix $[z_0(N_d, N_g)]$, and thus, more is difficult to perform the numerical identification. Instead of directly minimizing the error function $\mathrm{err}(N_d, N_g)$, it is more accurate to search for the optimal values of N_d and N_g that minimize the dimension of the projection basis, $(N_d + N_g)! \, / (N_d! \, N_g!)$.

□ *Notations.* In order to simplify the notations, in all the following of Section 10.4, N^{opt}, N_d^{opt}, and N_g^{opt} are rewritten as N, N_d, and N_g ("opt" is removed), and $[z_0(N_d^{\text{opt}}, N_g^{\text{opt}})]$ is simply rewritten as $[z_0]$.

□ *Representation of random field* $[\mathbf{K}^{\text{OAPSM}}]$. The representation $\{[\mathbf{K}^{\text{OAPSM}\,(m,N_d,N_g)}$ $(\mathbf{x})], \mathbf{x} \in \Omega\}$ of random field $\{[\mathbf{K}^{\text{OAPSM}}(\mathbf{x})], \mathbf{x} \in \Omega\}$ is written as

$$[\mathbf{K}^{\text{OAPSM}\,(m,N_d,N_g)}(\mathbf{x})] = \mathcal{K}^{(m,N_d,N_g)}(\mathbf{x}, \boldsymbol{\Xi}, [z_0]) \quad , \quad \forall \mathbf{x} \in \Omega. \tag{10.80}$$

Step 6: Identification of the Prior Stochastic Model $[\mathbf{K}^{\text{prior}}]$ of $[\mathbf{K}]$ in the General Class of the Non-Gaussian Random Fields

□ *Statistical inverse problem for the identification of the prior model.* The identification of the prior stochastic model $\{[\mathbf{K}^{\text{prior}}(\mathbf{x})], \mathbf{x} \in \Omega\}$ of $\{[\mathbf{K}(\mathbf{x})], \mathbf{x} \in \Omega\}$ is performed by using the maximum likelihood method (see Section 7.3.3) with the available data set, $\mathbf{w}_{\text{exp}}^1, \ldots, \mathbf{w}_{\text{exp}}^{\nu_{\text{exp}}}$ that are related to the random observation $\mathbf{W}$. We then have to solve the following optimization problem,

$$[z^{\text{prior}}] = \arg \max_{[z] \in \mathbb{V}_m(\mathbb{R}^N)} \sum_{\ell=1}^{\nu_{\text{exp}}} \log p_{\mathbf{W}^{(m,N_d,N_g)}} (\mathbf{w}_{\text{exp}}^\ell; [z]), \tag{10.81}$$

in which $p_{\mathbf{W}^{(m,N_d,N_g)}}$ is the pdf of random vector $\mathbf{W}^{(m,N_d,N_g)}$ defined by Equations (10.63) and Equation (10.64),

$$\mathbf{W}^{(m,N_d,N_g)} = \mathbf{h}(\mathcal{K}^{(m,N_d,N_g)}(\mathbf{x}^1, \boldsymbol{\xi}, [z]), \ldots, \mathcal{K}^{(m,N_d,N_g)}(\mathbf{x}^{N_p}, \boldsymbol{\xi}, [z])),$$

which is estimated for each point $\mathbf{w}_{\text{exp}}^\ell$ and for each value of $[z]$ by using the multidimensional kernel density estimation method (see Section 7.2) and the ν_{chaos} independent realizations $\boldsymbol{\xi}^{(1)}, \ldots, \boldsymbol{\xi}^{(\nu_{\text{chaos}})}$ of $\boldsymbol{\Xi}$.

□ *Method for solving the optimization problem.* Let us assume that $N > m$ that is generally the case. The parameterization $[z] = \mathcal{R}_{[z_0]}([v])$ of $\mathbb{V}_m(\mathbb{R}^N)$ is used for exploring, with a random search algorithm, the subset of $\mathbb{V}_m(\mathbb{R}^N)$, which is centered in $[z_0]$ computed in Step 5. The optimization problem is thus reformulated as follows:

$$[v^{\text{prior}}] = \arg \max_{[v] \in T_{[z_0]}} \sum_{\ell=1}^{\nu_{\text{exp}}} \log p_{\mathbf{W}^{(m,N_d,N_g)}} (\mathbf{w}_{\text{exp}}^\ell; \mathcal{R}_{[z_0]}([v])). \tag{10.82}$$

The corresponding prior model $[z^{\text{prior}}]$ can then be computed by

$$[z^{\text{prior}}] = \mathcal{R}_{[z_0]}([v^{\text{prior}}]). \tag{10.83}$$

Random search algorithm. For solving the optimization problem, $[v]$ is modeled by a random matrix $[\mathbf{V}] = \text{Proj}_{T_{[z_0]}}([\mathbf{\Lambda}])$ that is the projection on $T_{[z_0]}$ of a

$\mathbb{M}_{N,m}(\mathbb{R})$-valued random matrix $[\mathbf{\Lambda}]$ for which the entries are independent normalized Gaussian real-valued random variables, that is to say, $E\{[\mathbf{\Lambda}]_{ji}\} = 0$ and $E\{[\mathbf{\Lambda}]_{ji}^2\} = 1$. The positive parameter σ introduced in the parameterization of $\mathbb{V}_m(\mathbb{R}^N)$ allows for controlling the "diameter" of the subset (centered in $[z_0]$) that is explored by the random search algorithm.

□ *Representation of the prior stochastic model.* The representation of the prior stochastic model $\{[\mathbf{K}^{\text{prior}\,(m,N_d,N_g)}(\mathbf{x})], \mathbf{x} \in \Omega\}$ of random field $\{[\mathbf{K}(\mathbf{x})], \mathbf{x} \in \Omega\}$ is rewritten as

$$[\mathbf{K}^{\text{prior}\,(m,N_d,N_g)}(\mathbf{x})] = \mathcal{K}^{(m,N_d,N_g)}(\mathbf{x}, \mathbf{\Xi}, [z^{\text{prior}}]) \quad , \quad \forall \mathbf{x} \in \Omega, \quad (10.84)$$

in which the mapping $\mathcal{K}^{(m,N_d,N_g)}(\mathbf{x}, \mathbf{\Xi}, [z])$ is defined by Equation (10.62).

Step 7: Identification of a Posterior Stochastic Model $[\mathbf{K}^{\text{post}}]$ of $[\mathbf{K}]$

□ *Bayesian posterior.* A posterior stochastic model $\{[\mathbf{K}^{\text{post}}(\mathbf{x})], \mathbf{x} \in \Omega\}$ of random field $\{[\mathbf{K}(\mathbf{x})], \mathbf{x} \in \Omega\}$ is constructed by using the Bayes method (see Section 7.3.4). For that, in the PCE $\boldsymbol{\eta}^{\text{chaos}}(N_d, N_g) = [z]^T \mathbf{\Psi}(\mathbf{\Xi})$, the coefficients matrix $[z]$ is modeled [196] by a $\mathbb{V}_m(\mathbb{R}^N)$-valued random variable

$$[\mathbf{Z}] = \mathcal{R}_{[z^{\text{prior}}]}([\mathbf{V}]), \quad (10.85)$$

in which the posterior model $[\mathbf{V}^{\text{post}}]$ of the $T_{[z^{\text{prior}}]}$-valued random variable $[\mathbf{V}]$ is estimated by the Bayes method. We then obtain

$$\boldsymbol{\eta}^{\text{post}} = [\mathbf{Z}^{\text{post}}]^T \mathbf{\Psi}(\mathbf{\Xi}) \quad \text{with} \quad [\mathbf{Z}^{\text{post}}] = \mathcal{R}_{[z^{\text{prior}}]}([\mathbf{V}^{\text{post}}]). \quad (10.86)$$

□ *Prior model $[\mathbf{V}^{\text{prior}}]$ of $[\mathbf{V}]$.* For estimating the posterior model $[\mathbf{V}^{\text{post}}]$ of $[\mathbf{V}]$ using the Bayes method, we need to construct a prior model $[\mathbf{V}^{\text{prior}}]$ (see Section 7.3.4). The prior model $[\mathbf{V}^{\text{prior}}] = \text{Proj}_{T_{[z^{\text{prior}}]}}([\mathbf{\Lambda}^{\text{prior}}])$ is chosen as the projection on $T_{[z^{\text{prior}}]}$ of an $\mathbb{M}_{N,m}(\mathbb{R})$-valued random matrix $[\mathbf{\Lambda}^{\text{prior}}]$ for which the entries are independent normalized Gaussian real-valued random variables: $E\{[\mathbf{\Lambda}]_{ji}\} = 0$ and $E\{[\mathbf{\Lambda}]_{ji}^2\} = 1$.

□ *Posterior model $[\mathbf{V}^{\text{post}}]$ of $[\mathbf{V}]$ using the Bayes method.* For a sufficiently small value of σ, the statistical fluctuations of the $\mathbb{V}_m(\mathbb{R}^N)$-valued random matrix $[\mathbf{Z}^{\text{prior}}]$ are approximatively centered around $[z^{\text{prior}}]$. The Bayesian update (see Section 7.3.4) with the MCMC method (see Chapter 4) allows for constructing the posterior pdf of the $\mathbb{M}_{N,m}(\mathbb{R})$-valued random variable $[\mathbf{V}^{\text{post}}]$ and its generator by using the random solution

$$\mathbf{W}^{(m,N_d,N_g)} = \mathcal{B}^{(m,N_d,N_g)}(\mathbf{\Xi}, \mathcal{R}_{[z^{\text{prior}}]}([\mathbf{V}^{\text{prior}}])), \quad (10.87)$$

in which the mapping $\mathcal{B}^{(m,N_d,N_g)}$ is defined by Equation (10.63) with Equations (10.64) and (10.62), and by using the available data set,

$$\mathbf{w}_{\text{exp}}^1, \dots, \mathbf{w}_{\text{exp}}^{\nu_{\text{exp}}}. \tag{10.88}$$

□ *Representation of the posterior stochastic model.* The posterior stochastic model $\{[\mathbf{K}^{\text{post}\,(m,N_d,N_g)}(\mathbf{x})], \mathbf{x} \in \Omega\}$ of random field $\{[\mathbf{K}(\mathbf{x})], \mathbf{x} \in \Omega\}$ is rewritten as

$$[\mathbf{K}^{\text{post}\,(m,N_d,N_g)}(\mathbf{x})] = \mathcal{K}^{(m,N_d,N_g)}(\mathbf{x}, \mathbf{\Xi}, \mathcal{R}_{[z^{\text{prior}}]}([\mathbf{V}^{\text{post}}])) \,, \ \forall \mathbf{x} \in \Omega, \quad (10.89)$$

in which $\mathcal{K}^{(m,N_d,N_g)}(\mathbf{x}, \boldsymbol{\xi}, [z])$ is defined by Equation (10.62).

□ *Iterative procedure for a global optimization.* Once the probability distribution of $[\mathbf{V}^{\text{post}}]$ has been estimated with Step 7, by using the MCMC method (see Chapter 4), ν_{KL} independent realizations can be calculated for the random field $[\mathbf{G}^{\text{post}}(\mathbf{x})] = [G_0(\mathbf{x})] + \sum_{i=1}^{m} \sqrt{\sigma_i}\,[G_i(\mathbf{x})]\,\eta_i^{\text{post}}$ in which $\boldsymbol{\eta}^{\text{post}} = [\mathbf{Z}^{\text{post}}]^T \boldsymbol{\Psi}(\mathbf{\Xi})$ and where $[\mathbf{Z}^{\text{post}}] = \mathcal{R}_{[z^{\text{prior}}]}([\mathbf{V}^{\text{post}}])$. The identification procedure can then be restarted from Step 4 replacing $[\mathbf{G}^{\text{OAPSM}}]$ by $[\mathbf{G}^{\text{post}}]$.

10.5 Prior Stochastic Model for an Elastic Homogeneous Medium

We present the construction of a prior stochastic model of the (6 x 6) elasticity matrix in the framework of the three-dimensional linear elasticity for continuum media by using the random matrix theory presented in Section 5.4. An important extension is presented in order to be in capability to construct prior stochastic models in presence of material symmetries.

This section is limited to the case of homogeneous media and can also be viewed as a preparation of the tools for the construction of prior stochastic model of heterogeneous media by using the random fields theory that we will present in Section 10.6 and that corresponds to the synthesis presented in [207] of the works published in [94, 97, 98, 99]. This is the reason why, in this section, we use the notation $[\widetilde{\mathbf{C}}]$ for representing the random elasticity matrix (that is independent of $\mathbf{x}$ because the elastic medium is homogeneous) instead of using the notation $[\mathbf{K}(\mathbf{x})]$ for representing the matrix-valued random elasticity field as we have done in Sections 10.2.1, 10.3, and 10.4.

Hypotheses and Available Information Related to the Symmetry Class

□ *Hypotheses.*

1. The framework is the one of the 3D linear elasticity for a homogeneous random medium (material) at a given scale.
2. The random elasticity matrix $[\widetilde{\mathbf{C}}]$ is defined on $(\Theta, \mathcal{T}, \mathcal{P})$, is with values in $\mathbb{M}_n^+(\mathbb{R})$ with $n = 6$. The Kelvin matrix representation of the fourth-order symmetric elasticity tensor in 3D linear elasticity is used.

3. The symmetry classes (linear elastic symmetries) are one of the following ones: isotropic, cubic, transversely isotropic, trigonal, tetragonal, orthotropic, monoclinic, or anisotropic.

□ *Available information related to the symmetry class.*

1. A given symmetry class induced by a material symmetry is defined by a subset $\mathbb{M}_n^{\text{sym}}(\mathbb{R})$ of $\mathbb{M}_n^+(\mathbb{R})$.

2. Random matrix $[\widetilde{\mathbf{C}}]$ is assumed to have a mean value that belongs to $\mathbb{M}_n^+(\mathbb{R})$, but is, in mean, close to the given symmetry class defined by $\mathbb{M}_n^{\text{sym}}(\mathbb{R})$,

$$[\underline{\widetilde{C}}] = E\{[\widetilde{\mathbf{C}}]\} \in \mathbb{M}_n^+(\mathbb{R}) . \qquad (10.90)$$

3. Random matrix $[\widetilde{\mathbf{C}}]$ admits a positive-definite lower bound $[C_\ell] \in \mathbb{M}_n^+(\mathbb{R})$,

$$[\widetilde{\mathbf{C}}] - [C_\ell] > 0 \quad a.s. \qquad (10.91)$$

4. The statistical fluctuations of $[\widetilde{\mathbf{C}}]$ mainly belong to the symmetry class defined by $\mathbb{M}_n^{\text{sym}}(\mathbb{R})$, but can be more or less anisotropic with respect to this symmetry class.

5. The level of statistical fluctuations in the symmetry class must be controlled independently of the level of statistical anisotropic fluctuations.

Positive-Definite Matrices Having a Symmetry Class

□ *Definition of a symmetry class.*

1. A given symmetry class is defined by a subset $\mathbb{M}_n^{\text{sym}}(\mathbb{R}) \subset \mathbb{M}_n^+(\mathbb{R})$ such that any matrix $[M]$ in $\mathbb{M}_n^{\text{sym}}(\mathbb{R})$ exhibits the given symmetry and can be written as

$$[M] = \sum_{j=1}^N m_j [E_j^{\text{sym}}] \quad , \quad [E_j^{\text{sym}}] \in \mathbb{M}_n^S(\mathbb{R}) , \qquad (10.92)$$

in which $\{[E_j^{\text{sym}}], j = 1, \ldots, N\}$ is a matrix algebraic basis of $\mathbb{M}_n^{\text{sym}}(\mathbb{R})$ (see the Walpole tensor basis in [215, 216]), where $N \leq n(n+1)/2$, and where

$$\mathbf{m} = (m_1, \ldots, m_N) \in \mathcal{C}_{\mathbf{m}} \subset \mathbb{R}^N , \qquad (10.93)$$

in which the admissible subset $\mathcal{C}_{\mathbf{m}}$ of $\mathbb{R}^N$ is defined by

$$\mathcal{C}_{\mathbf{m}} = \{\mathbf{m} \in \mathbb{R}^N \mid \sum_{j=1}^N m_j [E_j^{\text{sym}}] \in \mathbb{M}_n^+(\mathbb{R})\} . \qquad (10.94)$$

2. The matrices $[E_j^{\text{sym}}]$ are symmetric (belong to $\mathbb{M}_n^S(\mathbb{R})$), but are not positive definite (do not belong to $\mathbb{M}_n^+(\mathbb{R})$).

3. The dimension N of a symmetry class is always such that $N \leq n(n+1)/2$ and is equal to: 2 for isotropic, 3 for cubic, 5 for transversely isotropic, 6 or 7 for trigonal, 6 or 7 for tetragonal, 9 for orthotropic, 13 for monoclinic, and 21 for anisotropic.

□ *Properties of matrices belonging to a given symmetry class.* If $[M]$ and $[M']$ belong to $\mathbb{M}_n^{\mathrm{sym}}(\mathbb{R})$, then

$$[M][M'] \in \mathbb{M}_n^{\mathrm{sym}}(\mathbb{R}) , \quad [M]^{-1} \in \mathbb{M}_n^{\mathrm{sym}}(\mathbb{R}) , \quad [M]^{1/2} \in \mathbb{M}_n^{\mathrm{sym}}(\mathbb{R}) . \quad (10.95)$$

Any matrix $[N]$ belonging to $\mathbb{M}_n^{\mathrm{sym}}(\mathbb{R})$ can be written as

$$[N] = \exp_{\mathrm{M}}([\mathcal{N}]) , \quad [\mathcal{N}] = \sum_{j=1}^{N} y_j \, [E_j^{\mathrm{sym}}] , \quad \mathbf{y} = (y_1, \ldots, y_N) \in \mathbb{R}^N , \quad (10.96)$$

in which $\exp_{\mathrm{M}}$ is the matrix exponential from $\mathbb{M}_n^S(\mathbb{R})$ into $\mathbb{M}_n^+(\mathbb{R})$. It should be noted that matrix $[\mathcal{N}]$ is a symmetric real matrix but does not belong to $\mathbb{M}_n^{\mathrm{sym}}(\mathbb{R})$ (because $\mathbf{y}$ is in $\mathbb{R}^N$ and therefore, $[\mathcal{N}]$ is not a positive-definite matrix).

Representation Introducing a Positive-Definite Lower Bound

The representation of random elasticity matrix $[\widetilde{\mathbf{C}}]$ is written as

$$[\widetilde{\mathbf{C}}] = [C_\ell] + [\mathbf{C}] , \quad (10.97)$$

in which the deterministic lower bound $[C_\ell]$ is given in $\mathbb{M}_n^+(\mathbb{R})$ and where the random matrix $[\mathbf{C}]$, which is such that

$$[\mathbf{C}] = [\widetilde{\mathbf{C}}] - [C_\ell] , \quad (10.98)$$

is constructed as a random matrix with values in $\mathbb{M}_n^+(\mathbb{R})$. The mean value $[\underline{C}] = E\{[\mathbf{C}]\}$ of $[\mathbf{C}]$ is written as

$$[\underline{C}] = [\underline{\widetilde{C}}] - [C_\ell] \in \mathbb{M}_n^+(\mathbb{R}) , \quad (10.99)$$

The lower bound $[C_\ell]$ can be in the symmetry class $\mathbb{M}_n^{\mathrm{sym}}(\mathbb{R})$ or not (if not $[C_\ell]$ can be close or can be not close to the symmetry class). This lower bound can be defined

– either by using some well-known micromechanics-based bounds,
– either by using the lower bound obtained from computational homogenization,
– or, in the absence of such information, by a simple expression such as $[C_\ell] = \epsilon[\underline{\widetilde{C}}]$ with $0 \leq \epsilon < 1$, from which it can be deduced that

$$[\underline{C}] = (1 - \epsilon)[\underline{\widetilde{C}}] > 0 . \quad (10.100)$$

Introducing Deterministic Matrices $[\underline{A}]$ and $[\underline{S}]$

☐ *Projection operator in the symmetry class.* Let $[\underline{A}]$ be the deterministic matrix in $\mathbb{M}_n^{\text{sym}}(\mathbb{R})$, which is defined by

$$[\underline{A}] = P^{\text{sym}}([\underline{C}]), \tag{10.101}$$

in which $[\underline{C}] \in \mathbb{M}_n^+(\mathbb{R})$ is defined by Equation (10.99) and where P^{sym} is the projection operator from $\mathbb{M}_n^+(\mathbb{R})$ onto $\mathbb{M}_n^{\text{sym}}(\mathbb{R})$. For a given symmetry class with $N < 21$, if there is no anisotropic statistical fluctuations, then the mean matrix $[\underline{C}]$ belongs to $\mathbb{M}_n^{\text{sym}}(\mathbb{R})$ and consequently, $[\underline{A}]$ is equal to $[\underline{C}]$.

☐ *Case of the anisotropic class (symmetry class with $N = 21$).* If the class of symmetry is anisotropic (thus $N = 21$), then $\mathbb{M}_n^{\text{sym}}(\mathbb{R})$ coincides with $\mathbb{M}_n^+(\mathbb{R})$ and again, $[\underline{A}]$ is equal to the mean matrix $[\underline{C}]$ that belongs to $\mathbb{M}_n^+(\mathbb{R})$.

☐ *Case of a given symmetry class with $N < 21$.* In general, for a given symmetry class with $N < 21$, and due to the presence of anisotropic statistical fluctuations, the mean matrix $[\underline{C}]$ of random matrix $[\mathbf{C}]$ belongs to $\mathbb{M}_n^+(\mathbb{R})$ but does not belong to $\mathbb{M}_n^{\text{sym}}(\mathbb{R})$. For this case, an invertible deterministic $(n \times n)$ real matrix $[\underline{S}]$ is introduced such that

$$[\underline{C}] = [\underline{S}]^T [\underline{A}] [\underline{S}]. \tag{10.102}$$

The construction of $[\underline{S}]$ is performed as follows. Let $[L_C]$ and $[L_A]$ be the upper triangular real matrices with positive diagonal entries resulting from the Cholesky factorization of matrices $[\underline{C}]$ and $[\underline{A}]$,

$$[\underline{C}] = [L_C]^T [L_C] \quad , \quad [\underline{A}] = [L_A]^T [L_A]. \tag{10.103}$$

Therefore, the matrix $[\underline{S}]$ is written as

$$[\underline{S}] = [L_A]^{-1} [L_C]. \tag{10.104}$$

If $N < 21$ with no anisotropic fluctuations, or, if $N = 21$, then $[\underline{S}] = [I_n]$.

Nonparametric Stochastic Model for $[\mathbf{C}]$

The random matrix $[\mathbf{C}]$ is written as

$$[\mathbf{C}] = [\underline{S}]^T [\mathbf{A}]^{1/2} [\mathbf{G}_0] [\mathbf{A}]^{1/2} [\underline{S}], \tag{10.105}$$

in which

− the deterministic $(n \times n)$ real matrix $[\underline{S}]$ is defined by Equation (10.104).
− the random matrix $[\mathbf{G}_0]$ belongs to ensemble SG_0^+ (see Section 5.4.7.1), which allows for modeling the anisotropic statistical fluctuations ($E\{[\mathbf{G}_0]\} = [I_n]$ and for which the level of statistical fluctuations is controlled by hyperparameter δ defined by Equation (5.59)).

– the random matrix $[\mathbf{A}]^{1/2}$ is the square root of a random matrix $[\mathbf{A}]$ with values in $\mathbb{M}_n^{\text{sym}}(\mathbb{R}) \subset \mathbb{M}_n^+(\mathbb{R})$, which models the statistical fluctuations in the given symmetry class, and which is statistically independent of random matrix $[\mathbf{G}_0]$,

$$E\{[\mathbf{A}]\} = [\underline{A}] \in \mathbb{M}_n^{\text{sym}}(\mathbb{R}) \subset \mathbb{M}_n^+(\mathbb{R}). \qquad (10.106)$$

Construction of the Random Matrix [A] Using the MaxEnt Principle

Random matrix $[\mathbf{A}]$ that allows for describing the statistical fluctuations in the class of symmetry $\mathbb{M}_n^{\text{sym}}(\mathbb{R})$ with $N < 21$, is constructed using the MaxEnt principle (see Sections 5.4.5 and 5.4.10).

☐ *Defining the available information.* Let $p_{[\mathbf{A}]}$ be the unknown pdf of random matrix $[\mathbf{A}]$, with respect to $d^S A$ on $\mathbb{M}_n^S(\mathbb{R})$ (see Equation (5.41)), with values in the given symmetry class $\mathbb{M}_n^{\text{sym}}(\mathbb{R}) \subset \mathbb{M}_n^+(\mathbb{R}) \subset \mathbb{M}_n^S(\mathbb{R})$ with $N < 21$. The available information is made up of

1. the support Supp $p_{[\mathbf{A}]}$ of the pdf that is the subset $\mathbb{S}_n = \mathbb{M}_n^{\text{sym}}(\mathbb{R})$.
2. the given mean value $E[\mathbf{A}] = [\underline{A}]$ and the constraint $E\{\log(\det[\mathbf{A}])\} = c_A$ with $|c_A| < +\infty$ that implies that $[\mathbf{A}]^{-1}$ is a second-order random variable. These two constraints define the vector $\mathbf{f}$ in $\mathbb{R}^\mu$ with $\mu = N + 1$ and the mapping $[A] \mapsto \mathcal{G}([A])$ from $\mathbb{S}_n$ into $\mathbb{R}^\mu$, such that

$$E\{\mathcal{G}([\mathbf{A}])\} = \mathbf{f}. \qquad (10.107)$$

☐ *Defining the parameterization.* The objective is to construct the parameterization of ensemble $\mathbb{S}_n = \mathbb{M}_n^{\text{sym}}(\mathbb{R})$, such that any matrix $[A]$ in $\mathbb{M}_n^{\text{sym}}(\mathbb{R})$ is written as

$$[A] = [\mathcal{A}(\mathbf{y})], \qquad (10.108)$$

in which $\mathbf{y}$ is a vector in $\mathbb{R}^N$ and where $\mathbf{y} \mapsto [\mathcal{A}(\mathbf{y})]$ is a given mapping from $\mathbb{R}^N$ into $\mathbb{M}_n^{\text{sym}}(\mathbb{R})$. Any matrix $[A]$ in $\mathbb{M}_n^{\text{sym}}(\mathbb{R})$ can then be written as

$$[A] = [\underline{A}]^{1/2} [N] [\underline{A}]^{1/2}, \qquad (10.109)$$

in which $[\underline{A}]^{1/2} \in \mathbb{M}_n^{\text{sym}}(\mathbb{R})$ is the square root of $[\underline{A}] \in \mathbb{M}_n^{\text{sym}}(\mathbb{R}) \subset \mathbb{M}_n^+(\mathbb{R})$ and, due to the invertibility of $[\underline{A}]^{1/2}$, where $[N]$ is a unique matrix belonging to $\mathbb{M}_n^{\text{sym}}(\mathbb{R})$, which admits the representation (see Equation (10.96)),

$$[N] = \exp_{\mathbb{M}}([\mathcal{N}(\mathbf{y})]) \quad, \quad [\mathcal{N}(\mathbf{y})] = \sum_{j=1}^{N} y_j [E_j^{\text{sym}}], \qquad (10.110)$$

in which $\mathbf{y} = (y_1, \dots, y_N) \in \mathbb{R}^N$.

☐ *Construction of [A] using the parameterization and generator of realizations.* The random matrix $[\mathbf{A}]$ with values in $\mathbb{M}_n^{\text{sym}}(\mathbb{R})$ is then written as

$$[\mathbf{A}] = [\underline{A}]^{1/2} [\mathbf{N}] [\underline{A}]^{1/2}, \tag{10.111}$$

where the random matrix $[\mathbf{N}]$ with values in $\mathbb{M}_n^{\text{sym}}(\mathbb{R})$ is such that

$$[\mathbf{N}] = \exp_{\mathbb{M}}([\mathcal{N}(\mathbf{Y})]) \quad , \quad [\mathcal{N}(\mathbf{Y})] = \sum_{j=1}^{N} Y_j [E_j^{\text{sym}}], \tag{10.112}$$

in which $\mathbf{Y} = (Y_1, \ldots, Y_N)$ is the random vector with values in $\mathbb{R}^N$ whose pdf $p_{\mathbf{Y}}$ on $\mathbb{R}^N$ and the generator of realizations have been detailed in Section 5.4.10. Since $[\mathbf{N}]$ can be written as $[\mathbf{N}] = [\underline{A}]^{-1/2} [\mathbf{A}] [\underline{A}]^{-1/2}$, and since $E[\mathbf{A}] = [\underline{A}]$, it can be deduced that

$$E\{[\mathbf{N}]\} = [I_n]. \tag{10.113}$$

10.6 Algebraic Prior Stochastic Model for a Heterogeneous Anisotropic Elastic Medium

An explicit construction of an algebraic prior stochastic model $\{[\mathbf{K}^{\text{APSM}}(\mathbf{x})], \mathbf{x} \in \Omega\}$ is presented for the elasticity random field of a heterogeneous anisotropic elastic medium that exhibits anisotropic statistical fluctuations (initially introduced in [191]), and for which the parameterization consists of the spatial-correlation lengths and of a positive-definite lower bound [193, 203]. An extension of this model can be found in [95] for some positive-definite lower and upper bounds introduced as constraints.

Introduction of an Adapted Representation

The random field $\{[\mathbf{K}^{\text{APSM}}(\mathbf{x})], \mathbf{x} \in \Omega\}$ is defined on $(\Theta, \mathcal{T}, \mathcal{P})$, is indexed by $\Omega \subset \mathbb{R}^d$ and with values in $\mathbb{M}_n^+(\mathbb{R})$, is homogeneous on $\mathbb{R}^d$, and is a second-order random field that is defined by

$$[\mathbf{K}^{\text{APSM}}(\mathbf{x})] = [C_\ell] + [\underline{C}]^{1/2} [\mathbf{G}_0(\mathbf{x})] [\underline{C}]^{1/2} \quad , \quad \forall \mathbf{x} \in \Omega. \tag{10.114}$$

The matrix $[\underline{C}]^{1/2}$ is the square root of the matrix $[\underline{C}]$ in $\mathbb{M}_n^+(\mathbb{R})$, independent of $\mathbf{x}$, such that

$$[\underline{C}] = [\underline{K}] - [C_\ell] \in \mathbb{M}_n^+(\mathbb{R}) \quad , \quad [\underline{K}] \in \mathbb{M}_n^+(\mathbb{R}). \tag{10.115}$$

The random field $\{[\mathbf{G}_0(\mathbf{x})], \mathbf{x} \in \mathbb{R}^d\}$ is defined on $(\Theta, \mathcal{T}, \mathcal{P})$, is indexed by $\mathbb{R}^d$ with values in $\mathbb{M}_n^+(\mathbb{R})$, is homogeneous on $\mathbb{R}^d$, and is a second-order random field such that, for all $\mathbf{x}$ in $\mathbb{R}^d$,

$$E\{[\mathbf{G}_0(\mathbf{x})]\} = [I_n] \quad , \quad [\mathbf{G}_0(\mathbf{x})] > 0 \quad \text{a.s.} \tag{10.116}$$

It can then be deduced that, for all $\mathbf{x}$ in Ω,

$$E\{[\mathbf{K}^{\text{APSM}}(\mathbf{x})]\} = [\underline{K}] \in \mathbb{M}_n^+(\mathbb{R}) \quad , \quad [\mathbf{K}^{\text{APSM}}(\mathbf{x})] - [C_\ell] > 0 \quad \text{a.s.} \quad (10.117)$$

Construction of Random Field $[\mathbf{G}_0]$ and Its Generator of Realizations

□ *Random fields* $\mathcal{U}_{jk}$ *as the stochastic germs of the random field* $[\mathbf{G}_0]$. Random field $\{[\mathbf{G}_0(\mathbf{x})], \mathbf{x} \in \mathbb{R}^d\}$ is constructed as a nonlinear transformation of $n(n+1)/2$ independent second-order, centered, homogeneous, Gaussian, and normalized random fields $\{\mathcal{U}_{jk}(\mathbf{x}), \mathbf{x} \in \mathbb{R}^d\}_{1 \leq j \leq k \leq n}$, defined on probability space $(\Theta, \mathcal{T}, \mathcal{P})$, indexed by $\mathbb{R}^d$, with values in $\mathbb{R}$, and named the stochastic germs of the non-Gaussian random field $[\mathbf{G}_0]$. We then have

$$E\{\mathcal{U}_{jk}(\mathbf{x})\} = 0 \quad , \quad E\{\mathcal{U}_{jk}(\mathbf{x})^2\} = 1 . \quad (10.118)$$

The random fields $\{\mathcal{U}_{jk}(\mathbf{x}), \mathbf{x} \in \mathbb{R}^d\}_{1 \leq j \leq k \leq n}$ are completely and uniquely defined by the $n(n+1)/2$ autocorrelation functions

$$\boldsymbol{\zeta} = (\zeta_1, \ldots, \zeta_d) \mapsto R_{\mathcal{U}_{jk}}(\boldsymbol{\zeta}) = E\{\mathcal{U}_{jk}(\mathbf{x}+\boldsymbol{\zeta})\mathcal{U}_{jk}(\mathbf{x})\}, \quad (10.119)$$

from $\mathbb{R}^d$ into $\mathbb{R}$, such that $R_{\mathcal{U}_{jk}}(0) = 1$. The spatial-correlation lengths $L_1^{jk}, \ldots,$ L_d^{jk} of $\{\mathcal{U}_{jk}(\mathbf{x}), \mathbf{x} \in \mathbb{R}^d\}$ are defined by

$$L_\alpha^{jk} = \int_0^{+\infty} |R_{\mathcal{U}_{jk}}(0, \ldots, \zeta_\alpha, \ldots, 0)| \, d\zeta_\alpha \quad , \quad \alpha = 1, \ldots d, \quad (10.120)$$

and are generally chosen as parameters for the parameterization. An example of an autocorrelation function with a small number of parameters is

$$R_{\mathcal{U}_{jk}}(\boldsymbol{\zeta}) = \rho_1^{jk}(\zeta_1) \times \ldots \times \rho_d^{jk}(\zeta_d), \quad (10.121)$$

$$\rho_\alpha^{jk}(\zeta_\alpha) = \{4(L_\alpha^{jk})^2/(\pi^2 \zeta_\alpha^2)\} \, \sin^2(\pi \zeta_\alpha/(2L_\alpha^{jk})) . \quad (10.122)$$

Each random field $\mathcal{U}_{jk}$ is mean-square continuous on $\mathbb{R}^d$ and its power spectral density function defined on $\mathbb{R}^d$ has a compact support that is written as

$$[-\pi/L_1^{jk}, \pi/L_1^{jk}] \times \ldots \times [-\pi/L_d^{jk}, \pi/L_d^{jk}]. \quad (10.123)$$

Such a model allows for sampling the random field as a function of the spatial-correlation lengths by using the Shannon theorem and has only $dn(n+1)/2$ real parameters. The details concerning the generation of realizations of random field $\mathcal{U}_{jk}$ can be found in [191, 193]. One possible method is based on the usual numerical simulation of homogeneous Gaussian vector-valued random field constructed with the stochastic integral representation of homogeneous stochastic fields (see [180, 164, 165]).

□ *Defining the random field* $\{[\mathbf{G}_0(\mathbf{x})], \mathbf{x} \in \mathbb{R}^d\}$ *and its generator of realizations.* For all $\mathbf{x}$ fixed in $\mathbb{R}^d$, the random matrix $[\mathbf{G}_0(\mathbf{x})]$ is chosen in ensemble SG_0^+ of positive-definite random matrices with a unit mean value (see Section 5.4.7.1).

The spatial-correlation structure of random field $[\mathbf{G}_0]$ is introduced in replacing the Gaussian random variables U_{jk} (see the end of Section 5.4.7.1)) by the Gaussian real-valued random fields $\{\mathcal{U}_{jk}(\mathbf{x}), \mathbf{x} \in \mathbb{R}^d\}$ for which the spatial-correlation structure is defined by the spatial-correlation lengths $\{L_\alpha^{jk}\}_{\alpha=1,\ldots,d}$ and where the statistical fluctuations of random field $[\mathbf{G}_0]$ are controlled by the hyperparameter δ of random matrix $[\mathbf{G}_0(\mathbf{x})]$, which is taken independent of $\mathbf{x}$ and such that

$$0 < \delta < \sqrt{(n+1)/(n+5)} < 1. \tag{10.124}$$

The random field $\{[\mathbf{G}_0(\mathbf{x})], \mathbf{x} \in \mathbb{R}^d\}$, is defined on $(\Theta, \mathcal{T}, \mathcal{P})$, is indexed by $\mathbb{R}^d$ with values in $\mathbb{M}_n^+(\mathbb{R})$, and is constructed as follows:

- For all $\mathbf{x}$ fixed in $\mathbb{R}^d$, the random matrix $[\mathbf{G}_0(\mathbf{x})]$ is written as

$$[\mathbf{G}_0(\mathbf{x})] = [\mathbf{L}(\mathbf{x})]^T [\mathbf{L}(\mathbf{x})], \tag{10.125}$$

  in which $[\mathbf{L}(\mathbf{x})]$ is the upper $(n \times n)$ real triangular random matrix.
- For $1 \leq j \leq k \leq n$, the random fields $\{[\mathbf{L}(\mathbf{x})]_{jk}, \mathbf{x} \in \Omega\}$ are independent.
- For $j < k$, the real-valued random field $\{[\mathbf{L}(\mathbf{x})]_{jk}, \mathbf{x} \in \Omega\}$ is defined by $[\mathbf{L}(\mathbf{x})]_{jk} = \sigma_n \mathcal{U}_{jk}(\mathbf{x})$ in which σ_n is such that $\sigma_n = \delta/\sqrt{n+1}$.
- For $j = k$, the positive-valued random field $\{[\mathbf{L}(\mathbf{x})]_{jj}, \mathbf{x} \in \Omega\}$ is defined by $[\mathbf{L}(\mathbf{x})]_{jj} = \sigma_n \sqrt{2 h(\mathcal{U}_{jj}(\mathbf{x}), a_j)}$ in which $a_j = (n+1)/(2\delta^2) + (1-j)/2$. Function $u \mapsto h(\alpha, u)$ is such that $\Gamma_\alpha = h(\alpha, U)$ is a gamma random variable with parameter α (U being a normalized Gaussian random variable).

□ *A few basic properties of random field* $[\mathbf{G}_0]$. The random field $\{[\mathbf{G}_0(\mathbf{x})], \mathbf{x} \in \Omega\}$, defined on $(\Theta, \mathcal{T}, \mathcal{P})$, indexed by $\mathbb{R}^d$, with values in $\mathbb{M}_n^+(\mathbb{R})$, is a homogeneous, second-order, and mean-square continuous random field. For all $\mathbf{x}$ in $\mathbb{R}^d$,

$$E\{\|\mathbf{G}_0(\mathbf{x})\|_F^2\} < +\infty \quad , \quad E\{[\mathbf{G}_0(\mathbf{x})]\} = [I_n]. \tag{10.126}$$

Since the hyperparameter δ is defined by $\delta = \{n^{-1} E\{\| [\mathbf{G}_0(\mathbf{x})] - [I_n] \|_F^2\}\}^{1/2}$, it can be deduced that

$$E\{\| \mathbf{G}_0(\mathbf{x}) \|_F^2\} = n (\delta^2 + 1). \tag{10.127}$$

For all $\mathbf{x}$ fixed in $\mathbb{R}^d$, the pdf (with respect to the measure $d^S G$) of random matrix $[\mathbf{G}_0(\mathbf{x})]$ is independent of $\mathbf{x}$ and is written (see Equation (5.58)) as

$$p_{[\mathbf{G}_0(\mathbf{x})]}([G]) = \mathbb{1}_{\mathbb{M}_n^+(\mathbb{R})}([G]) \times c_{G_0} \times \left(\det [G]\right)^{(n+1)\frac{(1-\delta^2)}{2\delta^2}} \times e^{-\frac{(n+1)}{2\delta^2} \operatorname{tr} [G]}.$$

For all $\mathbf{x}$ fixed in $\mathbb{R}^d$, the random variables $\{[\mathbf{G}_0(\mathbf{x})]_{jk}, 1 \leq j \leq k \leq 6\}$ are mutually dependent. The system of the marginal probability distributions of random field $\{[\mathbf{G}_0(\mathbf{x})], \mathbf{x} \in \Omega\}$ is completely defined and is not Gaussian. There exists a positive constant b_G independent of $\mathbf{x}$, but depending on δ, such that for all $\mathbf{x}$ in $\mathbb{R}^d$,

$$E\{\|[\mathbf{G}_0(\mathbf{x})]^{-1}\|^2\} \le b_G < +\infty . \tag{10.128}$$

Definition of the Hyperparameter s

The hyperparameter $\mathbf{s} \in \mathcal{C}_{ad} \subset \mathbb{R}^\mu$ of the algebraic prior stochastic model $\{[\mathbf{K}^{\text{APSM}}(\mathbf{x}; \mathbf{s})], \mathbf{x} \in \Omega\}$ that has been constructed for the anisotropic statistical fluctuations, is constituted of

1. the reshaping of $[C_\ell] \in \mathbb{M}_n^+(\mathbb{R})$ (the lower bound) and the reshaping of $[\underline{K}] \in \mathbb{M}_n^+(\mathbb{R})$ (the mean value),
2. the $d\,n(n+1)/2$ positive real numbers, $\{L_1^{jk}, \ldots, L_d^{jk}\}_{1 \le j \le k \le n}$ (the spatial-correlation lengths, for the parameterization given in the example) and δ (the dispersion) such that $0 < \delta < \sqrt{(n+1)/(n+5)}$.

Probabilistic Analysis of the Representative Volume Element Size in Stochastic Homogenization of a Heterogeneous Complex Microstructure

□ *Notation.* For writing the boundary value problem, instead of using the matrix notations introduced in Section 10.2.1, we use the classical notations for the elasticity problems in continuum mechanics. The convention of summation over the repeated Latin indices is used.

We consider a heterogeneous complex elastic microstructure that cannot be described in terms of constituents, for which the stochastic modeling is carried out at the mesoscale. The microstructure occupies the open bounded domain Ω^{meso} of $\mathbb{R}^3$, which is assumed to be a *representative volume element* (RVE). Let $\mathbf{x} = (x_1, x_2, x_3)$ be the generic point in Ω^{meso}. Let $\mathbf{U}^{\text{meso}}(\mathbf{x}) = (U_1^{\text{meso}}(\mathbf{x}), U_2^{\text{meso}}(\mathbf{x}), U_3^{\text{meso}}(\mathbf{x}))$ be the displacement field defined in Ω^{meso}. Let $\sigma^{\text{meso}}(\mathbf{x}) = \{\sigma_{ij}^{\text{meso}}(\mathbf{x})\}_{ij}$ be the stress tensor field and let $\varepsilon^{\text{meso}}(\mathbf{x}) = \{\varepsilon_{kh}^{\text{meso}}(\mathbf{x})\}_{kh}$ be the strain tensor. The constitutive equation is written as

$$\sigma^{\text{meso}}(\mathbf{x}) = \mathbb{C}^{\text{meso}}(\mathbf{x}) : \varepsilon^{\text{meso}}(\mathbf{x}) , \tag{10.129}$$

in which $\mathbb{C}^{\text{meso}}(\mathbf{x}) = \{\mathbb{C}_{ijkh}^{\text{meso}}(\mathbf{x})\}_{ijkh}$ is the fourth-order elasticity tensor at point $\mathbf{x}$, and which is written in terms of the components as

$$\sigma_{ij}^{\text{meso}}(\mathbf{x}) = \mathbb{C}_{ijkh}^{\text{meso}}(\mathbf{x})\, \varepsilon_{kh}^{\text{meso}}(\mathbf{x}) ,$$

In the framework of homogenization [146, 214], at the macroscale, the effective stress tensor σ^{e} and the effective strain tensor ε^{e} are defined by

$$\sigma^{\text{e}} = \frac{1}{|\Omega^{\text{meso}}|} \int_{\Omega^{\text{meso}}} \sigma^{\text{meso}}(\mathbf{x})\, d\mathbf{x} \ , \quad \varepsilon^{\text{e}} = \frac{1}{|\Omega^{\text{meso}}|} \int_{\Omega^{\text{meso}}} \varepsilon^{\text{meso}}(\mathbf{x})\, d\mathbf{x} . \tag{10.130}$$

□ *Change of scale: localization and random effective stiffness tensor.* The homogenization is done by using a homogeneous Dirichlet condition on the boundary of the RVE. The localization is then done with a given random effective strain

$\underline{\varepsilon}$ on the boundary $\partial \Omega^{\text{meso}}$ of the RVE which is independent of $\mathbf{x}$. We then have $\mathbf{U}^{\text{meso}}(\mathbf{x}) = \underline{\varepsilon}\mathbf{x}$ on $\partial \Omega^{\text{meso}}$ in which the given tensor $\underline{\varepsilon}$ is independent of $\mathbf{x}$ and is such that $\underline{\varepsilon} = \varepsilon^{\text{e}}$. For a given random effective strain $\underline{\varepsilon}$ on $\partial \Omega^{\text{meso}}$, the random local displacement field $\mathbf{U}^{\text{meso}}$ in the microstructure Ω^{meso} can be constructed by solving the following stochastic boundary value problem,

$$- \operatorname{div} \sigma^{\text{meso}} = \mathbf{0} \quad \text{in} \quad \Omega^{\text{meso}} , \tag{10.131}$$

$$\mathbf{U}^{\text{meso}}(\mathbf{x}) = \underline{\varepsilon}\mathbf{x} \quad \text{on} \quad \partial \Omega^{\text{meso}} , \tag{10.132}$$

$$\sigma^{\text{meso}} = \mathbb{C}^{\text{meso}}(\mathbf{x}) : \varepsilon(\mathbf{U}^{\text{meso}}(\mathbf{x})) , \tag{10.133}$$

in which the divergence of a second-order tensor $\text{\ss}(\mathbf{x}) = \{\text{\ss}(\mathbf{x})_{ij}\}_{ij}$ is defined by

$$\{\operatorname{div} \text{\ss}(\mathbf{x})\}_i = \partial \text{\ss}_{ij}(\mathbf{x})/\partial x_j , \tag{10.134}$$

and where the linear operator ε (strain tensor mapping) is defined, for $\mathbf{v}(\mathbf{x}) = (v_1(\mathbf{x}), v_2(\mathbf{x}), v_3(\mathbf{x}))$, by

$$\{\varepsilon(\mathbf{v}(\mathbf{x}))\}_{kh} = \frac{1}{2} \left(\frac{\partial v_k(\mathbf{x})}{\partial x_h} + \frac{\partial v_h(\mathbf{x})}{\partial x_k} \right) . \tag{10.135}$$

As the solution $\mathbf{U}^{\text{meso}}$ of Equations (10.131) to (10.133) depends linearly on $\underline{\varepsilon}$, the random local strain tensor $\varepsilon^{\text{meso}}(\mathbf{x}) = \varepsilon(\mathbf{U}^{\text{meso}}(\mathbf{x}))$ can be written as

$$\varepsilon^{\text{meso}}(\mathbf{x}) = \mathbb{H}(\mathbf{x}) : \underline{\varepsilon} , \tag{10.136}$$

in which the fourth-order tensor-valued random field $\mathbf{x} \mapsto \mathbb{H}(\mathbf{x})$ corresponds to the strain localization associated with the stochastic BVP. As $\varepsilon^{\text{meso}}(\mathbf{x})$ and $\underline{\varepsilon}$ are symmetric tensors, it can be deduced that

$$\mathbb{H}_{jk\ell m}(\mathbf{x}) = \mathbb{H}_{kj\ell m}(\mathbf{x}) = \mathbb{H}_{jkm\ell}(\mathbf{x}) . \tag{10.137}$$

The symmetric fourth-order random effective stiffness tensor $\mathbb{C}^{\text{e}}$ that is defined by $\sigma^{\text{e}} = \mathbb{C}^{\text{e}} : \varepsilon^{\text{e}}$ can then be written as

$$\mathbb{C}^{\text{e}} = \frac{1}{|\Omega^{\text{meso}}|} \int_{\Omega^{\text{meso}}} \mathbb{C}^{\text{meso}}(\mathbf{x}) : \mathbb{H}(\mathbf{x}) \, d\mathbf{x} . \tag{10.138}$$

□ *Modeling the random field* $\{\mathbb{C}^{\text{meso}}_{ijkh}(\mathbf{x})\}_{ijkh}, \mathbf{x} \in \Omega^{\text{meso}}\}$ *with the algebraic prior stochastic model* $\{[\mathbf{K}^{\text{APSM}}(\mathbf{x})], \mathbf{x} \in \Omega^{\text{meso}}\}$. The fourth-order elasticity tensor $\mathbb{C}^{\text{meso}}(\mathbf{x})$ is modeled as

$$\mathbb{C}^{\text{meso}}_{ijkh}(\mathbf{x}) = [\mathbf{K}^{\text{APSM}}(\mathbf{x})]_{IJ} \quad \text{with} \quad I = (i,j) \quad \text{and} \quad J = (k,h) , \tag{10.139}$$

in which indices I and J belong to $\{1, \ldots, 6\}$. With such a stochastic modeling of $\{\mathbb{C}^{\text{meso}}(\mathbf{x}), \mathbf{x} \in \Omega^{\text{meso}}\}$, the stochastic boundary value problem defined by

Equations (10.131) to (10.133), related to the microstructure at the mesoscale, has a unique second-order stochastic solution $\mathbf{U}^{\text{meso}}$.

☐ *FE discretization of the mean model of the RVE (microstructure).* The RVE of the microstructure is defined as the unit cube, $\Omega^{\text{meso}} = (]0, 1[)^3$. The mean model of the microstructure is a homogeneous anisotropic linear elastic medium such that

$$E\{[\mathbf{K}^{\text{APSM}}(\mathbf{x})]\} = [\underline{K}],\qquad(10.140)$$

$$[\underline{K}] = 10^{10} \times \begin{bmatrix} 3.3617 & 1.7027 & 1.3637 & -0.1049 & -0.2278 & 2.1013 \\ 1.7027 & 1.6092 & 0.7262 & 0.0437 & -0.1197 & 0.8612 \\ 1.3637 & 0.7262 & 1.4653 & -0.1174 & -0.1506 & 1.0587 \\ -0.1049 & 0.0437 & -0.1174 & 0.1319 & 0.0093 & -0.1574 \\ -0.2278 & -0.1197 & -0.1506 & 0.0093 & 0.1530 & -0.1303 \\ 2.1013 & 0.8612 & 1.0587 & -0.1574 & -0.1303 & 1.7446 \end{bmatrix}.$$

☐ *Finite element model of the Representative Volume Element.* The finite element model is constructed by using a regular mesh $12 \times 12 \times 12 = 1\,728$ nodes of domain Ω^{meso} yielding $5\,184$ DOFs. There are $11 \times 11 \times 11 = 1\,331$ finite elements, which are 8-nodes solid elements. For constructing the elementary matrices of the finite elements, the $2 \times 2 \times 2$ Gauss-Legendre quadrature points are used for each finite element, yielding $N_p = 10\,684$ points $\mathbf{x}^1, \dots, \mathbf{x}^{N_p}$ for the mesh.

☐ *Parameters of the mesoscale stochastic model.* The spatial correlation lengths L_1^K, L_2^K, and L_3^K of the random field $[\mathbf{K}^{\text{APSM}}]$ are defined, for $k = 1, 2, 3$, by

$$L_k^K = \int_0^{+\infty} |r^K(\zeta^k)|\, d\zeta_k \quad , \quad k = 1, 2, 3,\qquad(10.141)$$

in which $\zeta^1 = (\zeta_1, 0, 0)$, $\zeta^2 = (0, \zeta_2, 0)$, and $\zeta^3 = (0, 0, \zeta_3)$, and where, for $k = 1, 2, 3$, the function $\zeta^k \mapsto r^K(\zeta^k)$ is the correlation function defined, for all ζ_k in $\mathbb{R}$, by

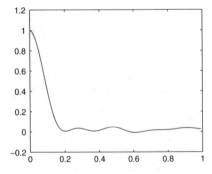

Fig. 10.2 Graph of the correlation function $\zeta \mapsto r^K(\zeta)$ for $L_d = 0.1$. Horizontal axis ζ. Vertical axis $r^K(\zeta)$. [Figure from [193]]

$$r^K(\zeta^k) = \frac{\text{tr } E\{([\mathbf{K}^{\text{APSM}}(\mathbf{x} + \zeta^k)] - [\underline{K}]) ([\mathbf{K}^{\text{APSM}}(\mathbf{x})] - [\underline{K}])\}}{E\{\|[\mathbf{K}^{\text{APSM}}(\mathbf{x})] - [\underline{K}]\|_F^2\}}. \quad (10.142)$$

From Equation (10.142), it can be deduced that $r^K(\mathbf{0}) = 1$. Let us assume that the spatial correlation lengths of the stochastic germ are chosen such that $L_1^{jk} = L_2^{jk} = L_3^{jk} = L_d$ for all j and k in $\{1, \ldots, 6\}$. Consequently, it can be seen that, for all real ζ, we have $r^K(\zeta, 0, 0) = r^K(0, \zeta, 0) = r^K(0, 0, \zeta)$ that we defined as $\text{r}^K(\zeta)$, and that $L_k^K = 1.113 L_d$. Figure 10.2 displays the graph of the correlation function $\zeta \mapsto \text{r}^K(\zeta)$ for $L_d = 0.1$.

□ *Mean-square convergence of the norm of the random effective stiffness matrix* $\|\mathbf{K}^e\|$. The fourth-order effective elasticity tensor $\mathbb{C}^e$ defined by Equation (10.138) is rewritten, by using the matrix notation, as

$$[\mathbf{K}^e]_{IJ} = \mathbb{C}^e_{ijkh} \quad \text{with} \quad I = (i, j) \quad \text{and} \quad J = (k, h), \quad (10.143)$$

in which indices I and J belong to $\{1, \ldots, 6\}$. A first convergence analysis has been performed with respect to the number n_s of realizations used by the stochastic solver based on the Monte Carlo numerical method. For $N_p = 10\,684$, $\delta = 0.4$, $L_d = 0.1$, and $m_{\text{DOF}} = 5\,184$ DOFs, the mean-square convergence with respect to n_s is reached for $n_s \geq 500$. A second convergence analysis has been carried out with respect to the mesh size. For $\delta = 0.4$, for the smallest value of the spatial correlation length corresponding to $L_d = 0.1$, and for $n_s = 900$, Figure 10.3 displays the mean-square convergence of the operator norm $\|\mathbf{K}^e\|$ of the random effective stiffness matrix $[\mathbf{K}^e]$ with respect to the number m_{DOF} of DOFs of the finite element model. It can be seen that the convergence is reached for $m_{\text{DOF}} = 5\,184$.

□ *Probabilistic analysis of the RVE size* $Z = \|\mathbf{K}^e\|/E\{\|\mathbf{K}^e\|\}$. A probabilistic analysis of the stochastic homogenization is performed with respect to the RVE size. It is recalled that, if the scales (microscale and macroscale) are separated, then the statistical fluctuations of the effective stiffness matrix are negligible. Such a separation can be obtained if the spatial correlation lengths of the random

Fig. 10.3 Mean-square convergence of the random effective stiffness matrix with respect to the number m_{DOF} of degrees of freedom of the finite element model for $\delta = 0.4$, $L_d = 0.1$, and for $n_s = 900$. Horizontal axis is m_{DOF}. Vertical axis is the relative error. [Figure from [193]]

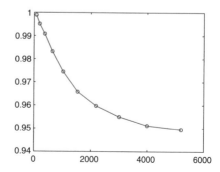

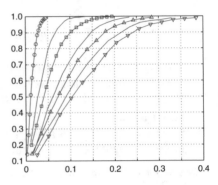

Fig. 10.4 Graph of $\beta \mapsto$ $\mathbb{P}(\beta) = P\{1 - \beta < Z \leq 1 + \beta\}$ for $\delta = 0.4$ and for $L_d = 0.1$ (circle), 0.2 (no marker), 0.3 (square), 0.4 (no marker), 0.5 (triangle-up), 0.6 (no marker), 0.7 (triangle-down). [Figure from [193]]

elasticity field are sufficiently small with respect to the size of the RVE. Since the side of the cube representing the domain of the RVE is 1, Figure 10.4 displays the evolution of the probability distribution of the normalized random variable $Z = \| \mathbf{K}^e \| / E\{\| \mathbf{K}^e \|\}$ as a function of L_d. For instance, this figure shows that, for $L_d = 0.2$, if we consider $\beta = 0.02, 0.04$, and 0.08, we have $P\{0.98 < Z \leq 1.02\} = 0.36$, $P\{0.96 < Z \leq 1.04\} = 0.65$, and $P\{0.92 < Z \leq 1.08\} = 0.95$. Clearly, the scales are badly separated for this value 0.2 of L_d. It can also be seen that the scales are reasonably well separated for $L_d = 0.1$.

10.7 Statistical Inverse Problem Related to the Elasticity Random Field for Heterogeneous Microstructures

This section is devoted to the identification of the elasticity random field for a heterogeneous microstructure by solving a statistical inverse problem for which two important aspects are developed on the basis of the methods presented in Sections 10.3 and 10.4. The first one is relative to a multiscale statistical inverse problem by using the multiscale experimental digital image correlation. The second one is relative to a complete identification of a posterior probabilistic model by using the Bayes method in high stochastic dimension. This section deals with the following items:

1. multiscale statistical inverse problem to be solved.
2. adaptation to the multiscale case of the first two steps of the methodology that has been presented in Section 10.4 for solving a statistical inverse problem in high dimension.
3. introduction of a family of prior stochastic models for the non-Gaussian tensor-valued random field at the mesoscale and its generator, which is derived from Sections 10.5 and 10.6.
4. multiscale identification of the prior stochastic model using a multiscale experimental digital image correlation, at the macroscale and at the mesoscale.

5. example of application of the method for multiscale experimental measurements of cortical bone in 2D plane stresses.
6. example of the construction of a Bayesian posterior of the elasticity random field for a heterogeneous anisotropic microstructure.

10.7.1 Multiscale Statistical Inverse Problem to Be Solved

We consider a material for which the elastic heterogeneous microstructure cannot be described in terms of its constituents such as the biological tissue of a cortical bone (see Figure 10.5). The objective is the identification of the apparent tensor-valued elasticity random field at the mesoscale, $\{\mathbb{C}^{\text{meso}}(\mathbf{x}), \mathbf{x} \in \Omega^{\text{meso}}\}$, by using multiscale experimental data.

10.7.2 Adaptation to the Multiscale Case of the First Two Steps of the General Methodology

In this section, we adapt the notations of the first two steps of the general methodology that has been presented in Section 10.4 for the statistical inverse problem in high dimension, to the multiscale case.

Step 1: Constructing a Prior Stochastic Model for $\mathbb{C}^{\text{meso}}$

The first step consists in introducing an adapted prior stochastic model, $\{\mathbb{C}^{\text{meso}}(\mathbf{x}; \mathbf{s})$, $\mathbf{x} \in \Omega^{\text{meso}}\}$, of the elasticity random field at the mesoscale, that is defined on the probability space $(\Theta, \mathcal{T}, \mathcal{P})$ and which depends on a vector-valued hyperparameter $\mathbf{s} \in \mathcal{C}_{\text{ad}}$. This hyperparameter is assumed to be in small dimension and will be constituted, for instance, of the statistical mean tensor, of the dispersion parameters, and of the spatial correlation lengths. Such a construction will be carried out by using the developments presented in Sections 10.5 and 10.6. As explained in Section 10.4, for the high dimensions, the real possibility to correctly identify the random field $\mathbb{C}^{\text{meso}}$, through a stochastic BVP, is directly related to the capability of the constructed prior stochastic model for representing fundamental properties such

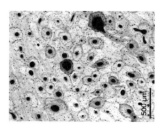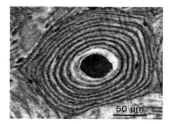

Fig. 10.5 Microstructure of a cortical bone at scale $5 \times 10^{-4}\ m$ (left) and one osteon at scale $5 \times 10^{-5}\ m$. [Photo from Julius Wolff Institute, Charité - Universitatsmedizin Berlin

as the lower bound, the positiveness, the invariance related to material symmetries, the mean value, the support of the spectrum, the spatial correlation lengths, the level of statistical fluctuations, etc. In order to simplify the notation, $\mathbb{C}^{\text{meso}}(\mathbf{x}; \mathbf{s})$ will be, sometimes, simply rewritten as $\mathbb{C}^{\text{meso}}(\mathbf{x})$, in particular, in Section 10.7.3.

Step 2: Identifying Hyperparameter s of the Prior Stochastic Model $\{\mathbb{C}^{\text{meso}}(\mathbf{x}; \mathbf{s}),$
$\mathbf{x} \in \Omega^{\text{meso}}\}$

The second step consists in identifying the hyperparameter $\mathbf{s}$ of the prior stochastic model $\{\mathbb{C}^{\text{meso}}(\mathbf{x}; \mathbf{s}), \mathbf{x} \in \Omega^{\text{meso}}\}$ at the mesoscale in the framework of a multiscale identification by using a multiscale experimental digital image correlation simultaneously at the macroscale and at the mesoscale.

10.7.3 Introduction of a Family of Prior Stochastic Models for the Non-Gaussian Tensor-Valued Random Field at the Mesoscale and Its Generator

In this section, the notations and the developments of Section 10.5 are reused and an advanced construction of a family of prior non-Gaussian tensor-valued random fields are presented for the mesoscale stochastic modeling of the apparent elasticity random field. The framework of this construction is defined as follows:

— We consider the 3D linear elasticity for the heterogeneous microstructure.

— We use the (6×6)-matrix notation for the fourth-order elasticity tensor, that is to say, we write

$$[\mathbf{C}^{\text{meso}}(\mathbf{x}; \mathbf{s})]_{IJ} = \mathbb{C}^{\text{meso}}_{ijkh}(\mathbf{x}; \mathbf{s}) \quad \text{with} \quad I = (i, j) \quad \text{and} \quad J = (k, h), \quad (10.144)$$

in which indices I and J belong to $\{1, \ldots, 6\}$.

— The apparent elasticity random field $\{[\mathbf{C}^{\text{meso}}(\mathbf{x}; \mathbf{s})], \mathbf{x} \in \Omega^{\text{meso}}\}$ of the microstructure Ω^{meso} at the mesoscale, depends on a hyperparameter $\mathbf{s}$ (that will be defined later and that is removed below for simplifying the notations).

— For all $\mathbf{x}$ fixed in Ω^{meso}, elasticity random matrix $[\mathbf{C}^{\text{meso}}(\mathbf{x}; \mathbf{s})]$,

  1. is, in mean, close to a given symmetry class (independent of $\mathbf{x}$), induced by a material symmetry.
  2. exhibits more or less anisotropic statistical fluctuations around this symmetry class.
  3. exhibits a level of statistical fluctuations in the symmetry class, which must be controlled independently of the level of anisotropic statistical fluctuations.

An Advanced Prior Stochastic Model of the Apparent Elasticity Random Field
$\{[\mathbf{C}^{\text{meso}}(\mathbf{x})], \mathbf{x} \in \Omega^{\text{meso}}\}$ **with a Given Symmetry Class**

In this section, we summarize the advanced prior stochastic model introduced and developed in [98] and more detailed in [208].

□ *Prior algebraic representation.* Similarly to Equation (10.97) a lower bound is introduced and the random field $\{[\mathbf{C}^{\text{meso}}(\mathbf{x})], \mathbf{x} \in \Omega^{\text{meso}}\}$ with values in $\mathbb{M}_n^+(\mathbb{R})$ is written as

$$\forall \mathbf{x} \in \Omega^{\text{meso}} \;,\; [\mathbf{C}^{\text{meso}}(\mathbf{x})] = [C_\ell(\mathbf{x})] + [\mathbf{C}(\mathbf{x})] \,, \qquad (10.145)$$

in which the lower bound $\{[C_\ell(\mathbf{x})], \mathbf{x} \in \Omega^{\text{meso}}\}$ is a deterministic field defined in Ω^{meso} with values in $\mathbb{M}_n^+(\mathbb{R})$, and where $\{[\mathbf{C}(\mathbf{x})], \mathbf{x} \in \Omega^{\text{meso}}\}$ is the random field with values in $\mathbb{M}_n^+(\mathbb{R})$, which is written (see Equation (10.105)) as

$$\forall \mathbf{x} \in \Omega^{\text{meso}} \;,\; [\mathbf{C}(\mathbf{x})] = [\underline{S}(\mathbf{x})]^T [\mathbf{A}(\mathbf{x})]^{1/2} [\mathbf{G}_0(\mathbf{x})] [\mathbf{A}(\mathbf{x})]^{1/2} [\underline{S}(\mathbf{x})] \,, \quad (10.146)$$

where

- the random field $\{[\mathbf{G}_0(\mathbf{x})], \mathbf{x} \in \Omega^{\text{meso}}\}$ with values in $\mathbb{M}_n^+(\mathbb{R})$ allows for representing the anisotropic statistical fluctuations around the symmetry class.
- the random field $\{[\mathbf{A}(\mathbf{x})], \mathbf{x} \in \Omega^{\text{meso}}\}$ with values in $\mathbb{M}_n^{\text{sym}}(\mathbb{R})$ is independent of random field $\{[\mathbf{G}_0(\mathbf{x})], \mathbf{x} \in \Omega^{\text{meso}}\}$ and allows for representing the statistical fluctuations in the given symmetry class defined by the set $\mathbb{M}^{\text{sym}}(\mathbb{R})$.
- the deterministic field $\{[\underline{S}(\mathbf{x})], \mathbf{x} \in \Omega^{\text{meso}}\}$ with values in $\mathbb{M}_n(\mathbb{R})$ is introduced for controlling the mean value of random field $[\mathbf{C}]$.

□ *Anisotropic statistical fluctuations.* The random field $\{[\mathbf{G}_0(\mathbf{x})], \mathbf{x} \in \Omega^{\text{meso}}\}$, which is a non-Gaussian $\mathbb{M}_n^+(\mathbb{R})$-valued random field, is defined in Section 10.6 and is such that $E\{[\mathbf{G}_0(\mathbf{x})]\} = [I_n]$. The hyperparameters of $\{[\mathbf{G}_0(\mathbf{x})], \mathbf{x} \in \Omega^{\text{meso}}\}$ are the $d \times n(n+1)/2$ spatial correlation lengths and a scalar dispersion parameter δ_G that controls the level of the anisotropic statistical fluctuations.

□ *Statistical fluctuations in the given symmetry class.* The random field $\{[\mathbf{A}(\mathbf{x})], \mathbf{x} \in \Omega^{\text{meso}}\}$ (that is independent of $[\mathbf{G}_0]$) is a non-Gaussian random field with values in $\mathbb{M}_n^{\text{sym}}(\mathbb{R})$, which is constructed by using Equation (10.101) and Equations (10.111) to (10.113). Consequently, we have

$$E\{[\mathbf{A}(\mathbf{x})]\} = [\underline{A}(\mathbf{x})] = P^{\text{sym}}([\underline{C}(\mathbf{x})]) \,, \qquad (10.147)$$

in which P^{sym} is the projection operator from $\mathbb{M}_n^+(\mathbb{R})$ on $\mathbb{M}_n^{\text{sym}}(\mathbb{R})$ and where

$$[\underline{C}(\mathbf{x})] = E\{[\mathbf{C}(\mathbf{x})]\} = E\{[\mathbf{C}^{\text{meso}}(\mathbf{x})]\} - [C_\ell(\mathbf{x})] \in \mathbb{M}_n^+(\mathbb{R}) \,. \qquad (10.148)$$

For all $\mathbf{x}$ fixed in Ω^{meso}, the random matrix $[\mathbf{A}(\mathbf{x})]$ with values in $\mathbb{M}_n^{\text{sym}}(\mathbb{R})$ is written as

$$[\mathbf{A}(\mathbf{x})] = [\underline{A}(\mathbf{x})]^{1/2} [\mathbf{N}(\mathbf{x})] [\underline{A}(\mathbf{x})]^{1/2} \,. \qquad (10.149)$$

The random field $\{[\mathbf{N}(\mathbf{x})], \mathbf{x} \in \Omega^{\text{meso}}\}$ with values in $\mathbb{M}_n^{\text{sym}}(\mathbb{R})$ is non-Gaussian, is such that

$$E\{[\mathbf{N}(\mathbf{x})]\} = [I_n], \tag{10.150}$$

and is written (see Equation (10.96)) as

$$[\mathbf{N}(\mathbf{x})] = \text{expm}([\boldsymbol{\mathcal{N}}(\mathbf{x})]) \quad , \quad [\boldsymbol{\mathcal{N}}(\mathbf{x})] = \sum_{j=1}^{N} Y_j(\mathbf{x})[E_j^{\text{sym}}], \tag{10.151}$$

where the $\mathbb{R}^N$-valued random field $\{\mathbf{Y}(\mathbf{x}), \mathbf{x} \in \mathbb{R}^d\}$ is non-Gaussian, homogeneous, second-order, mean-square continuous, and is constructed by using the MaxEnt principle (see Sections 5.3.8, 5.4.10, and 10.5). The hyperparameters of $\{[\mathbf{A}(\mathbf{x})], \mathbf{x} \in \Omega^{\text{meso}}\}$ are the $d \times N$ spatial correlation lengths and a scalar dispersion parameter δ_A that allows for controlling the statistical fluctuations in the symmetry class.

□ *Construction of the* $\mathbb{M}_n(\mathbb{R})$ *-valued deterministic field* $\{[\underline{S}(\mathbf{x})], \mathbf{x} \in \Omega^{\text{meso}}\}$. This construction is directly deduced from Equations (10.102) to (10.104). For $\mathbf{x}$ fixed in Ω^{meso}, the Cholesky factorization of the deterministic matrix $[\underline{C}(\mathbf{x})] \in \mathbb{M}_n^+(\mathbb{R})$, defined by Equation (10.148), yields the upper matrix $[L_{\underline{C}}(\mathbf{x})]$, while the Cholesky factorization of $[\underline{A}(\mathbf{x})] \in \mathbb{M}_n^{\text{sym}}(\mathbb{R})$, defined by Equation (10.147), yields the upper matrix $[L_{\underline{A}}(\mathbf{x})]$. As $[\underline{C}(\mathbf{x})] = [\underline{S}(\mathbf{x})]^T [\underline{A}(\mathbf{x})] [\underline{S}(\mathbf{x})]$, it can be deduced that

$$[\underline{S}(\mathbf{x})] = [L_{\underline{A}}(\mathbf{x})]^{-1} [L_{\underline{C}}(\mathbf{x})]. \tag{10.152}$$

Fully Anisotropic Case

The symmetry class is chosen as the anisotropic class that is thus defined by $N = 21$ and $\delta_A = 0$. Consequently, $[\mathbf{C}(\mathbf{x})] = [\underline{C}(\mathbf{x})]^{1/2}[\mathbf{G}_0(\mathbf{x})] [\underline{C}(\mathbf{x})]^{1/2}$, and we obtain (see Equation (10.114)),

$$\forall \mathbf{x} \in \Omega^{\text{meso}} \quad , \quad [\mathbf{C}^{\text{meso}}(\mathbf{x})] = [C_\ell(\mathbf{x})] + [\underline{C}(\mathbf{x})]^{1/2}[\mathbf{G}_0(\mathbf{x})] [\underline{C}(\mathbf{x})]^{1/2}. \tag{10.153}$$

Particular choice for the lower bound. If the lower bound is constructed as explained for obtaining Equation (10.40), then we can choose $[C_\ell(\mathbf{x})]$ as

$$[C_\ell(\mathbf{x})] = \frac{\varepsilon}{1+\varepsilon} E\{[\mathbf{C}^{\text{meso}}(\mathbf{x})]\} \quad \text{with} \quad 0 < \varepsilon \ll 1, \tag{10.154}$$

and then $[\underline{C}(\mathbf{x})]$ is written as

$$[\underline{C}(\mathbf{x})] = \frac{1}{1+\varepsilon} E\{[\mathbf{C}^{\text{meso}}(\mathbf{x})]\}. \tag{10.155}$$

Definition of the hyperparameter s for a homogeneous mean value. If the mean value is homogeneous, that is to say, if $[\underline{C}^{\text{meso}}] = E\{[\mathbf{C}^{\text{meso}}(\mathbf{x})]\}$ is independent of $\mathbf{x}$, then it can be seen that hyperparameter $\mathbf{s}$ is of dimension 25 and can be written as

$$\mathbf{s} = (\{[\underline{C}^{\text{meso}}]_{ij}\}_{i \geq j}, (L_1, L_2, L_3), \delta_G). \tag{10.156}$$

10.7.4 Multiscale Identification of the Prior Stochastic Model Using a Multiscale Experimental Digital Image Correlation, at the Macroscale and at the Mesoscale

This section deals with a new methodology recently developed [148, 149, 150] for the multiscale identification of the prior stochastic model of a complex heterogeneous microstructure made up of a biological tissue (cortical bone) using a multiscale experimental digital image correlation simultaneously measured at the macroscale and at the mesoscale.

Difficulties and Multiscale Identification

The problem concerns the experimental identification of the hyperparameter $\mathbf{s}$ of the prior stochastic model of the apparent elasticity random field $\{\mathbb{C}^{\mathrm{meso}}(\mathbf{x}; \mathbf{s})], \mathbf{x} \in \Omega^{\mathrm{meso}}\}$ defined at the mesoscale in the domain Ω^{meso}. Hyperparameter $\mathbf{s}$ is made up of the statistical mean tensor, $\underline{\mathbb{C}}^{\mathrm{meso}}$, and other parameters that allow for controlling the statistical fluctuations at the mesoscale. The difficulty of such an identification is due to the fact that $\underline{\mathbb{C}}^{\mathrm{meso}}$ cannot directly be identified using only the measurements of the displacement field $\mathbf{u}_{\mathrm{exp}}^{\mathrm{meso}}$ at the mesoscale in Ω^{meso}, and requires to simultaneously measure the displacement field $\mathbf{u}_{\mathrm{exp}}^{\mathrm{macro}}$ at the macroscale in Ω^{macro} (see Figure 10.6).

▷ *For the experimental identification of the prior stochastic model of the apparent elasticity random field at the mesoscale, some experimental multiscale field measurements are required and must be made simultaneously at the macroscale and at the mesoscale.*

Hypotheses Concerning the Experimental Digital Image Correlation at the Macro-Scale and at the Mesoscale

It is assumed that only a single specimen is measured, is submitted to a given load applied at the macroscale, and is tested. The displacement field is measured at the macroscale in Ω^{macro} (with a spatial resolution $10^{-3}\,m$, for instance), and simultaneously, is measured at the mesoscale in Ω^{meso} (with a spatial resolution $10^{-5}\,m$, for instance). The measured strain fields are directly deduced from the measured displacement fields.

Fig. 10.6 Scheme showing the displacement field $\mathbf{u}_{\mathrm{exp}}^{\mathrm{macro}}$ measured at the macroscale in Ω^{macro} and the displacement field $\mathbf{u}_{\mathrm{exp}}^{\mathrm{meso}}$ measured at the mesoscale in Ω^{meso}.

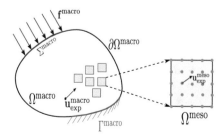

Hypotheses and Strategy for Solving the Statistical Inverse Problem

For the statistical inverse problem that allows for experimentally identifying the prior stochastic model of the apparent elasticity random field $\{\mathbb{C}^{\text{meso}}(\mathbf{x}; \mathbf{s}), \mathbf{x} \in \Omega^{\text{meso}}\}$ at the mesoscale, it is assumed that

- there is a separation of the macroscale from the mesoscale, which means that domain Ω^{meso} corresponds to an RVE.
- at the macroscale, the elasticity tensor is assumed to be independent of $\mathbf{x}$, which means that the material is assumed to be homogeneous at the macroscale.
- at the mesoscale, the apparent elasticity random field is statistically homogeneous (that is to say is a stationary stochastic process).

Under these hypotheses, we then have to construct

- a prior deterministic model of the elasticity tensor $\mathbb{C}^{\text{macro}}(\mathbf{a})$ at the macroscale, which depends on a vector-valued parameter $\mathbf{a}$ that belongs to an admissible set $\mathcal{A}^{\text{macro}}$.
- a prior stochastic model of the apparent elasticity random field $\{\mathbb{C}^{\text{meso}}(\mathbf{x}; \mathbf{s}), \mathbf{x} \in \Omega^{\text{meso}}\}$ at the mesoscale, which depends on a vector-valued hyperparameter $\mathbf{s}$ that belongs to an admissible set $\mathcal{S}^{\text{meso}}$.

Defining the Numerical Indicators for Solving the Statistical Inverse Problem

Three numerical indicators are introduced for solving the statistical inverse problem:

- a macroscopic numerical indicator $\mathcal{J}_1(\mathbf{a})$ that minimizes the distance between the experimental strain field and the computed strain field at the same macroscale.
- a mesoscopic numerical indicator $\mathcal{J}_2(\mathbf{s})$ that minimizes the distance between the experimental strain field and the statistical fluctuations of the computed strain random field at the mesoscale.
- a macroscopic-mesoscopic numerical indicator $\mathcal{J}_3(\mathbf{a}, \mathbf{s})$ that minimizes the distance between the elasticity tensor $\mathbb{C}^{\text{macro}}(\mathbf{a})$ at the macroscale and the effective elasticity tensor $\mathbb{C}^{\text{e}}(\mathbf{s})$ that is constructed by using a stochastic homogenization on the RVE, Ω^{meso}.

Computation of the Three Numerical Indicators

For the boundary value problems, the notations introduced in Section 10.6 are reused, in particular, for the divergence of a second-order tensor (see Equation (10.134)) and for the linear operator ε (see Equation (10.135)).

□ *Macroscopic numerical indicator.* The indicator $\mathcal{J}_1(\mathbf{a})$ is defined for minimizing the distance between the experimental strain field $\varepsilon_{\text{exp}}^{\text{macro}}(\mathbf{x})$ and the computed strain field $\varepsilon^{\text{macro}}(\mathbf{x}; \mathbf{a})$ at the macroscale,

$$\mathcal{J}_1(\mathbf{a}) = \int_{\Omega^{\text{macro}}} \|\varepsilon_{\text{exp}}^{\text{macro}}(\mathbf{x}) - \varepsilon^{\text{macro}}(\mathbf{x}; \mathbf{a})\|_F^2 \, d\mathbf{x}. \tag{10.157}$$

Let $\partial\Omega^{\text{macro}} = \Sigma^{\text{macro}} \cup \Gamma^{\text{macro}}$ be the boundary of domain Ω^{macro} (see Figure 10.7). The strain field $\varepsilon^{\text{macro}}(\mathbf{x}; \mathbf{a}) = \varepsilon(\mathbf{u}^{\text{macro}}(\mathbf{x}; \mathbf{a}))$ is computed by solving the following deterministic boundary value problem formulated with $\mathbf{u}^{\text{macro}}$ on domain Ω^{macro},

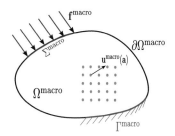

Fig. 10.7 Computational
model at the macroscale in
Ω^{macro}.

$$- \operatorname{div} \sigma^{\text{macro}} = \mathbf{0} \quad \text{in} \quad \Omega^{\text{macro}}, \tag{10.158}$$

$$\sigma^{\text{macro}} \mathbf{n}^{\text{macro}} = \mathbf{f}^{\text{macro}} \quad \text{on} \quad \Sigma^{\text{macro}}, \tag{10.159}$$

$$\mathbf{u}^{\text{macro}} = \mathbf{0} \quad \text{on} \quad \Gamma^{\text{macro}}, \tag{10.160}$$

$$\sigma^{\text{macro}} = \mathbb{C}^{\text{macro}}(\mathbf{a}) : \varepsilon(\mathbf{u}^{\text{macro}}) \quad , \quad \mathbf{a} \in \mathcal{A}^{\text{macro}}. \tag{10.161}$$

in which $\mathbf{n}^{\text{macro}}(\mathbf{x})$ is the outward unit normal to $\partial\Omega^{\text{macro}}$, where $\mathbf{f}^{\text{macro}}(\mathbf{x})$ is the given surface force field applied on Σ^{macro}.

☐ *Mesoscopic numerical indicator.* At the mesoscale, the indicator $\mathcal{J}_2(\mathbf{s})$ is defined for minimizing the distance between the normalized dispersion coefficient $\delta^{\text{meso}}(\mathbf{x}; \mathbf{s})$ that controls the level of the statistical fluctuations of the computed strain random field at the mesoscale, and the corresponding normalized dispersion coefficient $\delta^{\text{meso}}_{\text{exp}}$ of the experimental strain field at the mesoscale,

$$\mathcal{J}_2(\mathbf{s}) = \int_{\Omega^{\text{meso}}} \left(\delta^{\text{meso}}(\mathbf{x}; \mathbf{s}) - \delta^{\text{meso}}_{\text{exp}} \right)^2 d\mathbf{x}. \tag{10.162}$$

At the mesoscale (see Figure 10.8), the strain random field $\varepsilon^{\text{meso}}(\mathbf{x}) = \varepsilon(\mathbf{U}^{\text{meso}}(\mathbf{x}))$ is computed by solving, with the finite element method, the following stochastic

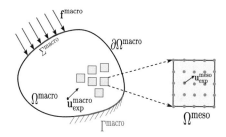

Fig. 10.8 Computational
model at the mesoscale in
Ω^{meso}.

boundary value problem in $\mathbf{U}^{\mathrm{meso}}$ on Ω^{meso},

$$- \operatorname{div} \sigma^{\mathrm{meso}} = \mathbf{0} \quad \text{in} \quad \Omega^{\mathrm{meso}}, \tag{10.163}$$

$$\mathbf{U}^{\mathrm{meso}} = \mathbf{u}_{\mathrm{exp}}^{\mathrm{meso}} \quad \text{on} \quad \partial\Omega^{\mathrm{meso}}, \tag{10.164}$$

$$\sigma^{\mathrm{meso}} = \mathbb{C}^{\mathrm{meso}}(\mathbf{x};\mathbf{s}) : \varepsilon(\mathbf{U}^{\mathrm{meso}}) \quad , \quad \mathbf{s} \in \mathcal{S}^{\mathrm{meso}}. \tag{10.165}$$

The average strain tensor is then defined by

$$\underline{\varepsilon}^{\mathrm{meso}}(\mathbf{s}) = \frac{1}{|\Omega^{\mathrm{meso}}|} \int_{\Omega^{\mathrm{meso}}} \varepsilon^{\mathrm{meso}}(\mathbf{x};\mathbf{s})\, d\mathbf{x}. \tag{10.166}$$

The corresponding experimental average strain tensor is written as

$$\underline{\varepsilon}_{\mathrm{exp}}^{\mathrm{meso}} = \frac{1}{|\Omega^{\mathrm{meso}}|} \int_{\Omega^{\mathrm{meso}}} \varepsilon_{\mathrm{exp}}^{\mathrm{meso}}(\mathbf{x})\, d\mathbf{x}. \tag{10.167}$$

Since Ω^{meso} is an RVE, then for all $\mathbf{s}$ in $\mathcal{S}^{\mathrm{meso}}$, we have $\underline{\varepsilon}^{\mathrm{meso}}(\mathbf{s}) = \underline{\varepsilon}_{\mathrm{exp}}^{\mathrm{meso}}$ a.s. For the computed strain random field at the mesoscale, the dispersion coefficients $\delta^{\mathrm{meso}}(\mathbf{x};\mathbf{s})$ is defined by

$$\delta^{\mathrm{meso}}(\mathbf{x};\mathbf{s}) = \frac{\sqrt{V^{\mathrm{meso}}(\mathbf{x};\mathbf{s})}}{\|\underline{\varepsilon}_{\mathrm{exp}}^{\mathrm{meso}}\|_F}, \tag{10.168}$$

in which $V^{\mathrm{meso}}(\mathbf{x};\mathbf{s})$ is given by

$$V^{\mathrm{meso}}(\mathbf{x};\mathbf{s}) = E\{\|\varepsilon^{\mathrm{meso}}(\mathbf{x};\mathbf{s}) - \underline{\varepsilon}^{\mathrm{meso}}(\mathbf{s})\|_F^2\}. \tag{10.169}$$

The dispersion coefficients $\delta_{\mathrm{exp}}^{\mathrm{meso}}$ for the experimental strain field at the mesoscale is defined by

$$\delta_{\mathrm{exp}}^{\mathrm{meso}} = \frac{\sqrt{V_{\mathrm{exp}}^{\mathrm{meso}}}}{\|\underline{\varepsilon}_{\mathrm{exp}}^{\mathrm{meso}}\|_F}, \tag{10.170}$$

in which $V_{\mathrm{exp}}^{\mathrm{meso}}$ is given by

$$V_{\mathrm{exp}}^{\mathrm{meso}} = \frac{1}{|\Omega^{\mathrm{meso}}|} \int_{\Omega^{\mathrm{meso}}} \|\varepsilon_{\mathrm{exp}}^{\mathrm{meso}}(\mathbf{x}) - \underline{\varepsilon}_{\mathrm{exp}}^{\mathrm{meso}}\|_F^2\, d\mathbf{x}. \tag{10.171}$$

☐ *Macroscopic-mesoscopic numerical indicator.* The indicator $\mathcal{J}_3(\mathbf{a},\mathbf{s})$ is defined for minimizing the distance between the macro elasticity tensor $\mathbb{C}^{\mathrm{macro}}(\mathbf{a})$ at the macroscale and the effective elasticity tensor $\mathbb{C}^{\mathrm{e}}(\mathbf{s})$ constructed by a stochastic homogenization using the RVE Ω^{meso},

$$\mathcal{J}_3(\mathbf{a},\mathbf{s}) = \|\mathbb{C}^{\mathrm{macro}}(\mathbf{a}) - E\{\mathbb{C}^{\mathrm{e}}(\mathbf{s})\}\|_F^2. \tag{10.172}$$

The stochastic homogenization (from the mesoscale to the macroscale) is performed as explained in Section 10.6 (see Equations (10.131) to (10.138)). It can also be performed in homogeneous stresses, for instance for a problem that is formulated in the 2D plane stresses. As Ω^{meso} is an RVE, theoretically, $\mathbb{C}^{\mathrm{e}}\,(\mathbf{s})$ should be deterministic. In fact, as we have explained in Section 10.6, $\mathbb{C}^{\mathrm{e}}\,(\mathbf{s})$ is an random quantity for which the statistical fluctuations are small if Ω^{meso} is a RVE, but are not exactly zero because the RVE is defined in a probabilistic framework. This is the reason why we have used $E\{\mathbb{C}^{\mathrm{e}}\,(\mathbf{s})\}$ instead of $\mathbb{C}^{\mathrm{e}}\,(\mathbf{s})$ in Equation (10.172).

Statistical Inverse Problem as a Multi-Objective Optimization Problem

As there are three numerical indicators, $\mathcal{J}_1(\mathbf{a})$ that depends on $\mathbf{a}$, $\mathcal{J}_2(\mathbf{s})$ that depends on $\mathbf{s}$, and $\mathcal{J}_3(\mathbf{a},\mathbf{s})$ that depends on $\mathbf{a}$ and on $\mathbf{s}$, the statistical inverse problem is formulated as the following multi-objective optimization problem,

$$\left(\mathbf{a}^{\mathrm{macro}},\mathbf{s}^{\mathrm{meso}}\right) = \arg \min_{\mathbf{a}\in\mathcal{A}^{\mathrm{macro}},\mathbf{s}\in\mathcal{S}^{\mathrm{meso}}} \mathcal{J}(\mathbf{a},\mathbf{s}), \qquad (10.173)$$

in which the notation $\min \mathcal{J}(\mathbf{a},\mathbf{s})$ is defined by

$$\min \mathcal{J}(\mathbf{a},\mathbf{s}) = \left(\min \mathcal{J}_1(\mathbf{a}), \min \mathcal{J}_2(\mathbf{s}), \min \mathcal{J}_3(\mathbf{a},\mathbf{s})\right). \qquad (10.174)$$

Solving the Multi-Objective Optimization Problem

☐ The deterministic BVP at the macroscale, which is defined by Equations (10.158) to (10.161), is discretized by using the finite element method.

☐ The stochastic BVP at the mesoscale, which is defined by Equations (10.163) to (10.165), is discretized by using the finite element method and is solved by using the Monte Carlo numerical simulation method (see Section 6.4).

☐ The multi-objective optimization problem can be solved by using a genetic algorithm and the Pareto front [44, 58]. The initial value $\mathbf{a}^{(0)}$ of $\mathbf{a} \in \mathcal{A}^{\mathrm{macro}}$ is computed by solving the optimization problem,

$$\mathbf{a}^{(0)} = \arg \min_{\mathbf{a}\in\mathcal{A}^{\mathrm{macro}}} \mathcal{J}_1(\mathbf{a}), \qquad (10.175)$$

by using, for instance, the simplex algorithm. The hyperparameter $\mathbf{s}^{\mathrm{meso}}$ is estimated in $\mathcal{S}^{\mathrm{meso}}$ as the point on the Pareto front that minimizes the distance between the Pareto front and the origin.

10.7.5 Example of Application of the Method for Multiscale Experimental Measurements of Cortical Bone in 2D Plane Stresses

☐ *Multiscale experimental measurements.*

The specimen is a cube with dimensions $0.01 \times 0.01 \times 0.01 \, m^3$ made up of a cortical bovine bone. The 2D optical measurements have been performed [148, 150]

Fig. 10.9 The speci-
men of cortical bone is
a cube with dimensions
$0.01 \times 0.01 \times 0.01 \, m^3$
(left figure). Measuring bench
at LMS of Ecole Polytech-
nique (right figure). [Figures
from [148]].

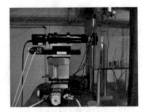

at LMS of Ecole Polytechnique, by using a multi-scale 2D-digital image correlation. Figure 10.9 displays the specimen of the cortical bone (left figure) and the
measuring bench (right figure). The experimental configuration of the specimen
is shown in Figure 10.10. The x_1 axis is horizontal and the x_2 axis is vertical.
A uniform surface force is applied following x_2 direction. The resultant force
that is applied is $9\,000 \, N$. The 2D domains Ω^{macro} and Ω^{meso}, and the 2D spatial
resolution of the multiscale 2D optical measurements are defined as follows:

- Macroscopic 2D domain Ω^{macro}: $0.01 \times 0.01 \, m^2$, meshed with a 10×10-points
  grid yielding a spatial resolution of $10^{-3} \times 10^{-3} \, m^2$.
- Mesoscopic 2D domain Ω^{meso}: $0.001 \times 0.001 \, m^2$, meshed with a 100×100-
  points grid yielding a spatial resolution of $10^{-5} \times 10^{-5} \, m^2$.

At the macroscale, the 2D optical measurements of the displacement field
$\mathbf{u}_{\text{exp}}^{\text{macro}} = \left(u_{\text{exp},1}^{\text{macro}}, u_{\text{exp},2}^{\text{macro}} \right)$ are shown in Figure 10.11. The left figure displays the
component $u_{\text{exp},1}^{\text{macro}}$ of the displacement field for direction x_1 while the right figure
displays the component $u_{\text{exp},2}^{\text{macro}}$ for direction x_2. At the mesoscale, the 2D optical
measurements of the displacement field $\mathbf{u}_{\text{exp}}^{\text{meso}} = \left(u_{\text{exp},1}^{\text{meso}}, u_{\text{exp},2}^{\text{meso}} \right)$ are shown in
Figure 10.12. The left figure displays the component $u_{\text{exp},1}^{\text{meso}}$ of the displacement
field for direction x_1 while the right figure displays the component $u_{\text{exp},2}^{\text{meso}}$ for
direction x_2.

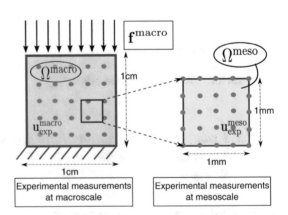

Fig. 10.10 2D scheme defining the experimental configuration of the specimen.

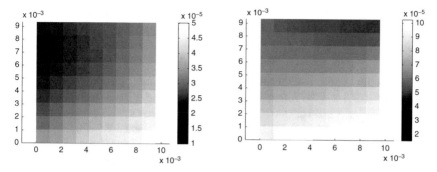

Fig. 10.11 Experimental 2D optical measurements of the displacement field $\mathbf{u}_{\exp}^{\text{macro}}$ at the macroscale. Left figure: component $u_{\exp,1}^{\text{macro}}$ (in m) of the displacement field for direction x_1 (horizontal direction). Right figure: component $u_{\exp,2}^{\text{macro}}$ (in m) of the displacement field for direction x_2 (vertical direction).

☐ *Hypotheses for the stochastic computational model.*

The computational mechanical model is based on the 2D modeling in plane stresses (see Figure 10.10). At the macroscale, Ω^{macro}, the material is assumed to be homogeneous, transverse isotropic, and linear elastic. The parameter $\mathbf{a}$ is such that $\mathbf{a} = (E_T^{\text{macro}}, \nu_T^{\text{macro}})$, in which E_T^{macro} is the transverse Young modulus and where ν_T^{macro} is the Poisson coefficient. At the mesoscale, the material is assumed to be heterogeneous, anisotropic, and linear elastic. The stochastic model of the apparent elasticity random field is deduced from the full anisotropic stochastic case (see Equations (10.153) to (10.156)), for which the statistical mean value is assumed to be transverse isotropic. The hyperparameter $\mathbf{s}$ is such that $\mathbf{s} = (\underline{E}_T, \underline{\nu}_T, L, \delta)$, in which $\underline{E}_T$ and $\underline{\nu}_T$ are the statistical mean values of the transverse Young modulus and of the Poisson coefficient, where L is the spatial correlation length such that $L_1 = L_2 = L_3 = L$, and where δ is the

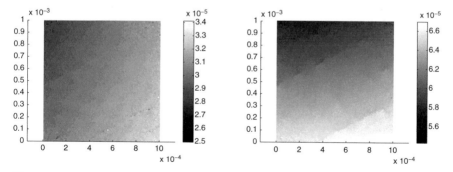

Fig. 10.12 Experimental 2D optical measurements of the displacement field $\mathbf{u}_{\exp}^{\text{meso}}$ at the mesoscale. Left figure: component $u_{\exp,1}^{\text{meso}}$ (in m) of the displacement field for direction x_1 (horizontal direction). Right figure: component $u_{\exp,2}^{\text{meso}}$ (in m) of the displacement field for direction x_2 (vertical direction).

dispersion parameter that allows for controlling the statistical fluctuations of the apparent elasticity random field.

□ *Results obtained with the multiscale identification procedure.*

The results obtained with the multiscale identification procedure presented in Section 10.7.4 are the following. The optimal value $\mathbf{a}^{\text{macro}}$ is such that $E_T^{\text{macro}} = 6.74 \times 10^9 \ Pa$ and $\nu_T^{\text{macro}} = 0.32$. The optimal value $\mathbf{s}^{\text{meso}}$ is such that $E_T^{\text{meso}} = 6.96 \times 10^9 \ Pa$, $\nu_T^{\text{meso}} = 0.37$, $L^{\text{meso}} = 5.06 \times 10^{-5} \ m$, and $\delta^{\text{meso}} = 0.28$. The identified spatial correlation length is in agreement with the assumption introduced concerning the separation of the scales and is of the same order of magnitude than the distance between adjacent lamellae of the osteons in cortical bovine femur. The identified values, $\mathbf{a}^{\text{macro}}$ and $\mathbf{s}^{\text{meso}}$, are in coherence with the values published in literature.

10.7.6 Example of the Construction of a Bayesian Posterior of the Elasticity Random Field for a Heterogeneous Anisotropic Microstructure

In this section, we present an example of the method presented in Section 10.4 related to the Bayesian posterior of the elasticity random field for a heterogeneous anisotropic microstructure (additional details concerning this example can be found in [201]). For the boundary value problem, the notations introduced in Section 10.6 are reused, in particular, for the divergence of a second-order tensor (see Equation (10.134)) and for the linear operator ε (see Equation (10.135)). Nevertheless, the matrix notation will be used for the elasticity tensor as in Section 10.4.

Mechanical Modeling

The microstructure occupies the domain $\Omega^{\text{meso}} = (]0, 1[)^3$ with generic point $\mathbf{x} = (x_1, x_2, x_3)$ and is made up of a heterogeneous complex material modeled, at the mesoscale, by a heterogeneous and anisotropic elastic random medium for which the elastic properties are defined by the fourth-order tensor-valued non-Gaussian random field $\mathbb{C}^{\text{meso}}(\mathbf{x}) = \{\mathbb{C}_{ijkh}^{\text{meso}}(\mathbf{x})\}_{ijkh}$, which is rewritten as

$$[\mathbf{K}(\mathbf{x})]_{IJ} = \mathbb{C}_{ijkh}^{\text{meso}}(\mathbf{x}) \quad \text{with} \quad I = (i, j) \quad \text{and} \quad J = (k, h), \tag{10.176}$$

in which indices I and J belong to $\{1, \ldots, 6\}$. The $\mathbb{R}^3$-valued displacement field $\mathbf{U}^{\text{meso}}$ is defined in Ω^{meso}. The boundary of Ω^{meso} is written as $\partial\Omega^{\text{meso}} = \Gamma_0 \cup \Gamma \cup \Gamma_{\text{obs}}$. The outward unit normal to $\partial\Omega^{\text{meso}}$ is denoted by $\mathbf{n}(\mathbf{x})$. A Dirichlet condition $\mathbf{U}^{\text{meso}} = \mathbf{0}$ is given on Γ_0 while a Neumann condition is given on Γ corresponding to the application of a given deterministic surface force field $\mathbf{g}^\Gamma$. There is no surface force field applied to Γ_{obs}, which is the part of the boundary for which field $\mathbf{U}^{\text{meso}}$ is observed (this corresponds to the hypothesis for which only partial experimental data are observed and then are available). The stochastic boundary value problem is written as

$$- \operatorname{div} \sigma^{\mathrm{meso}} = \mathbf{0} \quad \text{in} \quad \Omega^{\mathrm{meso}}, \tag{10.177}$$

$$\mathbf{U}^{\mathrm{meso}}(\mathbf{x}) = \mathbf{0} \quad \text{on} \quad \Gamma_0, \tag{10.178}$$

$$\sigma^{\mathrm{meso}}(\mathbf{x})\,\mathbf{n}(\mathbf{x}) = \mathbf{g}^{\Gamma}(\mathbf{x}) \quad \text{on} \quad \Gamma, \tag{10.179}$$

$$\sigma^{\mathrm{meso}}(\mathbf{x})\,\mathbf{n}(\mathbf{x}) = \mathbf{0} \quad \text{on} \quad \Gamma_{\mathrm{obs}}, \tag{10.180}$$

$$\sigma^{\mathrm{meso}} = \mathbb{C}^{\mathrm{meso}}(\mathbf{x}) : \varepsilon(\mathbf{U}^{\mathrm{meso}}). \tag{10.181}$$

The nominal model is chosen as a homogeneous material for which the elasticity tensor is $\mathbb{C}^{\mathrm{meso}}$ that is thus independent of $\mathbf{x}$. Let $[K]$ be the corresponding elasticity matrix such that $[K]_{IJ} = \mathbb{C}^{\mathrm{meso}}_{ijkh}$ with $I = (i, j)$ and $J = (k, h)$. It is assumed that matrix $[K]$ is written as

$$[K] = 10^{10} \times \begin{bmatrix} 3.361 & 1.702 & 1.363 & -0.104 & -0.227 & 2.101 \\ 1.702 & 1.609 & 0.726 & 0.043 & -0.119 & 0.861 \\ 1.363 & 0.726 & 1.465 & -0.117 & -0.150 & 1.058 \\ -0.104 & 0.043 & -0.117 & 0.131 & 0.009 & -0.157 \\ -0.227 & -0.119 & -0.150 & 0.009 & 0.153 & -0.130 \\ 2.101 & 0.861 & 1.058 & -0.157 & -0.130 & 1.744 \end{bmatrix}. \tag{10.182}$$

Finite Element Approximation of the Stochastic BVP

The 3D domain Ω^{meso} is meshed with $6 \times 6 \times 6 = 216$ finite elements using 8-nodes finite elements. There are 8 integration points in each finite element. The displacements are locked at points $(1, 0, 0)$, $(1, 1, 0)$, $(1, 1, 1)$, $(1, 0, 1)$ (Γ_0 is discrete). An External point load $(0, 1, 0)$ is applied to the node of coordinates $(0, 0, 1)$ (Γ is also discrete). There are: $N_p = 1\,728$ integration points in the finite element mesh, $m_{\mathrm{obs}} = 50$ observed degrees of freedom inside the face $x_1 = 0$, $m_{\mathrm{nobs}} = 967$ non-observed degrees of freedom, and the total number of degrees of freedom is $m_{\mathrm{DOF}} = 1\,017$.

Available Data Set

Let $\mathbf{W}$ be the random vector with values in $\mathbb{R}^{N_W}$ for which its components are constituted of $N_W = 50$ DOFs observed among the $1\,017$ DOFs of the finite element discretization of the displacement random field $\mathbf{U}^{\mathrm{meso}}$ (partial data). The $\nu_{\mathrm{exp}} = 200$ experimental realizations $\{\mathbf{w}_{\mathrm{exp}}^{\mathrm{meso},1}, \dots \mathbf{w}_{\mathrm{exp}}^{\mathrm{meso},\nu_{\mathrm{exp}}}\}$ of the random observation $\mathbf{W}^{\mathrm{meso}}$ are synthetically generated using the stochastic boundary value problem with a strong multiplicative stochastic perturbation of the APSM model. The data set $\{\mathbf{w}_{\mathrm{exp}}^{\mathrm{meso},1}, \dots \mathbf{w}_{\mathrm{exp}}^{\mathrm{meso},\nu_{\mathrm{exp}}}\}$ is considered as ν_{exp} independent realizations of the random vector $\mathbf{W}_{\mathrm{exp}}^{\mathrm{meso}}$.

Statistical Inverse Problem

☐ *Step 1: introduction of a family of algebraic prior stochastic models (APSM).*

The stochastic boundary value problem is elliptic. The algebraic prior stochastic model $\{[\mathbf{K}^{\text{APSM}}(\mathbf{x})], \mathbf{x} \in \Omega^{\text{meso}}\}$ of the random field $\{[\mathbf{K}(\mathbf{x})], \mathbf{x} \in \Omega^{\text{meso}}\}$ for the anisotropic case is the one defined in Section 10.6 (see Equations (10.114) to (10.128)). This APSM depends on the given mean value $[\underline{K}]$ and depends on the hyperparameter $\mathbf{s} = (\delta, L_d) \in \mathcal{C}_{\text{ad}} \subset \mathbb{R}^2$ in which δ is the dispersion parameter that allows for controlling the statistical fluctuations and where L_d is such that the spatial correlation lengths are written, for all j and k in $\{1, 2, 3\}$, as $L_1^{jk} = L_2^{jk} = L_3^{jk} = \delta_{jk} L_d$ with δ_{jk} the Kronecker symbol.

□ *Step 2: identification of an optimal APSM $[\mathbf{K}^{\text{OAPSM}}]$ for the non-Gaussian random field $[\mathbf{K}]$.*

The least-square method with the control of the statistical fluctuations is used (see Section 7.3.2). The optimization problem is solved by the trial method, and for $\mathbf{s}$ in $\mathcal{C}_{\text{ad}}$, the cost function $\mathcal{J}(\mathbf{s})$ is estimated by using the stochastic computational model for which the stochastic solver is the Monte Carlo numerical simulation method (see Section 6.4),

$$\mathbf{W}^{\text{meso}} = \mathbf{h}([\mathbf{K}^{\text{APSM}}(\mathbf{x}^1; \mathbf{s})], \dots, [\mathbf{K}^{\text{APSM}}(\mathbf{x}^{N_p}; \mathbf{s})]) . \tag{10.183}$$

Figure 10.13 displays the cost function $(\delta, L_c) \mapsto J(\delta, L_c)$ that is used for the identification of the optimal APSM with the experimental data set. The optimal value $\mathbf{s}^{\text{opt}} = (\delta^{\text{opt}}, L_d^{\text{opt}})$ is such that $\delta^{\text{opt}} = 0.42$ and $L_d^{\text{opt}} = 0.34$.

□ *Step 3: choice of an adapted representation for non-Gaussian random field $[\mathbf{K}]$ and optimal APSM for the non-Gaussian random field $[\mathbf{G}]$.*

Since the random field $[\mathbf{K}^{\text{APSM}}]$ is of square type (see Section 10.6), the corresponding optimal APSM $\{[\mathbf{G}^{\text{OAPSM}}(\mathbf{x})], \mathbf{x} \in \Omega^{\text{meso}}\}$ of the random field $[\mathbf{G}]$ is written as

$$[\mathbf{G}^{\text{OAPSM}}(\mathbf{x})] = \mathcal{G}^{-1}([\mathbf{K}^{\text{OAPSM}}(\mathbf{x})]) \quad , \quad \forall \mathbf{x} \in \Omega^{\text{meso}} , \tag{10.184}$$

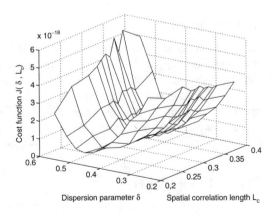

Fig. 10.13 Cost function $(\delta, L_c) \mapsto J(\delta, L_c)$ for the identification of the optimal APSM using experimental data set. [Figure from [201]].

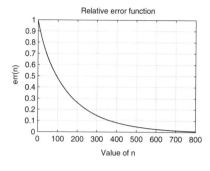
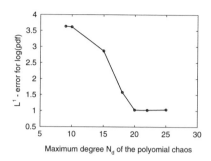

Fig. 10.14 Left figure: graph $m \mapsto \mathrm{err}(m)$ of the error function. Right figure: graph $N_d \mapsto \mathrm{err}(N_d, N_g)$ of the L^1-error function for $N_g = 4$. [Figure from [201]].

in which the mapping $\mathcal{G}^{-1}$ corresponds to the square-type representation that is completely defined in Section 10.3 (see Equations (10.46) to (10.51)). The representation defined by Equation (10.184) allows for generating $\nu_{\mathrm{KL}} = 1\,000$ independent realizations $[G^{(1)}], \dots, [G^{(\nu_{\mathrm{KL}})}]$ of random field $[\mathbf{G}^{\mathrm{OAPSM}}]$ such that

$$[G^{(\ell)}(\mathbf{x})] = \mathcal{G}^{-1}([K^{(\ell)}(\mathbf{x})]), \quad \forall \mathbf{x} \in \Omega^{\mathrm{meso}} \quad , \quad \ell = 1, \dots, \nu_{\mathrm{KL}}. \quad (10.185)$$

☐ *Step 4: construction of a truncated reduced representation of second-order random field* $[\mathbf{G}^{\mathrm{OAPSM}}]$.

The eigenvalue problem for the tensor-valued cross-covariance function of random field $\{[\mathbf{G}^{\mathrm{OAPSM}}(\mathbf{x})], \mathbf{x} \in \Omega^{\mathrm{meso}}\}$ is replaced by the eigenvalue problem for the finite family $\{[\mathbf{G}^{\mathrm{OAPSM}}(\mathbf{x}^1)], \dots, [\mathbf{G}^{\mathrm{OAPSM}}(\mathbf{x}^{N_p})]\}$, which is reshaped in a random vector $\mathbf{G}^{\mathrm{OAPSM}}_{\mathrm{resh}}$ with values in $\mathbb{R}^{36288}$ for which the mean vector and the covariance matrix $[C_{\mathbf{G}^{\mathrm{OAPSM}}_{\mathrm{resh}}}]$ are estimated by using the $1\,000$ independent realizations of $\mathbf{G}^{\mathrm{OAPSM}}_{\mathrm{resh}}$ that have been computed in Step 3. Figure 10.14 (left) displays the graph $m \mapsto \mathrm{err}(m)$ of the error function defined by

$$\mathrm{err}(m) = 1 - \frac{\sum_{i=1}^m \lambda_i}{\mathrm{tr}[C_{\mathbf{G}^{\mathrm{OAPSM}}_{\mathrm{resh}}}]}. \quad (10.186)$$

It can be seen that the convergence is reached for $m = 550$.

☐ *Step 5: construction of a truncated polynomial chaos expansion of* $\boldsymbol{\eta}^{\mathrm{OAPSM}}$ *and representation of random field* $[\mathbf{K}^{\mathrm{OAPSM}}]$.

The coefficients of truncated polynomial chaos expansion of the random vector $\boldsymbol{\eta}^{\mathrm{OAPSM}}$ are computed by using $\nu_{\mathrm{KL}} = 1\,000$ independent realizations. The L^1-error function $N_d \mapsto \mathrm{err}(N_d, N_g)$ is computed for $\boldsymbol{\eta}^{\mathrm{chaos}}(N_d, N_g)$ for several values of N_g. Figure 10.14 (right) displays the graph $N_d \mapsto \mathrm{err}(N_d, N_g)$ for $N_g = 4$. The convergence is obtained for $N_d^{\mathrm{opt}} = 20$ yielding $N^{\mathrm{opt}} = \mathbb{h}(N_d^{\mathrm{opt}}, N_g^{\mathrm{opt}}) = 10\,625$.

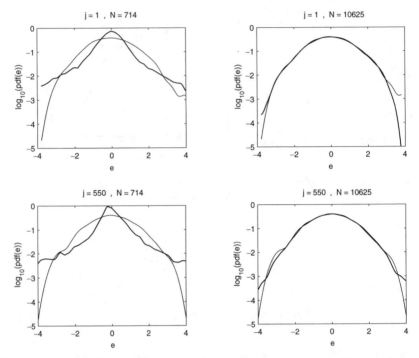

Fig. 10.15 Comparisons of the graph of the optimal APSM pdf $e \mapsto p_{\eta_j^{\text{OAPSM}}}(e)$ (thin solid line) with the graph of the pdf $e \mapsto p_{\eta_j^{\text{chaos}}}(e; [z_0])$ estimated using the truncated PCE with N chaos (thick solid line), for $j = 1$, $N = 714$ (up left), $j = 1$, $N = 10\,625$ (up right), $j = 550$, $N = 714$ (down left), $j = 550$, $N = 10\,625$ (down right). Vertical axis is the $\log_{10}$ of the pdf. Horizontal axis is the value e of η_j.[Figure from [201]].

The matrix $[z_0] \in \mathbb{M}_{N,m}(\mathbb{R})$ of the coefficients of the PCE of $\boldsymbol{\eta}^{\text{chaos}}(N_d^{\text{opt}}, N_g^{\text{opt}})$ is then represented by $5\,843\,750 = 10\,625 \times 550$ real coefficients. For $N_g = 4$, the convergence analysis of the truncated PCE $\boldsymbol{\eta}^{\text{chaos}}(N_d, N_g)$ with respect to the number $N = \mathrm{h}(N_d, N_g)$ of chaos is performed for the pdf of two coordinates, $j = 1$ and $j = 550$, and is illustrated in Figure 10.15.

☐ *Step 6: identification of the prior stochastic model* $[\mathbf{K}^{\text{prior}}]$ *of* $[\mathbf{K}]$ *in the general class of the non-Gaussian random fields*

The identification of the prior stochastic model $[\mathbf{K}^{\text{prior}}]$ of the random field $[\mathbf{K}]$ is performed in the general class of the non-Gaussian random fields. For the observed degrees of freedom k corresponding to the x_2-displacement of nodes 9 and 37 that belong to Γ_{obs}, Figure 10.16 displays the graphs of the functions $w_k \mapsto p_{W_{\text{exp},k}^{\text{meso}}}(w_k)$, $w_k \mapsto p_{W_k^{\text{OAPSM}}}(w_k)$, and $w_k \mapsto p_{W_k^{\text{prior}}}(w_k)$. It can be seen

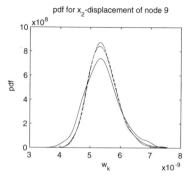
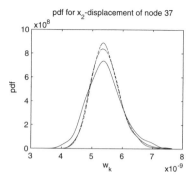

Fig. 10.16 For observed degrees of freedom k corresponding to the x_2-displacement of nodes 9 and 37, graphs of $w_k \mapsto p_{W_{\exp,k}^{\mathrm{meso}}}(w_k)$ (thin solid lines), $w_k \mapsto p_{W_k^{\mathrm{OAPSM}}}(w_k)$ (dashed lines), and $w_k \mapsto p_{W_k^{\mathrm{prior}}}(w_k)$ (dashed dotted lines). [Figure from [201]].

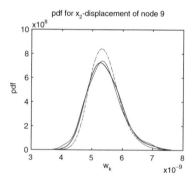
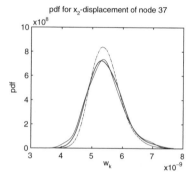

Fig. 10.17 For observed degrees of freedom k corresponding to the x_2-displacement of nodes 9 and 37, graphs of $w_k \mapsto p_{W_{\exp,k}^{\mathrm{meso}}}(w_k)$ (thin solid lines), $w_k \mapsto p_{W_k^{\mathrm{prior}}}(w_k)$ (dashed dotted lines), and $w_k \mapsto p_{W_k^{\mathrm{post}}}(w_k)$ (blue thick solid lines). [Figure from [201]].

that there still exists a deviation between the prior stochastic model and the reference. The posterior stochastic model estimated in Step 7 will allow for reducing this deviation.

□ *Step 7: identification of a posterior stochastic model* $[\mathbf{K}^{\mathrm{post}}]$ *of* $[\mathbf{K}]$.

The identification of the posterior stochastic model $[\mathbf{K}^{\mathrm{post}}]$ of the random field $[\mathbf{K}]$ is performed from the prior stochastic model estimated in Step 6. For the observed degrees of freedom k corresponding to the x_2-displacement of nodes 9 and 37 that belong to Γ_{obs}, Figure 10.17 displays the graphs of the functions $w_k \mapsto p_{W_{\exp,k}^{\mathrm{meso}}}(w_k)$, $w_k \mapsto p_{W_k^{\mathrm{prior}}}(w_k)$, and $w_k \mapsto p_{W_k^{\mathrm{post}}}(w_k)$. It can be seen that the posterior stochastic model gives an excellent prediction of the reference. However, the observations (for which experimental data are available) have been used for identifying the posterior stochastic model. In order to give a quality

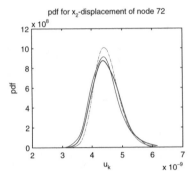 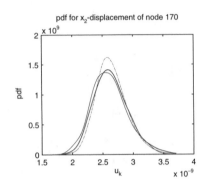

Fig. 10.18 For degrees of freedom k corresponding to the x_2-displacement of nodes 72 and 170 that are located inside domain Ω, graphs of $u_k \mapsto \widehat{p}_{U_{\exp,k}^{\mathrm{meso}}}(u_k)$ (thin solid lines), $u_k \mapsto p_{U_k^{\mathrm{prior}}}(u_k)$ (dashed dotted lines), and $u_k \mapsto p_{U_k^{\mathrm{post}}}(u_k)$ (blue thick solid lines). [Figure from [201]].

assessment of the posterior stochastic models, in Step 7, we present a comparison for degrees of freedom which are not observed (that is to say which are not used in the identification procedure of the posterior stochastic model).

Quality Assessment of the Posterior Stochastic Model

We present the quality assessment of the posterior stochastic model in presenting a comparison for degrees of freedom that are not used by the identification procedure of the posterior stochastic model. For degrees of freedom k that are not used for the identification of the posterior stochastic model and that correspond to the x_2-displacement of nodes 72 and 170 that are located inside domain Ω^{meso}, Figure 10.18 displays the graphs of $u_k \mapsto p_{U_{\exp,k}^{\mathrm{meso}}}(u_k)$, $u_k \mapsto p_{U_k^{\mathrm{prior}}}(u_k)$, and $u_k \mapsto p_{U_k^{\mathrm{post}}}(u_k)$. It can be seen that the prediction is very good.

10.8 Stochastic Continuum Modeling of Random Interphases from Atomistic Simulations for a Polymer Nanocomposite

This section is devoted to the stochastic modeling and to the inverse identification presented in [119, 120] related to the random field associated with the elastic properties in the interphase region between the polymer and a silicon nano-inclusion inserted in the polymer, for a polymer system reinforced by a Silica nanoscopic inclusion. This application gives an interesting illustration of the use of

— the advanced prior stochastic model of the apparent elasticity random field with a given symmetry class, presented in Section 10.7.3. The symmetry class is the

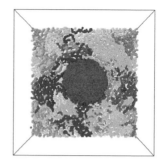

Fig. 10.19 Visualization of an instantaneous configuration of the polymer chains and of the atoms of the inclusion. [Figure from [119]].

transversely isotropic class, regardless of the unit normal vector to the sphere defined by the boundary of the inclusion.
- the molecular dynamics for generating atomistic simulated data.
- the methodology presented in Section 10.4 (the first two steps) concerning the identification of the prior stochastic model by solving a statistical inverse problem.

Objective

In the framework of the continuum mechanics, a prior stochastic model of the matrix-valued random elasticity field $\mathbf{x} \mapsto [\mathbf{C}^{meso}(\mathbf{x})]$ is constructed for describing the elastic behavior of the interphase between an amorphous polymer and a silicon nano-inclusion inserted in the polymer, which constitutes a polymer nanocomposite. The identification of the hyperparameters of the prior stochastic model of $[\mathbf{C}^{meso}]$ is performed by using atomistic simulations as experiments.

Simulation Procedure for the Inverse Statistical Problem

For solving the inverse statistical problem, the simulation procedure is carried out with two main steps.

1. The first one is an atomistic simulation. Figure 10.19 shows a visualization of an instantaneous configuration of the polymer chains and of the atoms of the inclusion.
2. The second step concerns the statistical inverse problem for carrying out the identification of the prior stochastic model of the random elasticity field describing the behavior of the interphase.

Physical Description and Molecular Dynamics Modeling

□ *Polymer nanocomposite.* The amorphous polymer is made up of long chains with CH_2 sites, represented through a coarse graining with harmonic potentials and

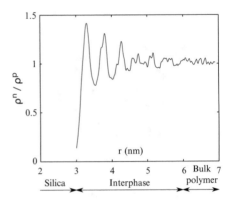

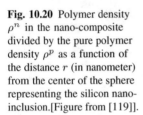

Fig. 10.20 Polymer density ρ^n in the nano-composite divided by the pure polymer density ρ^p as a function of the distance r (in nanometer) from the center of the sphere representing the silicon nano-inclusion.[Figure from [119]].

the Lennard Jones potential. The silicon nano-inclusion is made up of amorphous bulk of SiO_2 molecules described in terms of Si and O atoms, with Coulomb potential. The interaction CH_2 - SiO_2 is modeled by a Lennard Jones potential.

☐ *Simulations with a target volume fraction of* 4.7%. The atomistic simulation domain is a cube of $6.8 \times 10^{-9}\ m$ side, which contains 10 polymer chains. Each polymer chain is made up of $1\,000\ CH_2$ sites yielding a total of $10\,000\ CH_2$. The SiO_2 nano-inclusion is a sphere of $3 \times 10^{-9}\ m$ diameter with 275 Si atoms and 644 O atoms.

Atomistic Simulations
The atomistic simulations have been performed with 10, 80, and 320 chains [120]. The results presented after are limited to the atomistic simulations that have been performed with 10 chains in the following conditions. The temperature is $T = 100\ ^\circ K$ and the pressure P is the control variable. Six mechanical tests in traction and shear are simulated. A time-spatial averaging is performed for estimating the apparent strain that allows for deducing the components of the apparent elasticity matrix in the sense of the continuum mechanics. As an illustration of results of the atomistic simulation, Figure 10.20 displays the polymer density ρ^n in the nano-composite divided by the pure polymer density ρ^p as a function of the distance r from the center of the sphere representing the silicon nano-inclusion. This figure shows that the interphase thickness e is between $2 \times 10^{-9}\ m$ and $3 \times 10^{-9}\ m$. The atomistic simulations have shown that this interphase thickness is independent of the diameter of the silicon nano-inclusion (for the class of diameters: 3, 6, and $9.6 \times 10^{-9}\ m$).

Statistical Inverse Problem for the Identification of a Prior Stochastic Model of the Elasticity Field Describing the Interphase

☐ *Prior stochastic model of the elasticity field describing the interphase.* The prior non-Gaussian stochastic model $\mathbf{x} \mapsto [\mathbf{C}^{\mathrm{meso}}(\mathbf{x})]$ in the interphase is constructed as explained in Section 10.7.3 for which there are no anisotropic statistical

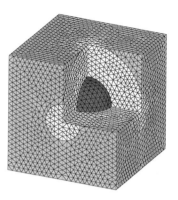

Fig. 10.21 3D view of the meshed continuum model (the inclusion appears in magenta (or in dark gray), the interphase in white and the polymer matrix in turquoise (or in gray)). [Figure from [120]].

fluctuations but there are only statistical fluctuations in the class of transversally isotropic material symmetries in the spherical coordinates (r, φ, ψ) in the orthonormal spherical frame (e_r, e_φ, e_ψ). The hyperparameters are the dispersion parameter related to the statistical fluctuations in the symmetry class, the spatial-correlation length L_r along the radial direction e_r, and the two spatial-correlation lengths L_φ and L_ψ along the two directions e_φ and e_ψ. For the identification, it has been assumed that $L_\varphi = L_\psi$ that is denoted by L_a.

☐ *Continuum mechanics modeling.* The finite element method is used for solving the 6 stochastic BVP corresponding to the 6 mechanical tests. The elasticity field of the linear elastic interphase region is modeled by the prior stochastic model for the class of transversally isotropic material symmetries in the spherical coordinates. The Silica inclusion and the polymer bulk are assumed to be linear elastic isotropic and to be homogeneous and deterministic media, for which the elastic properties (bulk and shear moduli) have been estimated from the MD simulations. The final computational model is made up of 190 310 elements (corresponding to 102 561 degrees of freedom). It should be noted that such a mesh density provides at least four integration points per correlation length (in mean), regardless of the direction or the configuration tested while solving the statistical inverse problem, hence ensuring a good sampling of the correlation structure. Figure 10.21 shows a 3D view of the meshed continuum model.

Likelihood Method for Solving the Statistical Inverse Problem

The methodology presented in Section 10.4 (the first two steps), based on the maximum likelihood method, is used for estimating the optimal values of the hyperparameters of the prior stochastic model of the non-Gaussian matrix-valued random field $\mathbf{x} \mapsto [\mathbf{C}^{\text{meso}}(\mathbf{x})]$ defined in the interphase domain. The observed random quantity corresponds to the random apparent elasticity matrix $[\mathbf{C}^{\text{app}}]$ related to the domain. The probability density function of $[\mathbf{C}^{\text{app}}]$ is estimated by using with 200 realizations that are computed by stochastic homogenization, using the finite element solutions of the 6 stochastic BVP with the prior stochastic model $\mathbf{x} \mapsto [\mathbf{C}^{\text{meso}}(\mathbf{x})]$

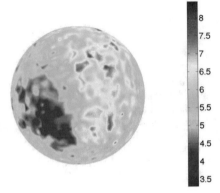

Fig. 10.22 Plot of a realization for elasticity random field C_{11}^{meso} (in GPa) in the interphase domain (relatively close to the exterior boundary surface of the interphase domain) and computed with the optimal values of the hyperparameters). [Figure from [120]].

in the interphase region. The likelihood function is estimated for $[\mathbf{C}^{\text{app,MD}}]$ that has been computed with the molecular dynamics simulations (considered as the experiments).

Result of the Identification

The optimal values of the hyperparameters are 0.2 for the dispersion parameter (20% of statistical fluctuations), $e/4$ for the radial spatial correlation length (e is the interphase thickness), $3.5 \times 10^{-9}\ m$ for the (average) angular spatial correlation length. Figure 10.22 displays a realization for elasticity random field C_{11}^{meso} (in GPa) inside the interphase domain (relatively close to the exterior boundary surface of the interphase domain) and computed with the optimal values of the hyperparameters).

In conclusion, the stochastic field thus defined exhibits transverse isotropy in spherical coordinates (in accordance with the MD-predicted geometrical configuration of the polymer chain segments in the neighborhood of the nano-inclusion), and interestingly depends on a low-dimensional hyperparameter. It is shown that the elasticity random field exhibits nonnegligible fluctuations, and that the estimates of the spatial correlation lengths are actually consistent with characteristic lengths of the atomistic model.

References

[1] Agmon N, Alhassid Y, Levine RD. An algorithm for finding the distribution of maximal entropy, *Journal of Computational Physics*, **30**(2), 250–258 (1979) doi:10.1016/0021-9991(79)90102-5.

[2] Amabili M, Sarkar A, Paidoussis MP. Reduced-order models for nonlinear vibrations of cylindrical shells via the proper orthogonal decomposition method, *Journal of Fluids and Structures*, **18**(2), 227–250 (2003) doi:10.1016/j.jfluidstructs.2003.06.002.

[3] Amsallem D, Farhat C. An online method for interpolating linear parametric reduced-order models, *SIAM Journal on Scientific Computing*, **33**(5), 2169–2198 (2011) doi:10.1137/100813051.

[4] Amsallem D, Zahr MJ, Farhat C. Nonlinear model order reduction based on local reduced-order bases, *International Journal for Numerical Methods in Engineering* 2012; **92**(10), 891–916 (2012) doi:10.1002/nme.4371.

[5] Amsallem D, Zahr M, Choi Y, Farhat C. Design optimization using hyper-reduced-order models, *Structural and Multidisciplinary Optimization*, **51**(4), 919–940 (2015) doi:0.1007/s00158-014-1183-y.

[6] Anderson TW. *An Introduction to Multivariate Statistical Analysis, Third Edition*, John Wiley & Sons, New York, 2003.

[7] Andrews HC, Patterson CL. Singular value decomposition and digital image processing, *Transactions on Acoustics, Speech, and Signal Processing - IEEE*, **24**, 26–53 (1976).

[8] Argyris JH, Kelsey S. The analysis of fuselages of arbitrary cross-section and taper, *Aircraft Engineering and Aerospace technology* **31**, 62–74 (1959) doi:10.1108/eb033088.

[9] Arnoux A, Batou A, Soize C, Gagliardini L. Stochastic reduced order computational model of structures having numerous local elastic modes in low frequency dynamics, *Journal of Sound and Vibration*, **332**(16), 3667–3680 (2013) doi:10.1016/j.jsv.2013.02.019.

© Springer International Publishing AG 2017
C. Soize, *Uncertainty Quantification*, Interdisciplinary Applied Mathematics 47,
DOI 10.1007/978-3-319-54339-0

[10] Arnst M, Ghanem R, Soize C. Identification of Bayesian posteriors for co-efficients of chaos expansion, *Journal of Computational Physics*, **229**(9), 3134–3154 (2010) doi:10.1016/j.jcp.2009.12.033.

[11] Arnst M, Soize C, Ghanem R. Hybrid sampling/spectral method for solving stochastic coupled problems, *SIAM/ASA Journal on Uncertainty Quantification*, **1**(1), 218–243 (2013) doi:10.1137/120894403.

[12] Arnst M, Ghanem R, Phipps E, Red-Horse J. Reduced chaos expansions with random coefficients in reduced-dimensional stochastic modeling of coupled problems, *International Journal for Numerical Methods in Engineering*, **29**(5), 352–376 (2014) doi:10.1002/nme.4595.

[13] Au SK, Beck JL. Subset simulation and its application to seismic risk based on dynamic analysis, *Journal of Engineering Mechanics - ASCE*, **129**(8), 901–917 (2003) doi:10.1061/(ASCE)0733-9399(2003)129:8(901).

[14] Au SK, Beck JL. Important sampling in high dimensions, *Structural Safety*, **25**(2), 139–163 (2003) doi:10.1016/S0167-4730(02)00047-4.

[15] Babuska I, Nobile F, Tempone R. A stochastic collocation method for elliptic partial differential equations with random input data, *SIAM Journal on Numerical Analysis*, **45**(3), 1005–1034 (2007) doi:10.1137/050645142.

[16] Batou A, Soize C. Identification of stochastic loads applied to a non-linear dynamical system using an uncertain computational model and experimental responses, *Computational Mechanics*, **43**(4), 559–571 (2009).

[17] Batou A, Soize C, Corus M. Experimental identification of an uncertain computational dynamical model representing a family of structures, *Computer and Structures*, **89**(13–14), 1440–1448 (2011) doi:10.1016/j.compstruc.2011.03.004.

[18] Batou A, Soize C. Rigid multibody system dynamics with uncertain rigid bodies, *Multibody System Dynamics*, **27**(3), 285–319 (2012) doi:10.1007/s11044-011-9279-2.

[19] Batou A, Soize C. Calculation of Lagrange multipliers in the construction of maximum entropy distributions in high stochastic dimension, *SIAM/ASA Journal on Uncertainty Quantification*, **1**(1), 431–451 (2013) doi:10.1137/120901386.

[20] Batou A, Soize C. Stochastic modeling and identification of an uncertain computational dynamical model with random fields properties and model uncertainties, *Archive of Applied Mechanics*, **83**(6), 831–848 (2013) doi:10.1007/s00419-012-0720-7.

[21] Batou A, Soize C, Audebert S. Model identification in computational stochastic dynamics using experimental modal data, *Mechanical Systems and Signal Processing*, **50–51**, 307–322 (2014) doi:10.1016/j.ymssp.2014.05.010.

[22] Benfield WA, Hruda RF. Vibration analysis of structures by component mode substitution, *AIAA Journal*, **9**, 1255–1261 (1971) doi:10.2514/3.49936.

[23] Berveiller M, Sudret B, Lemaire M. Stochastic finite elements: A non intrusive approach by regression, *European Journal of Computational Mechanics*, **15**(1–3), 81–92 (2006) doi:10.3166/remn.15.81–92.

[24] Blatman G, Sudret B. Adaptive sparse polynomial chaos expansion based on least angle regression, *Journal of Computational Physics*, **230**(6), 2345–2367 (2011) doi:10.1016/j.jcp.2010.12.021.

[25] Bowman AW, Azzalini A. *Applied Smoothing Techniques for Data Analysis: The Kernel Approach with S-Plus Illustrations*, Oxford University Press, 1997.

[26] Burrage K, Lenane I, Lythe G. Numerical methods for second-order stochastic differential equations, *SIAM Journal of Scientific Computing*, **29**(1), 245–264 (2007) doi:10.1137/050646032.

[27] Cameron RH, Martin WT. The orthogonal development of non-linear functionals in series of Fourier-Hermite functionals, *Annals of Mathematics*, Second Series, **48**(2), 385–392 (1947) doi:10.2307/1969178.

[28] Capiez-Lernout E, Soize C. Nonparametric modeling of random uncertainties for dynamic response of mistuned bladed disks, *Journal of Engineering for Gas Turbines and Power-Transactions of the ASME*, **126**(3), 610–618 (2004) doi:10.1115/1.1760527.

[29] Capiez-Lernout E, Soize C., Lombard JP, Dupont C, Seinturier E. Blade manufacturing tolerances definition for a mistuned industrial bladed disk, *Journal of Engineering for Gas Turbines and Power-Transactions of the ASME*, **127**(3), 621–628 (2005) doi:10.1115/1.1850497.

[30] Capiez-Lernout E, Pellissetti M, Pradlwarter H, Schueller GI, Soize C. Data and model uncertainties in complex aerospace engineering systems, *Journal of Sound and Vibration*, **295**(3–5), 923–938 (2006) doi:10.1016/j.jsv.2006.01.056.

[31] Capiez-Lernout E, Soize C. Robust design optimization in computational mechanics, *Journal of Applied Mechanics - Transactions of the ASME*, **75**(2), 1–11 (2008) doi:10.1115/1.2775493.

[32] Capiez-Lernout E, Soize C. Robust updating of uncertain damping models in structural dynamics for low- and medium-frequency ranges, *Mechanical Systems and Signal Processing*, **22**(8), 1774–1792 (2008) doi:10.1016/j.ymssp.2008.02.005.

[33] Capiez-Lernout E, Soize C. Design optimization with an uncertain vibroacoustic model, *Journal of Vibration and Acoustics*, **130**(2), 1–8 (2008) doi:10.1115/1.2827988.

[34] Capiez-Lernout E, Soize C, Mignolet MP. Computational stochastic statics of an uncertain curved structure with geometrical nonlinearity in three-dimensional elasticity, *Computational Mechanics*, **49**(1), 87–97 (2012) doi:10.1007/s00466-011-0629-y.

[35] Capiez-Lernout E, Soize C, Mignolet MP. Post-buckling nonlinear static and dynamical analyses of uncertain cylindrical shells and experimental validation, *Computer Methods in Applied Mechanics and Engineering*, **271**(1), 210–230 (2014) doi:10.1016/j.cma.2013.12.011.

[36] Capiez-Lernout E, Soize C, Mbaye M. Mistuning analysis and uncertainty quantification of an industrial bladed disk with geometrical nonlinearity, *Journal of Sound and Vibration*, **356**, 124–143 (2015) doi:10.1016/j.jsv.2015.07.006.

[37] Capillon R, Desceliers C, Soize C. Uncertainty quantification in computational linear structural dynamics for viscoelastic composite structures, *Computer Methods in Applied Mechanics and Engineering*, **305**, 154–172 (2016) doi:10.1016/j.cma.2016.03.012.

[38] Carlberg K, Bou-Mosleh C, Farhat C. Efficient non-linear model reduction via a least-squares Petrov-Galerkin projection and compressive tensor approximations, *International Journal for Numerical Methods in Engineering*, **86**(2), 155–181 (2011) doi:10.1002/nme.3050.

[39] Carlberg K, Farhat C. A low-cost, goal-oriented compact proper orthogonal decomposition basis for model reduction of static systems, *International Journal for Numerical Methods in Engineering*, **86**(3), 381–402 (2011) doi:10.1002/nme.3074.

[40] Carlberg K, Farhat C, Cortial J, Amsallem D. The GNAT method for nonlinear model reduction: effective implementation and application to computational fluid dynamics and turbulent flows, *Journal of Computational Physics*, **242**, 623–647 (2013) doi:10.1016/j.jcp.2013.02.028.

[41] Cataldo E, Soize C, Sampaio R, Desceliers C. Probabilistic modeling of a nonlinear dynamical system used for producing voice, *Computational Mechanics*, **43**(2), 265–275 (2009).

[42] Cataldo E, Soize C, Sampaio R. Uncertainty quantification of voice signal production mechanical model and experimental updating, *Mechanical Systems and Signal Processing*, **40**(2), 718–726 (2013) doi:10.1016/j.ymssp.2013.06.036.

[43] Cataldo E, Soize C. Jitter generation in voice signals produced by a two-mass stochastic mechanical model, *Biomedical Signal Processing and Control*, **27**, 87–95 (2016) doi:10.1016/j.bspc.2016.02.003.

[44] Censor Y. Pareto optimality in multiobjective problems, *Applied Mathematics and Optimization*, **4**(1), 41–59 (1977) doi:10.1007/BF01442131.

[45] Chaturantabut S, Sorensen DC. Nonlinear model reduction via discrete empirical interpolation, *SIAM Journal on Scientific and Statistical Computing*, **32**(5), 2737–2764 (2010) doi:10.1137/090766498.

[46] Chebli H, Soize C. Experimental validation of a nonparametric probabilistic model of non homogeneous uncertainties for dynamical systems, *The Journal of the Acoustical Society of America*, **115**(2), 697–705 (2004) doi:10.1121/1.1639335.

[47] Chen C, Duhamel D, Soize C. Uncertainties model and its experimental identification in structural dynamics for composite sandwich panels, *Journal of Sound and Vibration*, **294**(1–2), 64–81 (2006) doi:10.1016/j.jsv.2005.10.013.

[48] Chevalier L, Cloupet S, Soize C. Probabilistic model for random uncertainties in steady state rolling contact, *Wear Journal*, **258**(10), 1543–1554 (2005).

[49] Coifman RR, Lafon S, Lee AB, Maggioni M, Nadler B, Warner F, Zucker SW. Geometric diffusions as a tool for harmonic analysis and structure definition of data: Diffusion maps, *PNAS*, **102**(21), 7426–7431 (2005).

[50] Clément A, Soize C, Yvonnet J. Computational nonlinear stochastic homogenization using a non-concurrent multiscale approach for hyperelastic heterogenous microstructures analysis, *International Journal for Numerical Methods in Engineering*, **91**(8), 799–824 (2012) doi:10.1002/nme.4293.

[51] Clément A, Soize C, Yvonnet J. Uncertainty quantification in computational stochastic multiscale analysis of nonlinear elastic materials, *Computer Methods in Applied Mechanics and Engineering*, **254**, 61–82 (2013) doi:10.1016/j.cma.2012.10.016.

[52] Clough RW, Penzien J. *Dynamics of Structures*, McGraw-Hill, New York, 1975.

[53] Cottereau R, Clouteau D, Soize C. Construction of a probabilistic model for impedance matrices, *Computer Methods in Applied Mechanics and Engineering*, **196**(17–20), 2252–2268 (2007).

[54] Craig RR, Bampton MCC. Coupling of substructures for dynamic analyses, *AIAA Journal*, **6**, 1313–1322 (1968) doi:10.2514/3.4741.

[55] Craig RR. A review of time domain and frequency domain component mode synthesis method, *Combined Experimental-Analytical Modeling of Dynamic Structural Systems*, edited by D.R. Martinez and A.K. Miller, **67**, ASME-AMD, New York, 1985.

[56] Das S, Ghanem R, Spall J. Asymptotic sampling distribution for polynomial chaos representation of data: A maximum-entropy and fisher information approach, *SIAM Journal on Scientific Computing*, **30**(5), 2207–2234 (2008) doi:10.1137/060652105.

[57] Das S, Ghanem R, Finette S. Polynomial chaos representation of spatio-temporal random field from experimental measurements, *Journal of Computational Physics*, **228**, 8726–8751 (2009) doi:10.1016/j.jcp.2009.08.025.

[58] Deb K. *Multi-Objective Optimization using Evolutionary Algorithms*, John Wiley & Sons, Chichester, 2001.

[59] Debusschere BJ, Najm HN, Pebay PP, Knio OM, Ghanem R, Le Maitre OP. Numerical challenges in the use of polynomial chaos representations for stochastic processes, *SIAM Journal on Scientific Computing*, **26**(2), 698–719 (2004) doi:10.1137/S1064827503427741.

[60] Degroote J, Virendeels J, Willcox K. Interpolation among reduced-order matrices to obtain parameterized models for design, optimization and probabilistic analysis, *International Journal for Numerical Methods in Fluids*, **63**, 207–230 (2010) doi:10.1002/fld.2089.

[61] de Klerk D, Rixen DJ, Voormeeren SN. General framework for dynamic substructuring: History, review, and classification of techniques, *AIAA Journal* **46**, 1169–1181 (2008) doi:10.2514/1.33274.

[62] Desceliers C, Soize C, Cambier S. Non-parametric-parametric model for random uncertainties in non-linear structural dynamics: Application to earthquake engineering, *Earthquake Engineering & Structural Dynamics*, **33**(3), 315–327 (2004) doi:10.1002/eqe.352.

[63] Desceliers C, Ghanem R., Soize C. Maximum likelihood estimation of stochastic chaos representations from experimental data, *International Journal for Numerical Methods in Engineering*, **66**(6), 978–1001 (2006) doi:10.1002/nme.1576.

[64] Desceliers C, Soize C, Ghanem R. Identification of chaos representations of elastic properties of random media using experimental vibration tests, *Computational Mechanics*, **39**(6), 831–838 (2007) doi:10.1007/s00466-006-0072-7.

[65] Desceliers C, Soize C, Grimal Q, Talmant M, Naili S. Determination of the random anisotropic elasticity layer using transient wave propagation in a fluid-solid multilayer: Model and experiments, *The Journal of the Acoustical Society of America*, **125**(4), 2027–2034 (2009) doi:10.1121/1.3087428.

[66] Desceliers C, Soize C, Naili S, Haiat G. Probabilistic model of the human cortical bone with mechanical alterations in ultrasonic range, *Mechanical Systems and Signal Processing*, **32**, 170–177 (2012) doi:10.1016/j.ymssp.2012.03.008.

[67] Desceliers C, Soize C, Zarroug M. Computational strategy for the crash design analysis using an uncertain computational mechanical model, *Computational Mechanics*, **52**(2), 453–462 (2013) doi:10.1007/s00466-012-0822-7.

[68] Desceliers C, Soize C, Yanez-Godoy H, Houdu E, Poupard O. Robustness analysis of an uncertain computational model to predict well integrity for geologic CO2 sequestration, *Computational Geosciences*, **17**(2), 307–323 (2013) doi:10.1007/s10596-012-9332-0.

[69] Doostan A, Ghanem R, Red-Horse J. Stochastic model reduction for chaos representations, *Computer Methods in Applied Mechanics and Engineering*, **196**(37–40), 3951–3966 (2007) doi:10.1016/j.cma.2006.10.047.

[70] Doostan A, Owhadi H. A non-adapted sparse approximation of PDEs with stochastic inputs, *Journal of Computational Physics*, **230**(8), 3015–3034 (2011) doi:10.1016/j.jcp.2011.01.002.

[71] Duchereau J, Soize C. Transient dynamics in structures with nonhomogeneous uncertainties induced by complex joints, *Mechanical Systems and Signal Processing*, **20**(4), 854–867 (2006) doi:10.1016/j.ymssp.2004.11.003.

[72] Durand JF, Soize C, Gagliardini L. Structural-acoustic modeling of automotive vehicles in presence of uncertainties and experimental identification and validation, *Journal of the Acoustical Society of America*, **124**(3), 1513–1525 (2008) doi:10.1121/1.2953316.

[73] Dyson FJ, Mehta ML. Statistical theory of the energy levels of complex systems. Parts IV, V. *Journal of Mathematical Physics*, **4**, 701–719 (1963) doi:10.1063/1.1704008.

[74] Ernst OG, Mugler A, Starkloff HJ, Ullmann E. On the convergence of generalized polynomial chaos expansions, *ESAIM: Mathematical Modelling and Numerical Analysis*, **46**(2), 317–339 (2012) doi:10.1051/m2an/2011045.

[75] Ewins DJ. The effects of detuning upon the forced vibrations of bladed disks, *Journal of Sound and Vibration*,**9**(1), 65–79 (1969) doi:10.1016/0022-460X(69)90264-8.

[76] Ezvan O, Batou A, Soize C. Multi-scale reduced-order computational model in structural dynamics for the low- and medium-frequency ranges, *Computer and Structures*, **160**, 111–125 (2015) doi:10.1016/j.compstruc.2015.08.007.

[77] Ezvan O, Batou A, Soize C, Gagliardini L. Multilevel model reduction for uncertainty quantification in computational structural dynamics, *Computational Mechanics*, Submitted July 2nd (2016).

[78] Farhat C, Avery P, Chapman T, Cortial J. Dimensional reduction of nonlinear finite element dynamic models with finite rotations and energy-based mesh sampling and weighting for computational efficiency, *International Journal for Numerical Methods in Engineering*, **98**(9), 625–662 (2014) doi:10.1002/nme.4668.

[79] Farhat C, Chapman T, Avery P. Structure-preserving, stability, and accuracy properties of the Energy-Conserving Sampling and Weighting (ECSW) method for the hyper reduction of nonlinear finite element dynamic models, *International Journal for Numerical Methods in Engineering*, **102**(5), 1077–1110 (2015) doi:10.1002/nme.4820.

[80] Fernandez C, Soize C, Gagliardini L. Sound-insulation layer modelling in car computational vibroacoustics in the medium-frequency range, *Acta Acustica United with Acustica (AAUWA)*, **96**(3), 437–444 (2010) doi:10.3813/AAA.918296.

[81] Frauenfelder P, Schwab C, Todor RA. Finite elements for elliptic problems with stochastic coefficients, *Computer Methods in Applied Mechanics and Engineering*, **194**(2–5), 205–228 (2005) doi:10.1016/j.cma.2004.04.008.

[82] Ganapathysubramanian B, Zabaras N. Sparse grid collocation schemes for stochastic natural convection problems, *Journal of Computational Physics*, **25**(1), 652–685 (2007) doi:10.1016/j.jcp.2006.12.014.

[83] Gerbrands JJ. On the relationships between SVD, KLT and PCA, *Pattern Recognition*, **14**(1), 3756381 (1981).

[84] Ghanem R, Spanos PD. Polynomial chaos in stochastic finite elements, *Journal of Applied Mechanics - Transactions of the ASME*, **57**(1), 197–202 (1990) doi:10.1115/1.2888303.

[85] Ghanem R, Spanos PD. *Stochastic Finite Elements: A spectral Approach*, Springer-Verlag, New-York, 1991 (revised edition, Dover Publications, New York, 2003).

[86] Ghanem R, Kruger RM. Numerical solution of spectral stochastic finite element systems, *Computer Methods in Applied Mechanics and Engineering*, **129**(3), 289–303 (1996) doi:10.1016/0045-7825(95)00909-4.

[87] Ghanem R, Doostan R, Red-Horse J. A probabilistic construction of model validation, *Computer Methods in Applied Mechanics and Engineering*, **197**(29–32), 2585–2595 (2008) doi:10.1016/j.cma.2007.08.029.

[88] Ghosh D, Ghanem R. Stochastic convergence acceleration through basis enrichment of polynomial chaos expansions, *International Journal for Numerical Methods in Engineering*, **73**(2), 162–184 (2008) doi:10.1002/nme.2066.

[89] Givens GH, Hoeting JA. *Computational Statistics*, 2nd edition, John Wiley & Sons, Hoboken, New Jersey, 2013.

[90] Golub GH, Van Loan CF. Matrix Computations, Fourth Edition, The Johns Hopkins University Press, Baltimore, 2013.

[91] Grepl MA, Maday Y, Nguyen NC, Patera A. Efficient reduced-basis treatment of nonaffine and nonlinear partial differential equations, *ESAIM: Mathematical Modelling and Numerical Analysis*, **41**(03), 575–605 (2007) doi:10.1051/m2an:2007031.

[92] Guilleminot J, Soize C, Kondo D, Binetruy C. Theoretical framework and experimental procedure for modelling volume fraction stochastic fluctuations in fiber reinforced composites, *International Journal of Solid and Structures*, **45**(21), 5567–5583 (2008) doi:10.1016/j.ijsolstr.2008.06.002.

[93] Guilleminot J, Soize C, Kondo D. Mesoscale probabilistic models for the elasticity tensor of fiber reinforced composites: experimental identification and numerical aspects, *Mechanics of Materials*, **41**(12), 1309–1322 (2009) doi:10.1016/j.mechmat.2009. 08.004.

[94] Guilleminot J, Soize C. A stochastic model for elasticity tensors with uncertain material symmetries, *International Journal of Solids and Structures*, **47**(22–23), 3121–3130 (2010) doi:10.1016/j.ijsolstr.2010.07.013.

[95] Guilleminot J, Noshadravan A, Soize C, Ghanem R. A probabilistic model for bounded elasticity tensor random fields with application to polycrystalline microstructures, *Computer Methods in Applied Mechanics and Engineering*, **200**(17–20), 1637–1648 (2011) doi:10.1016/j.cma.2011.01.016.

[96] Guilleminot J, Soize C. Probabilistic modeling of apparent tensors in elastostatics: a MaxEnt approach under material symmetry and stochastic boundedness constraints, *Probabilistic Engineering Mechanics*, **28**(SI), 118–124 (2012) doi:10.1016/j.prob engmech.2011.07.004.

[97] Guilleminot J, Soize C. Generalized stochastic approach for constitutive equation in linear elasticity: A random matrix model, *International Journal for Numerical Methods in Engineering*, **90**(5),613–635 (2012) doi:10.1002/nme.3338.

[98] Guilleminot J, Soize C. Stochastic model and generator for random fields with symmetry properties: application to the mesoscopic modeling of elastic random media, *Multiscale Modeling and Simulation (A SIAM Interdisciplinary Journal)*, **11**(3), 840–870 (2013) doi:10.1137/120898346.

[99] Guilleminot J, Soize C. On the statistical dependence for the components of random elasticity tensors exhibiting material symmetry properties, *Journal of Elasticity*, **111**(2), 109–130 (2013) doi 10.1007/s10659-012-9396-z.

[100] Guilleminot J, Soize C. ISDE-based generator for a class of non-gaussian vector-valued random fields in uncertainty quantification, *SIAM Journal on Scientific Computing*, **36**(6), A2763–A2786 (2014) doi:10.1137/130948586.

[101] Gupta AK, Nagar DK. *Matrix Variate Distributions*, Chapman & Hall/CRC, Boca Raton, 2000.

[102] Guyan RJ. Reduction of stiffness and mass matrices, *AIAA Journal*, **3**, 380–380 (1965) doi:10.2514/3.2874.

[103] Hairer E, Lubich C, Wanner G. *Geometric Numerical Integration: Structure-Preserving Algorithms for Ordinary Differential Equations*, Springer-Verlag, Heidelberg, 2002.

[104] Han S, Feeny BF. Enhanced proper orthogonal decomposition for the modal analysis of homogeneous structures, *Journal of Vibration and Control*, **8**(1), 19–40 (2002) doi:10.1177/1077546302008001518.

[105] Hastings WK. Monte Carlo sampling methods using Markov chains and their applications, *Biometrika*, **57**(1), 97–109 (1970).

[106] Horova I, Kolacek J, Zelinka J. *Kernel Smoothing in Matlab*, World Scientific, Singapore, 2012.

[107] Huang TS. *Picture Processing and Digital Filtering*, Springer, Berlin, 1975.

[108] Hurty, WC. Vibrations of structural systems by component mode synthesis, *Journal of Engineering Mechanics - ASCE*, **86**, 51–69 (1960).

[109] Hurty WC. Dynamic analysis of structural systems using component modes, *AIAA Journal*, **3**, 678–685 (1965) doi:10.2514/3.2947.

[110] Ikeda N, Watanabe S. *Stochastic Differential Equations and Diffusion Processes*, North Holland, Amsterdam, 1981.

[111] Irons B. Structural eigenvalue problems - elimination of unwanted variables, *AIAA Journal* **3**, 961–962 (1965) doi:10.2514/3.3027.

[112] Jaynes ET. Information theory and statistical mechanics, *Physical Review*, **106**(4), 620–630 and **108**(2), 171–190 (1957).

[113] Kassem M, Soize C, Gagliardini L. Structural partitioning of complex structures in the medium-frequency range. An application to an automotive vehicle, *Journal of Sound and Vibration*, **330**(5), 937–946 (2011) doi:10.1016/j.jsv.2010.09.008.

[114] Keshavarzzadeh V, Ghanem R, Masri SF, Aldraihem OJ. Convergence acceleration of polynomial chaos solutions via sequence transformation, *Computer Methods in Applied Mechanics and Engineering*, **271**, 167–184 (2014) doi:10.1016/j.cma.2013.12.003.

[115] Khasminskii R. *Stochastic Stability of Differential Equations*, 2nd edition, Springer, 2012.

[116] Kerschen G, Golinval JC, Vakakis AF, Bergman LA. The method of proper orthogonal decomposition for dynamical characterization and order reduction of mechanical systems: an overview, *Nonlinear Dynamics*, **41**, 147–169 (2005) doi:10.1007/s11071-005-2803-2.

[117] Kim K, Wang XQ, Mignolet MP. Nonlinear reduced order modeling of isotropic and functionally graded plates, Proceedings of the 49th Structures, Structural Dynamics, and Materials Conference, *AIAA Paper*, AIAA-2008-1873 (2008) doi:10.2514/6.2008-1873.

[118] Krée P, Soize C. *Mathematics of Random Phenomena*, D. Reidel Publishing Company, Dordrecht, 1986 (Revised edition of the French edition *Mécanique aléatoire*, Dunod, Paris, 1983).

[119] Le TT. *Stochastic modeling in continuum mechanics of the inclusion-matrix interphase from molecular dynamics simulations* (in French: Modélisation stochastique, en mécanique des milieux continus, de l'interphase inclusion-matrice à partir de simulations en dynamique moléculaire), Thèse de doctorat de l'Université Paris-Est, Marne-la-Vallée, France, 2015.

[120] Le TT, Guilleminot J, Soize C. Stochastic continuum modeling of random interphases from atomistic simulations. Application to a polymer nanocomposite, *Computer Methods in Applied Mechanics and Engineering*, **303**, 430–449 (2016) doi:10.1016/j.cma.2015.10.006.

[121] Leissing T, Soize C, Jean P, Defrance J. Computational model for long-range non-linear propagation over urban cities, *Acta Acustica united with Acustica (AAUWA)*, **96**(5), 884–898 (2010) doi:10.3813/AAA.918347.

[122] Le Maitre OP, Knio OM, Najm HN. Uncertainty propagation using Wiener-Haar expansions, *Journal of Computational Physics*, **197**(1), 28–57 (2004) doi:10.1016/j.jcp.2003.11.033.

[123] Le Maitre OP, Knio OM. *Spectral Methods for Uncertainty Quantification with Applications to Computational Fluid Dynamics*, Springer, Heidelberg, 2010.

[124] Lestoille N, Soize C, Funfschilling C. Stochastic prediction of high-speed train dynamics to long-time evolution of track irregularities, *Mechanics Research Communications*, **75**, 29–39 (2016) doi:10.1016/j.mechrescom.2016.05.007.

[125] Leung AYT. *Dynamic Stiffness and Substructures*, Springer-Verlag, Berlin, 1993.

[126] Lucor D, Su CH, Karniadakis GE. Generalized polynomial chaos and random oscillators, *International Journal for Numerical Methods in Engineering*, **60**(3), 571–596 (2004) doi:10.1002/nme.976.

[127] Macocco K, Grimal Q, Naili S, Soize C. Elastoacoustic model with uncertain mechanical properties for ultrasonic wave velocity prediction; application to cortical bone evaluation, *Journal of the Acoustical Society of America*, **119**(2), 729–740 (2006).

[128] MacNeal RH. A hybrid method of component mode synthesis, *Computers and Structures*, **1**, 581–601 (1971).

[129] Marzouk YM, Najm HN. Dimensionality reduction and polynomial chaos acceleration of Bayesian inference in inverse problems, *Journal of Computational Physics*, **228**(6), 1862–1902 (2009) doi:10.1016/j.jcp.2008.11.024.

[130] Matthies HG, Keese A. Galerkin methods for linear and nonlinear elliptic stochastic partial differential equations, *Computer Methods in Applied Mechanics and Engineering*, **194**(12–16), 1295–1331 (2005) doi:10.1016/j.cma.2004.05.027.

[131] Mbaye M, Soize C, Ousty JP. A reduced-order model of detuned cyclic dynamical systems with geometric modifications using a basis of cyclic modes, *ASME Journal of Engineering for Gas Turbines and Power*, **132**(11), 112502-1-9 (2010) doi:10.1115/1.4000805.

[132] Mbaye M, Soize C, Ousty JP, Capiez-Lernout E. Robust analysis of design in vibration of turbomachines, *ASME Journal of Turbomachinery*, **35**(2), 021008-1-8 (2013) doi:10.1115/1.4007442.

[133] Mehta ML. *Random Matrices and the Statisticals Theory of Energy Levels*, Academic Press, New York, 1967.

[134] Mehta ML. *Random Matrices, Revised and Enlarged Second Edition*, Academic Press, San Diego, 1991.

[135] Mehta ML. *Random Matrices, Third Edition*, Elsevier, San Diego, 2014.

[136] Metropolis N, Ulam S. The Monte Carlo method, *Journal of the American Statistical Association*, **44**(247), 335–341 (1949).

[137] Metropolis N, Rosenbluth AW, Rosenbluth MN, Teller AH, Teller E. Equations of state calculations by fast computing machine, *The Journal of Chemical Physics*, **21**(6),1087–1092 (1953).

[138] Michel G. *Buckling of cylindrical thin shells under a shear dynamic loading* (in French: Flambage de coques minces cylindriques sous un chargement dynamique de cisaillement), Thèse de Doctorat, INSA Lyon, 1997.

[139] Mignolet MP, Soize C. Stochastic reduced order models for uncertain nonlinear dynamical systems, *Computer Methods in Applied Mechanics and Engineering*, **197**(45–48), 3951–3963 (2008) doi:10.1016/j.cma.2008.03.032.

[140] Mignolet MP, Soize C. Nonparametric stochastic modeling of linear systems with prescribed variance of several natural frequencies, *Probabilistic Engineering Mechanics*, **23**(2–3), 267–278 (2008) doi:10.1016/j.probengmech.2007.12.027.

[141] Mignolet MP, Soize C., Avalos J. Nonparametric stochastic modeling of structures with uncertain boundary conditions / coupling between substructures, *AIAA Journal*, **51**(6), 1296–1308 (2013) doi:10.2514/1.J051555.

[142] Mignolet MP, Przekop A, Rizzi SA, Spottswood SM. A review of indirect/non-intrusive reduced order modeling of nonlinear geometric structures, *Journal of Sound and Vibration*, **332**(10), 2437–2460 (2013) doi:doi:10.1016/j.jsv.2012.10.017.

[143] Morand HJP, Ohayon R. *Fluid Structure Interaction*, John Wiley & Sons, Hoboken, New Jersey, 1995.

[144] Muravyov AA, Rizzi SA. Determination of nonlinear stiffness with application to random vibration of geometrically nonlinear structures, *Computers and Structures*, **81**(15), 1513–1523 (2003) doi:10.1016/S0045-7949(03)00145-7.

[145] Najm HN. Uncertainty quantification and polynomial chaos techniques in computational fluid dynamics, *Journal Review of Fluid Mechanics*, **41**, 35–52 (2009) doi:10.1146/annurev.fluid.010908.165248.

[146] Nemat-Nasser S, Hori M. *Micromechanics: Overall Properties of Heterogeneous Materials*, Second revised edition, Elsevier, Amsterdam, 1999.

[147] Nguyen N, Peraire J. An efficient reduced-order modeling approach for non-linear parametrized partial differential equations, *International Journal for Numerical Methods in Engineering*, **76**(1), 27–55 (2008) doi:10.1002/nme.2309.

[148] Nguyen MT. *Multiscale identification of the apparent random elasticity field of heterogeneous microstructures. Application to a biological tissue* (in French: Identification multi-échelle du champ d'élasticité apparent stochastique de microstructures hétérogènes. Application à un tissu biologique), Thèse de doctorat de l'Université Paris-Est, Marne-la-Vallée, France, 2013.

[149] Nguyen MT, Desceliers C, Soize C, Allain JM, Gharbi H. Multiscale identification of the random elasticity field at mesoscale of a heterogeneous microstructure using multiscale experimental observations, *International Journal for Multiscale Computational Engineering*, **13**(4), 281–295 (2015) doi:10.1615/IntJMultCompEng.2015011435.

[150] Nguyen MT, Allain JM, Gharbi H, Desceliers C, Soize C. Experimental multiscale measurements for the mechanical identification of a cortical bone by digital image correlation, *Journal of the Mechanical Behavior of Biomedical Materials*, **63**, 125–133 (2016) doi:10.1016/j.jmbbm.2016.06.011.

[151] Nouy A. A generalized spectral decomposition technique to solve a class of linear stochastic partial differential equations, *Computer Methods in Applied Mechanics and Engineering*, **196**(45–48), 4521–4537 (2007) doi:10.1016/j.cma.2007.05.016.

[152] Nouy A. Proper Generalized Decomposition and separated representations for the numerical solution of high dimensional stochastic problems, *Archives of Computational Methods in Engineering*, **17**(4), 403–434 (2010) 10.1007/s11831-010-9054-1.

[153] Nouy A, Soize C. Random fields representations for stochastic elliptic boundary value problems and statistical inverse problems, *European Journal of Applied Mathematics*, **25**(3), 339–373 (2014) doi:10.1017/S0956792514000072.

[154] Ohayon R, Soize C. *Structural Acoustics and Vibration*, Academic Press, San Diego, 1998.

[155] Ohayon R, Soize C. Advanced computational dissipative structural acoustics and fluid-structure interaction in low- and medium-frequency domains. Reduced-order models and uncertainty quantification, *International Journal of Aeronautical and Space Sciences*, **13**(2), 127–153 (2012) doi:10.5139/IJASS.2012.13.2.127.

[156] Ohayon R, Soize C. *Advanced Computational Vibroacoustics - Reduced-Order Models and Uncertainty Quantification*, Cambridge University Press, New York, 2014.

[157] Ohayon R, Soize C, Sampaio R. Variational-based reduced-order model in dynamic substructuring of coupled structures through a dissipative physical interface: Recent advances, *Archives of Computational Methods in Engineering*, **21**(3), 321–329 (2014) doi:10.1007/s11831-014-9107-y.

[158] Paidoussis MP, Issid NT. Dynamic stability of pipes conveying fluid, *Journal of Sound and Vibration*, **33**(3), 267–294 (1974) doi:10.1016/S0022-460X(74)80002-7.

[159] Paidoussis MP. *Fluid-Structure Interactions: Slender Structures and Axial Flow*, Vol 1, Academic Press, San Diego, California, 1998.

[160] Paul-Dubois-Taine A, Amsallem D. An adaptive and efficient greedy procedure for the optimal training of parametric reduced-order models, *International Journal for Numerical Methods in Engineering*, **102**(5), 1262–1292 (2015) doi:10.1002/nme.4759.

[161] Pellissetti M, E. Capiez-Lernout E, Pradlwarter H, Soize C, Schueller GI, Reliability analysis of a satellite structure with a parametric and a non-parametric probabilistic model, *Computer Methods in Applied Mechanics and Engineering*, **198**(2), 344–357 (2008).

[162] Perrin G, Soize C, Duhamel D, Funfschilling C. Identification of polynomial chaos representations in high dimension from a set of realizations, *SIAM Journal on Scientific Computing*, **34**(6), A2917–A2945 (2012) doi:10.1137/11084950X.

[163] Perrin G, Soize C, Duhamel D, Funfschilling C. Karhunen-Loève expansion revisited for vector-valued random fields: scaling, errors and optimal basis, *Journal of Computational Physics*, **242**(1), 607–622 (2013) doi:10.1016/j.jcp.2013.02.036.

[164] Poirion F, Soize C. Numerical simulation of homogeneous and inhomogeneous Gaussian stochastic vector fields, *La Recherche Aérospatiale* (English edition), **1**, 41–61 (1989).

[165] Poirion F, Soize C. Numerical methods and mathematical aspects for simulation of homogeneous and non homogeneous Gaussian vector fields, pp. 17–53, in *Probabilistic Methods in Applied Physics*, edited by P. Krée and W. Wedig, Springer-Verlag, Berlin, 1995.

[166] Pradlwarter HJ, Schueller GI. Local domain Monte Carlo simulation, *Structural Safety*, **32**(5), 275–280 (2010) doi:10.1016/j.strusafe.2010.03.009.

[167] Puig B, Poirion F, Soize C. Non-Gaussian simulation using Hermite polynomial expansion: Convergences and algorithms, *Probabilistic Engineering Mechanics*, **17**(3), 253–264 (2002) doi:10.1016/S0266-8920(02)00010-3.

[168] Ritto TG, Soize C, Sampaio R. Nonlinear dynamics of a drill-string with uncertainty model of the bit-rock interaction, *International Journal of Non-Linear Mechanics*, **44**(8), 865–876 (2009) doi:10.1016/j.ijnonlinmec.2009.06.003.

[169] Ritto TG, Soize C, Sampaio R. Robust optimization of the rate of penetration of a drill-string using a stochastic nonlinear dynamical model, *Computational Mechanics*, **45**(5), 415–427 (2010) doi:10.1007/s00466-009-0462-8.

[170] Ritto TG, Soize C, Rochinha FA, Sampaio R. Dynamic stability of a pipe conveying fluid with an uncertain computational model, *Journal of Fluid and Structures*, **49**, 412–426 (2014) doi:10.1016/j.jfluidstructs.2014.05.003.

[171] Rubin S. Improved component-mode representation for structural dynamic analysis, *AIAA Journal*, **13**, 995–1006 (1975) doi:10.2514/3.60497.

[172] Rubinstein RY, Kroese DP. *Simulation and Monte Carlo method*, Second Edition, John Wiley & Sons, Hoboken, New Jersey, 2008.

[173] Ryckelynck D. A priori hyperreduction method: an adaptive approach, *Journal of Computational Physics*, **202**, 346–366 (2005) doi:10.1016/j.jcp.2004.07.015.

[174] Sakji S, Soize C, Heck JV. Probabilistic uncertainties modeling for thermomechanical analysis of plasterboard submitted to fire load, *Journal of Structural Engineering, ASCE*, **134**(10), 1611–1618 (2008) doi:10.1061/(ASCE)0733-9445(2008)134:10(1611).

[175] Sakji S, Soize C, Heck JV. Computational stochastic heat transfer with model uncertainties in a plasterboard submitted to fire load and experimental validation, *Fire and Materials Journal*, **33**(3), 109–127 (2009) doi:10.1002/fam.982.

[176] Sampaio R, Soize C. Remarks on the efficiency of POD for model reduction in nonlinear dynamics of continuous elastic systems, *International Journal for Numerical Methods in Engineering*, **72**(1), 22–45 (2007) doi:10.1002/nme.1991.

[177] Schueller GI. Efficient Monte Carlo simulation procedures in structural uncertainty and reliability analysis, recent advances, *Structural Engineering and Mechanics*, **32**(1), 1–20 (2009) doi:10.12989/sem.2009.32.1.001.

[178] Serfling RJ. *Approximation Theorems of Mathematical Statistics*, John Wiley & Sons, Hoboken, New Jersey, 1980 (Paperback edition published in 2002).

[179] Shannon CE. A mathematical theory of communication. *The Bell System Technical Journal*, **27**, 379–423 and 623–659 (1948).

[180] Shinozuka M. Simulations of multivariate and multidimensional random processes. *Journal of Acoustical Society of America*, **39**(1), 357–367 (1971) doi:10.1121/1.1912338.

[181] Smith RC. *Uncertainty Quantification: Theory, Implementation, and Applications*, SIAM, Philadelphia, 2014.

[182] Soize C. Oscillators submitted to squared gaussian processes, *Journal of Mathematical Physics*, **21**(10), 2500–2507 (1980) doi:10.1063/ 1.524356.

[183] Soize C. *The Fokker-Planck Equation for Stochastic Dynamical Systems and its Explicit Steady State Solutions*, World Scientific Publishing Co Pte Ltd, Singapore, 1994.

[184] Soize C. A nonparametric model of random uncertainties for reduced matrix models in structural dynamics, *Probabilistic Engineering Mechanics*, **15**(3), 277–294 (2000) doi:10.1016/S0266-8920(99)00028-4.

[185] Soize C. Maximum entropy approach for modeling random uncertainties in transient elastodynamics, *The Journal of the Acoustical Society of America*, **109**(5), 1979–1996 (2001) doi:10.1121/1.1360716.

[186] Soize C. Random matrix theory and non-parametric model of random uncertainties, *Journal of Sound ·and Vibration*, **263**(4), 893–916 (2003) doi:10.1016/S0022-460X(02)01170-7.

[187] Soize C, Chebli H. Random uncertainties model in dynamic substructuring using a nonparametric probabilistic model, *Journal of Engineering Mechanics*, **129**(4), 449–457 (2003).

[188] Soize C, Ghanem R. Physical systems with random uncertainties: Chaos representation with arbitrary probability measure, *SIAM Journal on Scientific Computing*, **26**(2), 395–410 (2004) doi:10.1137/S1064827503424505.

[189] Soize C. A comprehensive overview of a non-parametric probabilistic approach of model uncertainties for predictive models in structural dynamics, *Journal of Sound and Vibration*, **288**(3), 623–652 (2005) doi:10.1016/j.jsv.2005.07.009.

[190] Soize C. Random matrix theory for modeling uncertainties in computational mechanics, *Computer Methods in Applied Mechanics and Engineering*, **194**(12–16), 1333–1366 (2005) doi:10.1016/j.cma.2004.06.038.

[191] Soize C. Non Gaussian positive-definite matrix-valued random fields for elliptic stochastic partial differential operators, *Computer Methods in Applied Mechanics and Engineering*, **195**(1–3), 26–64 (2006) doi:10.1016/j.cma.2004.12.014.

[192] Soize C. Construction of probability distributions in high dimension using the maximum entropy principle. Applications to stochastic processes, random

fields and random matrices, *International Journal for Numerical Methods in Engineering*, **76**(10), 1583–1611 (2008) doi:10.1002/nme.2385.

[193] Soize C. Tensor-valued random fields for meso-scale stochastic model of anisotropic elastic microstructure and probabilistic analysis of representative volume element size, *Probabilistic Engineering Mechanics*, **23**(2–3), 307–323 (2008) doi:10.1016/j.probengmech.2007.12.019.

[194] Soize C, Capiez-Lernout E, Ohayon R. Robust updating of uncertain computational models using experimental modal analysis, *AIAA Journal*, **46**(11), 2955–2965 (2008) doi:10.2514/1.38115.

[195] Soize C, Capiez-Lernout E, Durand JF, Fernandez C, Gagliardini L. Probabilistic model identification of uncertainties in computational models for dynamical systems and experimental validation, *Computer Methods in Applied Mechanics and Engineering*, **198**(1), 150–163, (2008) doi:10.1016/j.cma.2008.04.007.

[196] Soize C, Ghanem R. Reduced chaos decomposition with random coefficients of vector-valued random variables and random fields, *Computer Methods in Applied Mechanics and Engineering*, **198**(21–26), 1926–1934 (2009) doi:10.1016/j.cma.2008.12.035.

[197] Soize C. Generalized Probabilistic approach of uncertainties in computational dynamics using random matrices and polynomial chaos decompositions, *International Journal for Numerical Methods in Engineering*, **81**(8), 939–970 (2010) doi:10.1002/nme.2712.

[198] Soize C. Information theory for generation of accelerograms associated with SRS, *Computer-Aided Civil and Infrastructure Engineering*, **25**(5), 334–347 (2010) doi:10.1111/j.1467-8667.2009.00643.x.

[199] Soize C, Desceliers C. Computational aspects for constructing realizations of polynomial chaos in high dimension, *SIAM Journal On Scientific Computing*, **32**(5), 2820–2831 (2010) doi:10.1137/100787830.

[200] Soize C. Identification of high-dimension polynomial chaos expansions with random coefficients for non-Gaussian tensor-valued random fields using partial and limited experimental data, *Computer Methods in Applied Mechanics and Engineering*, **199**(33–36), 2150–2164 (2010) doi:10.1016/j.cma.2010.03.013.

[201] Soize C. A computational inverse method for identification of non-Gaussian random fields using the Bayesian approach in very high dimension, *Computer Methods in Applied Mechanics and Engineering*, **200**(45–46), 3083–3099 (2011) doi:10.1016/j.cma.2011. 07.005.

[202] Soize C, Batou A. Stochastic reduced-order model in low-frequency dynamics in presence of numerous local elastic modes, *Journal of Applied Mechanics - Transactions of the ASME*, **78**(6), 061003-1 to 9 (2011) doi:10.1115/1.4002593.

[203] Soize C. *Stochastic Models of Uncertainties in Computational Mechanics*, American Society of Civil Engineers (ASCE), Reston, 2012.

[204] Soize C., Poloskov IE. Time-domain formulation in computational dynamics for linear viscoelastic media with model uncertainties and stochastic exci-

tation, *Computers and Mathematics with Applications*, **64**(11), 3594–3612 (2012) doi:10.1016/j.camwa. 2012.09.010.

[205] Soize C. Bayesian posteriors of uncertainty quantification in computational structural dynamics for low- and medium-frequency ranges, *Computers and Structures*, **126**, 41–55 (2013) doi:10.1016/j.compstruc.2013.03.020.

[206] Soize C. Polynomial chaos expansion of a multimodal random vector, *SIAM/ASA Journal on Uncertainty Quantification*, **3**(1), 34–60 (2015) doi:10.1137/140968495.

[207] Soize C. Random Matrix Models and Nonparametric Method for Uncertainty Quantification, pp. 1–69, in *Handbook for Uncertainty Quantification*, edited by R. Ghanem, D. Higdon, and H. Owhadi, doi:10.1007/978-3-319-11259-6_5-1, SpringerReference, Springer, 2016.

[208] Soize C. Random vectors and random fields in high dimension - Parametric model-based representation, identification from data, and inverse problems, pp. 1–53, in *Handbook for Uncertainty Quantification*, edited by R. Ghanem, D. Higdon, and H. Owhadi, doi:10.1007/978-3-319-11259-6_30-1, Springer-Reference, Springer, 2016.

[209] Soize C, Farhat C. Uncertainty quantification of modeling errors for nonlinear reduced-order computational models using a nonparametric probabilistic approach, *International Journal for Numerical Methods in Engineering*, on line, 2016, doi:10.1016/j.jmbbm.2016.06.011.

[210] Soize C, Ghanem R. Data-driven probability concentration and sampling on manifold, *Journal of Computational Physics*, **321**, 242–258 (2016) doi:10.1016/j.jcp.2016.05.044.

[211] Sudret B. Global sensitivity analysis using polynomial chaos expansions, *Reliability Engineering & System Safety*, **93**(7), 964–979 (2008) doi:10.1016/j.ress.2007.04.002.

[212] Sullivan TJ. *Introduction to Uncertainty Quantification*, Springer, 2015.

[213] Tipireddy R, Ghanem R. Basis adaptation in homogeneous chaos spaces, *Journal of Computational Physics*, **259**, 304–317 (2014) doi:10.1016/j.jcp.2013.12.009.

[214] Torquato S. *Random Heterogeneous Materials, Microstructure and Macroscopic Properties*, Springer-Verlag, New York, 2002.

[215] Walpole LJ. Elastic behavior of composite materials: theoretical foundations, *Advances in Applied Mechanics*, **21**, 169–242 (1981) doi:10.1016/S0065-2156(08)70332-6.

[216] Walpole LJ. Fourth-rank tensors of the thirty-two crystal classes: Multiplication Tables, *Proceedings of the Royal Society A*, 391, 149–179 (1984) doi:10.1098/rspa.1984.0008. (1984)

[217] Wan XL, Karniadakis GE. Multi-element generalized polynomial chaos for arbitrary probability measures, *SIAM Journal on Scientific Computing*, **28**(3), 901–928 (2006) doi:10.1137/050627630.

[218] Whitehead DS. Effects of mistuning on the vibration of turbomachine blades induced by wakes, *Journal of Mechanical Engineering Science*, **8**(1), 15–21. (1966) doi:10.1243/JMES-JOUR-1966-008-004-02.

[219] Wigner EP. On the statistical distribution of the widths and spacings of nuclear resonance levels, *Mathematical Proceedings of the Cambridge Philosophical Society*, **47**(04), 790–798 (1951) doi:10.1017/S0305004100027237.

[220] Wigner EP. Distribution laws for the roots of a random Hermitian matrix, pp 446–461, in *Statistical Theories of Spectra: Fluctuations*, edited by C.E. Poter, Academic Press, New York, 1965.

[221] Willcox K, Peraire J. Balanced model reduction via the proper orthogonal decomposition, *AIAA Journal*, **40**(11), 2323–2330 (2002).

[222] Xiu DB, Karniadakis GE. Wiener-Askey polynomial chaos for stochastic differential equations, *SIAM Journal on Scientific Computing*, **24**(2), 619–644 (2002) doi:10.1137/S1064827501387826.

[223] Xiu DB. *Numerical Methods for Stochastic Computations: A Spectral Method Approach*, Princeton University Press, Princeton, 2010.

[224] Zahr M, Farhat C. Progressive construction of a parametric reduced-order model for PDE-constrained optimization, *International Journal for Numerical Methods in Engineering*, **102**(5), 1077–1110 (2015) doi:0.1002/nme.4770.

Glossary

Deterministic variable: A deterministic scalar variable is denoted by a lowercase letter such as x.

Deterministic vector: A deterministic vector is denoted by a boldface, lowercase letter such as in $\mathbf{x} = (x_1, \ldots, x_n)$.

Random variable: A random scalar variable is denoted by an uppercase letter such as X.

Random vector: A random vector is denoted by a boldface, uppercase letter such as in $\mathbf{X} = (X_1, \ldots, X_n)$.

Deterministic matrix: A deterministic matrix is denoted by an uppercase (or lowercase) letter between brackets such as $[A]$ (or $[a]$).

Random matrix: A random matrix is denoted by a boldface, uppercase letter between brackets such as $[\mathbf{A}]$.

$\mathbf{x} = (x_1, \ldots, x_n)$: Vector in $\mathbb{K}^n$.

$\overline{x}_j$: complex conjugate of the complex number x_j.

A_{jk}: entry $[A]_{jk}$ of matrix $[A]$.

$\mathrm{tr}\{[A]\}$: trace of matrix $[A]$.

$[A]^T$: transpose of matrix $[A]$.

$[I_n]$: identity (or unit) identity matrix in $\mathbb{M}_n$.

$[0_{N,n}]$: zero matrix in $\mathbb{M}_{N,n}$.

δ_{jk}: Kronecker's symbol such that $\delta_{jk} = 0$ if $j \neq k$ and $\delta_{jk} = 1$ if $j = k$.

© Springer International Publishing AG 2017

C. Soize, *Uncertainty Quantification*, Interdisciplinary Applied Mathematics 47,

DOI 10.1007/978-3-319-54339-0

$1_B(\mathbf{x})$: indicator function of a set B defined by $1_B(\mathbf{x}) = 1$ if $\mathbf{x} \in B$ and $1_B(\mathbf{x}) = 0$ if $\mathbf{x} \notin B$.

i: pure imaginary complex number satisfying $i^2 = -1$.

$\mathbb{C}$: set of all the complex numbers.

$\mathbb{C}^n$: Hermitian vector space of dimension n.

$\mathbb{K}$: set $\mathbb{R}$ or $\mathbb{C}$.

$\mathbb{M}_{n,m}(\mathbb{R})$: set of all the $(n \times m)$ real matrices,

$\mathbb{M}_n(\mathbb{R})$: set of all the $(n \times n)$ real matrices,

$\mathbb{M}_n(\mathbb{C})$: set of all the $(n \times n)$ complex matrices,

$\mathbb{M}_n^S(\mathbb{R})$: set of all the $(n \times n)$ real symmetric matrices.

$\mathbb{M}_n^U(\mathbb{R})$: set of all the upper triangular real $(n \times n)$ matrices with positive diagonal entries.

$\mathbb{M}_n^+(\mathbb{R})$: set of all the $(n \times n)$ real symmetric positive-definite matrices.

$\mathbb{M}_n^{+0}(\mathbb{R})$: set of all the $(n \times n)$ real symmetric semipositive-definite matrices.

$\mathbb{N}$: set of all the integers $0, 1, 2, \ldots$.

$\mathbb{N}^*$: set of all the positive integers $1, 2, \ldots$.

$\mathbb{R}$: set of all the real numbers (real line).

$\mathbb{R}^+$: set $[0, +\infty[$ of all the positive and zero real numbers.

$\mathbb{R}^n$: Euclidean vector space of dimension n.

E: mathematical expectation.

$(\Theta, \mathcal{T}, \mathcal{P})$: probability space.

a.s.: almost surely.

l.i.m: limit in mean square.

m. − s.: mean-square.

$\mathbb{H}$: Hilbert space of the $\mathbb{R}$-valued functions that are square integrable on $\mathbb{R}^{N_g}$ with respect to the probability distribution $p_\Xi(\boldsymbol{\xi}) \, d\boldsymbol{\xi}$.

$L^1(\Omega, \mathbb{R}^n)$: vector space (Banach) of all the integrable functions defined on a subset Ω of $\mathbb{R}^d$ with values in $\mathbb{R}^n$.

$L^2(\Omega, \mathbb{R}^n)$: vector space (Hilbert space) of all the square integrable functions defined on a subset Ω of $\mathbb{R}^d$ with values in $\mathbb{R}^n$.

$L^0(\Theta, \mathbb{R}^n)$: vector space of all the $\mathbb{R}^n$-valued random variable defined on $(\Theta, \mathcal{T}, \mathcal{P})$.

$L^2(\Theta, \mathbb{R}^n)$: subspace (Hilbert space) of $L^0(\Theta, \mathbb{R}^n)$ of the second-order random variables.

$< \mathbf{x}, \mathbf{y} >$: Euclidean inner product $\sum_{j=1}^n x_j y_j$ in $\mathbb{R}^n$.

$< \mathbf{x}, \mathbf{y} >$: Hermitian inner product $\sum_{j=1}^n x_j \overline{y}_j$ in $\mathbb{C}^n$.

$\ll [G], [H] \gg = \text{tr}\{[G]^T [H]\}$: inner product in $\mathbb{M}_n(\mathbb{R})$.

$< g, h >_{\mathbb{H}}$: inner product in Hilbert space $\mathbb{H}$.

$< \mathbf{X}, \mathbf{Y} >_{\Theta}$: inner product $E\{<\mathbf{X},\mathbf{Y}>\}$ in $L^2(\Theta, \mathbb{R}^n)$.

$\|\mathbf{x}\|$: Euclidean norm $<\mathbf{x},\mathbf{x}>^{1/2}$ of vector $\mathbf{x}$ in $\mathbb{R}^n$.

$\|\mathbf{x}\|$: Hermitian norm $<\mathbf{x},\mathbf{x}>^{1/2}$ of vector $\mathbf{x}$ in $\mathbb{C}^n$.

$\|g\|_{\mathbb{H}}$: Hilbert norm $< g, g >_{\mathbb{H}}^{1/2}$ of a function g in Hilbert space $\mathbb{H}$.

$\|A\|$: operator norm in $\mathbb{M}_n(\mathbb{R})$ such that $\|A\| = \sup_{\|\mathbf{x}\| \le 1} \|[A]\mathbf{x}\|$ with $\mathbf{x} \in \mathbb{R}^n$.

$\|\mathbf{X}\|_{\Theta}$: norm $E\{<\mathbf{X},\mathbf{X}>\}^{1/2}$ in $L^2(\Theta, \mathbb{R}^n)$.

$\exp_{\mathbb{M}}$: matrix exponential defined on set $\mathbb{M}_n^S(\mathbb{R})$.

$\log_{\mathbb{M}}$: principal matrix logarithm defined on set $\mathbb{M}_n^+(\mathbb{R})$, the inverse of $\exp_{\mathbb{M}}$.

$[A] > 0$: means that matrix $[A]$ is positive definite.

$[A] \ge 0$: means that matrix $[A]$ is semipositive definite.

$[A] > [B]$: means that matrix $[A] - [B]$ is positive definite.

$[A] \ge [B]$: means that matrix $[A] - [B]$ is semipositive definite.

Index

© Springer International Publishing AG 2017
C. Soize, *Uncertainty Quantification*, Interdisciplinary Applied Mathematics 47,
DOI 10.1007/978-3-319-54339-0

Printed in the United States
By Bookmasters